D1531896

ENVIRONMENT AND SOLID WASTES

Characterization, Treatment, and Disposal

Edited by

CHESTER W. FRANCIS
STANLEY I. AUERBACH
Environmental Sciences Division
Oak Ridge National Laboratory
Oak Ridge, Tennessee

Technical Editor
VIVIAN A. JACOBS
Oak Ridge National Laboratory
Oak Ridge, Tennessee

BUTTERWORTHS
Boston • London
Sydney • Wellington • Durban • Toronto

An Ann Arbor Science Book

Ann Arbor Science is an imprint of Butterworth Publishers

Library of Congress Catalog Card Number 82-71528
ISBN 0-250-40583-0

Published by Butterworth Publishers
10 Tower Office Park
Woburn, MA 01801

Printed in the United States of America

ENVIRONMENT AND SOLID WASTES

Characterization, Treatment, and Disposal

Proceedings of the Fourth Life Sciences Symposium
Environment and Solid Wastes
Gatlinburg, Tennessee
October 4-8, 1981

Sponsored by

Oak Ridge National Laboratory
U.S. Department of Energy
U.S. Environmental Protection Agency
Electric Power Research Institute

Francis **Auerbach**

Chester W. Francis is a group leader in the Earth Sciences Section, Environmental Sciences Division, Oak Ridge National Laboratory. He earned his MS and PhD in soil science from the University of Wisconsin, and his BS in agronomy from Iowa State University. Dr. Francis is a principal investigator on a number of research projects evaluating the fate of contaminants in the environment from disposal of solid and hazardous wastes. He has published more than 40 technical papers (including one patent) dealing with movement of radionuclides and trace elements in soils and uptake in plants, biological denitrification waste treatment processes, and solid waste management practices. He has served as Associate Editor for the *Journal of Environmental Quality* and is a member of the Soil Science Society of America, the American Association for the Advancement of Science, and Sigma Xi.

Stanley I. Auerbach is Director of the Environmental Sciences Division at Oak Ridge National Laboratory, Oak Ridge, Tennessee. He earned his PhD in zoology from Northwestern University and his BS and MS in zoology (animal ecology) from the University of Illinois. Dr. Auerbach is an Adjunct Research Professor at the University of Georgia and the University of Tennessee. He has served as president of the Ecological Society of America, has published a number of papers, is a Fellow of the American Association for the Advancement of Science, and is a member of several other professional societies that reflect his interests in ecology and environmental sciences. Dr. Auerbach is also a member of the Advisory Research Committee, Resources for the Future; the Commission on Physical Sciences, Mathematics, and Resources for the National Research Council, National Academy of Sciences; and the Board of Environmental Consultants for the Tennessee Tombigbee Waterway Project. He is also Ecology Editor for *Environment International* and is on the Advisory Board of *Environmental and Experimental Botany.*

CONTENTS

Section 3
Transport and Transformation

Section 4
Environmental and Health Consequences

Section 5
Needs and Challenges of Monitoring

Section 6
Environmental Control Systems: Techniques and Technology

PREFACE

The purpose of the Fourth Life Sciences Symposium, *Environment and Solid Wastes: Characterization, Treatment, and Disposal,* was to evaluate the impact of solid waste management practices in the utility, chemical, mining, and municipal communities on human health and the environment. The multiagency and multidisciplinary nature of the program allowed the participants to identify relevant problems and provide practical solutions in solid waste management and environmental science. Invited speakers included representatives from major industrial communities, state and federal research and regulatory agencies, leading universities, and community action groups. The objective was to improve communication between the wide range of scientific disciplines, decision-makers, and members of the industrial community and interested public relative to the implications of solid and hazardous wastes on human health and the environment. These published proceedings of the sessions provide a compilation of these concepts and discussions for the scientific community and interested public.

We wish to express our thanks to the members of the Symposium committee: W. J. Boegly, Jr., Environmental Sciences Division, Oak Ridge National Laboratory; S. R. Cordle, Office of Environmental Processes and Effects Research, U.S. Environmental Protection Agency, Washington, DC; S. L. Daniels, Dow Chemical Company, Midland, Michigan; R. E. Franklin, Ecological Research Division, U.S. Department of Energy, Washington, DC; R. A. Griffin, Illinois State Geological Survey, Champaign, Illinois; and P. A. Vesilind, Department of Civil Engineering, Duke University, Durham, North Carolina. Special thanks are extended to those members of the committee that served as session chairmen and wrote the overviews for their sessions: Problems in Characterization, R. A. Griffin; Transport and Transformation, S. R. Cordle; Environmental and Health Consequences, R. E. Franklin; Needs and Challenges in Monitoring, I. P. Murarka; and Environmental Control Systems: Techniques and Technology, P. A. Vesilind. T. Tamura, Head, Earth Sciences Section, Environmental Sciences Division, Oak Ridge National Laboratory, served as a special advisor to the symposium committee.

We very much appreciate the thoughtfulness and consideration given by Bonnie S. Reesor, Conference Coordinator for Oak Ridge National Laboratory, in making the conference, hotel, and special dinner arrangements. Also, thanks to Joy Simmons for keeping the financial accounting up to date. The symposium would not have run so smoothly without the excellent support staff: Frank S. Brinkley, Philip Lowry, Ivy M. Munro, and Pam Valliant.

Special thanks are in order for Carolyn Henley and Linda Croff for their secretarial assistance before, during, and after the symposium. Preparation of these proceedings has been significantly easier because of the able assistance provided by Vivian Jacobs, Joyce Francis, and Carol Johnson technical editors; for this, we are grateful. We hope these proceedings will aid the reader in understanding more clearly the potential health and environmental effects of solid wastes in the environment.

C. W. Francis
S. I. Auerbach

WELCOME AND INTRODUCTION

Stanley I. Auerbach
Oak Ridge National Laboratory
Oak Ridge, Tennessee

On behalf of the Oak Ridge National Laboratory (ORNL) it is a pleasure to welcome you to the fourth symposium in the annual Life Sciences Symposium Series. This symposium is another in the ORNL series of symposia devoted to subjects that involve a broad area of research and development, and encompass multidisciplinary programmatic efforts that are typical of much of what we do at ORNL. Interdisciplinary research activities at large, multipurpose research and development institutions play a key role in the science and technology of today. We feel that we can organize our total resources to address broad problems that require the application of numerous techniques that vary from engineering to the biological, social, and physical sciences.

In 1980 ORNL established a program in nonnuclear waste disposal because we became aware of the many important technical and policy issues involved in land disposal of wastes. Resolving the uncertainties surrounding these issues is a goal to which all of us are dedicated and can be achieved through professional exchange of information. It is this organized, professional exchange to which this symposium is dedicated.

The principal objectives of the symposium are to evaluate how the management of solid wastes from the utility, chemical, mining and municipal communities influences the quality of the environment and, potentially, human health, and to elicit some of the more useful and practical technological solutions to the relevant problems associated with their management. The multiagency, multidisciplinary nature of the program will provide such a forum to evaluate the current concepts of solid-waste management and the relevant environmental sciences. This symposium has been structured to facilitate discussions that will assess the environmental and health consequences of solid wastes in the environment. The published proceedings of the sessions will document these concepts and discussions for the scientific community and the interested public.

Through the dedicated efforts of our organizing committee, a distinguished group of speakers representing the broad aspects of this environmental issue have been brought together. The diversity of viewpoints, the differences in background, and the differences in responsibility will be reflected in their presentations and will add immeasurably to our understanding of a problem area that has received, and is continuing to receive, national attention and concern.

The problems of groundwater contamination, spoilage of drinking wells, and increased shortage of space for landfills are now widely recognized and attest to the fact that we are dealing with an environmental–technology problem of considerable magnitude. It is somewhat ironic that it was in the latter part of the era of the environment, the decade of the 1970s, that this problem became evident to the broad public sector. That is perhaps understandable, because the immediate concerns of the public in the early part of the 1970s were with what were then the dominant issues of water and air pollution. As we all know, however, these challenges are still with us, and some of our major environmental problems are associated with those two media. Now we have added a third medium of concern—the lithosphere in the broadest sense. It is part of the total system that supports the biosphere. It is, therefore, a considerable challenge to start to apply the means whereby we can handle the increased volume and chemical complexity of our waste by-products in a manner that is commensurate with public need, one that reflects the best efforts of our technology and will be an appropriate balance between costs and benefits.

We are adjacent to the Great Smoky Mountains National Park; from an ecological viewpoint, it is one of the most interesting and unusual of the parks in the United States. The deciduous forests in the eastern part of the park are still representative of the magnificent untouched forests that existed in the United States before the arrival of European man. Its scientific importance is reflected in its designation as an International Biosphere Reserve.

The town of Gatlinburg, the gateway to the Smokies, still reflects some aspects of its original mountain culture tempered with the modern concept of a tourist mecca. Again, before I turn the podium over to my co-chairman, Chester W. Francis, I want to welcome you on behalf of ORNL to this symposium, to Gatlinburg, and to the Great Smoky Mountains.

SECTION 1

SOURCES AND MANAGEMENT OF WASTES

CHAPTER 1

OVERVIEW

C. W. Francis

Oak Ridge National Laboratory
P.O. Box X
Oak Ridge, Tennessee 37830

Recent publicity received by incidents such as polychlorinated biphenyls in Michigan, Kepone in Virginia, and Love Canal in New York has increased public awareness relative to how the management of solid and hazardous waste can influence public health and the environment. These events have, in turn, resulted in increased emphasis on the regulatory aspects of managing wastes [e.g., recent regulations promulgated under the Resource Conservation and Recovery Act (RCRA)]. Some of these regulations will change significantly the manner in which wastes are managed and, in most instances, the attendant costs will be increased appreciably. These costs, however, may be small relative to the costs imposed on human health and environment or even the costs involved to remedy a waste management practice at a later date. Thus, in-depth evaluations relative to how various waste management practices affect the quality of human health and environment are badly needed. Such evaluations encompass a variety of wastes and management schemes. They also require a wide range in scientific and engineering talents to assess the magnitude of the impacts relative to the disposal costs involved. This section reviews the various health and environmental problems and addresses the objectives and strategies relative to the management of wastes from the chemical, municipal, mining, utility, and emerging-energy communities.

Joseph Mayhew presents data from a recent survey conducted by the Chemical Manufacturers Association among its membership describing how the new regulations under RCRA have altered the manner in which the chemical industry manages its hazardous waste. Among the interesting observations are: (1) the chemical industry only discards about 10% of its hazardous waste by using offsite commercial disposal facilities; (2) many companies are increasing their incineration capacity; (3) disposal in waste lagoons is phasing out; and (4) the use of landfills is either declining or remaining steady (depending on the waste's physicochemical characteristics) because of the lack of federal standards relative to construction of landfills.

The management alternatives applicable to the utilization/disposal of municipal sludges are addressed by David Zenz. In 1977, approximately one-third of the sludge produced by U.S. wastewater municipalities was used in so-called sale or give-away programs, primarily as a soil amendment. A surprisingly high percentage (30%) was disposed of in the ocean. Future disposal options will depend on federal and state environmental regulations; however, it appears that most municipal sludges can be disposed of as a nonhazardous waste, because data have indicated that it is highly unusual to find a sludge classified as hazardous using the U.S. Environmental Protection Agency extraction procedure toxicity test.

The management of coal-fired electrical utility wastes in the United States and the United Kingdom is described by John Maulbetsch and John Brown respectively. In the United Kingdom, a significantly greater proportion of the fly ash (approximately 40%) has been used in the construction and associated industries relative to that used for similar purposes in the United States (approximately 25%). However, disposal of ash in the two countries is very similar, predominantly by landfills and lagoons. Utilities in the United Kingdom appear to have more experience than utilities in the United States in handling large volumes of ash at specific sites (volume of ash in excess of 3500 tons/d) and possibly have conducted more research relative to restoration of landfills for agricultural use. This probably reflects the greater demand for land in the United Kingdom than in the United States. Both representatives of the utilities indicated that waste management of ash from coal will continue to be a major environmental issue as oil is replaced by coal as an energy source. However, environmental research combined with careful waste management planning is important in assuring that impacts to health and environment are kept to an acceptable risk.

Lawrence Kronenberger describes some of the tentative solid waste management plans for the emerging energy technologies, oil shale retorting, and coal conversion to liquid and gaseous fuels. Oil shale retorting will produce extremely large volumes of solid wastes, quantities in excess of 50,000 tons/d at an individual facility producing 50,000 bbl of crude oil per day. Commercial-sized coal conversion facilities will produce significantly less solid wastes (approximately 4000 to 5000 tons/d), which is still a large volume relative to conventional coal-fired electrical utilities. In both cases, oil shale retorting and coal conversion, the solid waste management plans will be centered on how to dispose of the large volumes of nonhazardous (relatively nontoxic) wastes in an economical yet environmentally acceptable manner.

This section indicates that the major quantity of waste will be that classified as nonhazardous or exempt from hazardous classification (e.g., coal processing wastes). The volume of wastes classified as hazardous is declining as conservation and alternative management practices are implemented. Also, the regulations recently enacted by RCRA have been directed at the establishment of procedures to avoid the mismanagement of hazardous types of waste; for example, identification and listing of wastes, standards applicable to transporters of hazardous wastes, as well as standards for owners and operators of hazardous waste treatment, storage, and disposal facilities. Little effort has been directed at the large volumes of nonhazardous waste. Apparently, the likely impacts on health and environment will result from the noncompliance under RCRA (assuming current RCRA regulations are adequate) and disposal of nonhazardous wastes. The impact on land disposal of the large volumes of nonhazardous wastes will vary greatly, depending on the physicochemical characteristics of the waste and the geochemical and hydrologic conditions of the waste site. The major concern here is that long-term releases of contaminants in the solid waste leachates (which are not highly toxic) will overtax the natural attenuation or disposal mechanisms that may result in accumulation of harmful quantities of these contaminants in surface and groundwaters as well as other sensitive areas of the environment.

CHAPTER 2

EFFECT OF RCRA ON THE CHEMICAL INDUSTRY

Joseph J. Mayhew
Chemical Manufacturing Association
Washington, DC

ABSTRACT

The Resource Conservation and Recovery Act (RCRA) of 1976 has had a major effect on the way the chemical industry treats and disposes of its wastes. In 1980 and 1981 the U.S. Environmental Protection Agency (EPA) promulgated regulations that place stringent controls on the handling of hazardous wastes. These regulations have both favored certain types of treatment and disposal and provided the economic incentive to modify manufacturing processes that minimize the production of residuals that must be discarded.

This chapter describes the results of an informal survey of several major chemical companies to determine what specific changes in hazardous waste handling practices have occurred since the promulgation of the regulations, and what long-term trends in generation and disposal are expected.

The survey includes information on research needs, shifts in methods of disposal, and short-term solutions in response to regulatory and scientific uncertainty concerning acceptable methods of disposal.

The Chemical Manufacturers Association represents approximately 200 firms that are engaged in the manufacture of basic chemical products. Our industry has a major interest and plays an important role in the proper management of solid wastes. In this chapter, I will describe how the chemical industry, as represented by my association, has been affected by the Resource Conservation Recovery Act and how that may have altered the way the chemical industry manages its hazardous waste.

In 1976, Congress passed the Resource Conservation Recovery Act. Up to that time, hazardous waste had been controlled primarily by the states or local governments. In many cases the states were doing an adequate job; in others, however, hazardous materials were being improperly disposed of or were not being given the special handling that is required. The act, as passed, covered several different aspects of the management of hazardous waste; it also covered more generally the disposal of solid waste. In December 1978, the U.S. Environmental Protection Agency proposed regulations for the generation, transportation, and disposal of hazardous waste. Numerous comments were received concerning those proposed regulations, and in May 1980 the EPA put many aspects of them into final form. The regulations that went into effect on November 19, 1980, governed the generation, transportation, and disposal or treatment of hazardous waste. The treatment and disposal portion of the regulations provided only guidance for interim-status facilities (i.e., facilities that were in place as of November 19, 1980). In 1981, as the second phase of its programs, EPA issued regulations for new and existing facilities. Many of these regulations are yet to be made final by EPA, so we are dealing with a set of regulations that probably will be fluid for the next two to three years.

Given the definition of hazardous waste of May 1980, it has been estimated that from 40 to 60 million metric tons of hazardous waste are generated in the United States each year. This range is based on surveys of various sectors of industry rather than a total survey; therefore, we may expect to see the number vary considerably from the prediction. The definition of hazardous waste has a major effect on the estimate, and changes in the definition by EPA or by individual states will have a major effect on this estimate. The EPA estimates that as much as 62% of this hazardous waste is generated by the chemical industry. In recent surveys by the Eckhardt Committee, it has been estimated that our industry disposes of most of its own materials, as much as 90% on its own property, and this practice will continue.

I have conducted an informal survey of about 16% of the industry. The purpose of the survey was to find out what changes the regulations had brought about and what research the industry felt would be needed to answer the challenge of these new regulations. The first thing the survey showed was that the effect of the regulations after their effective date in November 1980 was not dramatic; the major effect had come earlier, subsequent to the passing of a law in 1976. This is because the industry believed the intent of the government was to control these activities closely, and disposal of hazardous waste would be given all the priority afforded other emissions into the environment. The effect that occurred in November 1980 (the first effective date of the hazardous waste regulations) was primarily in the administration or management of wastes. There was only a minor effect on disposal methods at that time, primarily in increased reporting requirements. These included the manifesting requirements, the filing of contingency plans, and the filing of training plans and other documents to assure EPA that management was being conducted in a proper manner.

Other than these paperwork requirements, what were the major effects of RCRA? There were two major aspects: first, changes in generation and, second, changes in disposal or treatment practices.

The amount of wastes being generated have not changed markedly since 1976. There have been some estimates of a 2 to 3% growth per year based simply on the growth of materials being produced. However, this growth may be counteracted by other changes such as increased recycling of materials and process design changes to reduce the amount of waste produced. Some recycling efforts had been occurring for several years; however, with the increased emphasis on control of the disposal of hazardous waste, the cost of disposal also rose. This increased disposal cost allowed management to consider the reuse of some materials that had once been relatively inexpensive to discard. For example, spent sulfuric acid has been recycled in this country for a number of years because it was economic to do so. It could be regenerated and resold to the petroleum industry at costs equivalent to the generation of virgin sulfuric acid. Other materials will now fall into that category. Process design changes are also a factor in reducing the amount of wastes being generated. This factor will take some years to manifest itself because of the length of time it takes to replace or modify facilities.

What then are the changes in the disposal practices of the industry? First, many companies are increasing their incinerator capacity because of the certainty of the regulations in this area and the indication by EPA that they prefer this treatment option. There is also some indication that the public in general finds this method of treatment or disposal to be more acceptable. However, not all hazardous wastes can be incinerated; metal sludges, for example, cannot be treated by incineration.

The second major change in methods of disposal is in the area of waste lagoons. Many companies are phasing out these facilities except as storage facilities. With respect to landfills, the industry's use of these facilities is either declining or remaining steady, depending on the mix of waste that the company is handling. No standards for construction of new landfills currently exist; that is, there are no federal standards. This lack of standards has caused a lapse in the construction of landfills.

As was mentioned earlier, the chemical industry discards only about 10% of its hazardous waste by the use of offsite or commercial disposal facilities. The recently completed survey of the industry indicates this use of commercial facilities will remain at about the same level in the short term and decline in the long term. There is an important distinction here: those industries with a number of small units, such as distribution and marketing installations, will continue to use a larger percentage of the offsite disposal capacity, whereas large, integrated facilities will continue to rely on their own disposal facilities. A factor influencing an increased use of company-owned treatment and disposal-facilities modification is in the area of legal liability. A significant portion of the chemical industry believes that the best way to control the long-term care of materials placed in or on the land is by ownership of these facilities. Because a new Superfund legislation would always make a waste-generating company liable for disposal of such wastes, the trend will be definitely toward continued ownership and care of the waste by the company that is generating it.

The survey also covered some other areas such as the use of waste exchanges. Of those companies surveyed, only one was using a waste exchange and only for a minor portion of its waste. The other companies reported they had established an internal waste exchange because the companies are large and can either trade materials within the company for reuse opportunities or can find markets for these byproducts through the regular marketing system.

In addition to the questions about the effect of the regulations on waste disposal in the industry, the respondents were asked to indicate what they felt would be the long-term trend in waste management over the next 20 to 30 years within their companies. The most common response was that new manufacturing facilities would be designed with in-process changes that would eliminate many of the wastes that are currently being produced. This is seen as a significant trend and one that is contrary to government predictions of increased generation of hazardous waste.

The chemical industry has made some major changes in the way it has handled its waste. These changes came about primarily with the passage of the RCRA. By requiring more reporting and planning, the recent hazardous waste regulations have affected the way the companies have been doing business. The long-term trends in waste generation cannot be predicted at this time, but no major increases in waste generation are expected. If process design changes are successful in eliminating wastes, new production units will produce fewer wastes than older or existing production units. Also, the industry will continue to treat or dispose of most of its own waste, and will increase this practice in the years to come. Incineration and land disposal techniques will continue to be the mainstay of hazardous waste management practices for the foreseeable future. The new technologies that are being demonstrated today will not, in the near future, account for a significant portion of the waste that is disposed of or treated.

DISCUSSION

***R. A. Griffin*, Illinois Geological Survey:** Would you comment on the impact on RCRA on monitoring requirements of the chemical industry, particularly relative to groundwater.

J. Mayhew: I'll start with the easy part. Groundwater monitoring requirement for land treatment facilities have not gone into effect yet; they start in November. Many people are now putting in their wells and that sort of thing. We'll know a little more in 2 or 3 months about how that's working out. In general, the paperwork requirements of the manifest in the annual reports have allowed them to keep much better track of where their wastes are going. I know that's a fairly general response.

***David Friedman*, EPA:** Two comments relative to RCRA are that (1) increased cost of petrochemical feedstocks has resulted in increased recycle and reuse of residuals, and (2) RCRA has caused companies to learn much more about their residuals (composition, properties, etc.) which have resulted in utilization and less waste for disposal.

J. Mayhew: I might just add something about energy recovery. In a similar survey we found that a company is now recovering about $80 million worth of fuel value per year by burning some of these wastes in heat-recovery equipment.

***Ed Portner*, Johns Hopkins University Applied Physics Laboratory:** I'm interested in your survey. Did you get any comments from the various companies about what the regulatory environment is in the different states where, ultimately, many of the regulations are going to be enforced? What is the technical depth of the various state regulatory agencies? Also, are they doing anything to provide central facilities within the state, etc.?

J. Mayhew: We didn't really cover that in the survey. In general conversations with the membership, a lot of things are going on in the states that are very interesting in terms of more stringent regulations; different definitions of hazardous wastes have caused some confusion in the regulations. Some of the states in some regional areas are pressing for regional facilities, but again as I pointed out, that only affects a minor 10% of our wastes anyway.

***Arthur W. Hounslow*, Oklahoma State University:** I was a little confused on your comments on onsite and offsite disposal. Does this include your own landfill operations? If it does, I can see you have good control over soil contamination and surface runoff, but you still don't have as much control probably over groundwater contamination and movement as a commercial offsite disposal operation.

J. Mayhew: The 90 and 10% figures include landfills. I guess one of the big problems with putting your landfill on your own facility is that selection of a manufacturing facility site is usually based on transportation and things like that, whereas that selection of landfill site is done for other reasons. There are other criteria for the selection of a landfill site and that certainly complicates the picture.

***S. Cordle*, EPA:** I have a two-part question. First, were your research needs generated from the survey or are these just your general knowledge of the discussion with members?

J. Mayhew: There were two sources, really—the survey in which I asked the questions, and we do have a task force that addresses such things. There's a lot of correspondence from other sources than the survey.

S. Cordle: The second part of my question is self-serving depending on how you answer it: Who do you think should be doing this research and how do you think it should be paid for?

J. Mayhew: I think everybody should be doing it, and we all should pay for it. We all will anyway.

CHAPTER 3

MUNICIPAL SLUDGE MANAGEMENT

David R. Zenz and Cecil Lue-Hing

Metropolitan Sanitary District of Greater Chicago
100 East Erie
Chicago, Illinois 60611

ABSTRACT

Municipal sludge utilization and disposal is one of the most difficult problems facing municipalities today. In the United States, about 18,000 dry tons per day of municipal sludge is projected to be produced by 1985. Municipal sludge is normally handled by incineration, land application, ocean disposal, or landfilling. Incineration may not be a feasible alternative in the future unless methods can be found to keep auxiliary fuel requirements to a minimum. Land application has become increasingly popular with municipalities, because the fertilizer value of the sludge can be used. Although ocean disposal offers a simple, low-cost solution to municipal sludge agencies, its use remains uncertain until the U.S. Environmental Protection Agency (EPA) issues regulations or appeals a recent federal court ruling that found ocean disposal cannot be completely terminated by an EPA mandate. Landfilling can be a low-cost disposal method if land space is available in urban areas. Most municipal sludge can be disposed of in nonhazardous landfills because they would not be classified as hazardous under the EPA extraction procedure toxicity test. A survey of municipal sludge practices revealed five sludge management options in order of preference: (1) sale and giveaway, (2) ocean disposal, (3) landfilling, (4) land application, and (5) lagooning.

INTRODUCTION

The utilization and/or disposal of the solid residues from wastewater treatment has been and remains one of the most difficult problems that municipal wastewater treatment facilities must resolve. Although the technology available for treating municipal wastewater is sufficient, a solution to the management of the solids produced from wastewater treatment remains a perplexing problem.

MUNICIPAL SLUDGE MANAGEMENT

Table I contains estimates of municipal sludge production in dry tons per day from three sources.

Table I. Municipal Wastewater Treatment Plant Sludge Production (dry tons/d)

	Year			
Data Source	**1970**	**1972**	**1985**	**1990**
Dean[a]		10,000		13,000
CAST[b]	8,945		18,140	
Patterson[c]	10,849		19,332	

[a]Dean, R. B. 1973. "Disposal and Reuse of Sludge and Sewage: What Are the Options?" Proceedings of the Conference on Land Disposal of Municipal Effluents and Sludges, Rutgers University, NJ.

[b]Council for Agricultural Science and Technology. 1976. "Application of Sewage Sludge to Cropland: Appraisal of Potential Hazards of the Heavy Metals to Plants and Animals," Report No. 64.

[c]Patterson, J. W. 1979. "Hazardous Waste Management in Illinois," Ill. Inst. Nat. Res. Doc. No. 79/32.

Dean (1973) estimated that the national production of sludge in 1972 was 10,000 dry tons/d. By 1990, he estimated that the quantity of sludge will increase to 13,000 tons/d. His estimate, which is based on sewered population figures and per capita solids production, includes the effect of process variations such as digestion and incineration.

The Council of Agricultural Science and Technology (CAST) (1976) also prepared estimates for municipal sludge production. CAST estimated daily production in dry tons per day for 1970 and 1985 to be 8945 and 18,140 dry tons/d respectively.

Patterson (1979) also calculated sludge production using estimates of sewered population and per capita sludge productions for various wastewater treatment methods. His figures, which closely parallel those of CAST, are 10,849 and 17,332 dry tons/d for 1970 and 1985 respectively.

SLUDGE MANAGEMENT ALTERNATIVES

Only four basic types of sludge management options are available to municipal agencies: (1) incineration, (2) ocean disposal, (3) landfill operations, and (4) land spreading.

Incineration

Incineration includes all processes in which the organic solids in municipal sludge are combusted at high temperatures. About 35% (dry weight) of the sludge remains as an ash requiring disposal. The primary advantage of incineration is volume reduction.

The incineration process requires large amounts of energy unless the sludge can be dewatered sufficiently to make combustion self-sustaining without the addition of auxiliary fuel. However, existing dewatering processes are often unable to produce the solids concentration of 25 to 35% required for self-sustaining combustion.

Table II presents data collected by Olexsey and Farrell (1974) on auxiliary fuel consumption at seven large incineration facilities in the United States. The weighted-average consumption of No. 2 fuel oil needed to combust one ton of dry solids is 51.6 gal. With the present high cost of fuel and the lack of a firm supply, this option will be difficult to implement unless methods can be found to produce self-sustaining combustion.

Incineration facilities are not actually sludge disposal operations because the facility itself does not provide a final resting place for the ash produced. Landfills are usually used for disposal. In large urban centers, sufficient landfill space is becoming increasingly difficult to find, and many facilities are required to transport ash to remote sites.

New incineration facilities are faced with the problem of complying with provisions of the Clean Air Act. Under this law, new air emission sources that are placed in nonattainment areas, where the air quality standards are not being met, must first receive an "offset." A municipality must show a corresponding removal of an existing emission to

Table II. Auxiliary Fuel Consumption for Sludge Incineration in Cities During 1972[a]

City	Sludge Production (dry tons/d)	Fuel/Dry Tons of Solids[b]
Jersey City, NJ	15	28
Providence, RI	30	28
Tonawanda, NY	6	17
Rochester, NY	17	42
Hartford, CT	25	160
St. Louis, MO	55	22
St. Paul, MN	275	45

[a]Olexsey, R. A., and J. B. Farrell. 1974. "Sludge Incineration and Fuel Conservation," *News Environ. Res.* EPA, May 3.
[b]Gallons of No. 2 oil.

justify the new emission. In other words, a municipality must offset the new source by reducing or eliminating an existing source. Because municipalities usually cannot comply with this requirement, few new municipal sludge incinerators will be approved by EPA.

Ocean Disposal

Ocean disposal offers coastal cities a readily available repository for municipal sludge. This method has been practiced by many U.S. cities, including New York and Los Angeles.

Ocean disposal was scheduled for complete elimination in the United States by January 1, 1981, as a result of regulations issued by EPA under the Marine Protection Research and Sanctuaries Act (MPRSA). However, the city of New York sued the EPA over these regulations. A recent ruling by a New York federal court on the city of New York suit found that EPA had incorrectly interpreted its authority under the MPRSA and must issue regulations stating those conditions under which ocean disposal will be acceptable. The court clearly stated that EPA cannot arbitrarily prohibit all ocean disposal. As of September, 1981, EPA has not attempted to appeal the federal court ruling nor has it issued regulations. The status of ocean disposal remains uncertain until new developments occur.

Landfill Operations

Landfill operations are practiced by many municipalities. Included under this term is sludge lagooning when sludge is not intended for removal. Landfilling operations are generally very economical and, when properly operated, can be an environmentally acceptable sludge management scheme. The chief difficulties are the lack of available land in urban areas and the general public reluctance to accept landfill sites near residential areas.

In Rule 261.24 (Hazardous Waste Management System, *Fed. Regist.*, May 19, 1980), the EPA classified a solid waste as hazardous if its extract contains any of the contaminants at a concentration equal to or greater than the respective value shown in Table III. Municipal sludge could be classified as hazardous if it failed to pass the extraction procedure toxicity test.

Table III. Limits of Contaminants in Extracts from Sludges According to EPA Rule 261.24

Contaminant	Maximum Concentration (mg/L)
Arsenic	5.0
Barium	100.0
Cadmium	1.0
Chromium	5.0
Lead	5.0
Mercury	0.2
Selenium	1.0
Silver	5.0
Endrin	0.02
Lindane	0.4
Methoxychlor	10.0
Toxaphene	0.5
2,4-D	10.0
Silvex	1.0

The Metropolitan Sanitary District of Greater Chicago (District) has performed the extraction procedure on its various sludges to determine whether the resulting extracts exceeded the levels shown in Table III. The range of metals found is shown in Table IV, and that of pesticides is shown in Table V.

No extract exceeded the levels given in Table III. Therefore, no District sludge would be classified as hazardous under the EPA Interim Final Rule 261.25. From the number and type of sludges tested in large areas such as metropolitan Chicago, where industrial inputs are strong, the implication is that most municipal sludges would be classified as nonhazardous and would not need to be placed in hazardous waste landfills.

Table IV. Ranges of Metal Concentrations in Extracts from Metropolitan Sanitary District of Greater Chicago Sludges[a]

Sludge Source (No. of Samples)	Ranges of Metal Concentration (mg/L)							
	As	Ba	Cd	Cr	Pb	Hg[b]	Se	Ag
Heat Dried Sludge (21)	<0.2	<0.2	0.14-0.59	0.09-0.65	<0.02-0.23	<0.2-2.4	<0.2	<0.02
Nu Earth (6)	<0.2	<0.2	<0.02-0.22	0.19-0.42	0.06-0.23	<0.2-1.2	<0.2	<0.02
WSW Digester Drawoff (30)	<0.2	<0.2	<0.02-0.08	0.06-0.23	<0.02-0.12	<0.2-5.8	<0.2	<0.02
Calumet Digester Drawoff (30)	<0.2	<0.2	<0.02	<0.02-0.39	<0.02-0.14	<0.2-3.7	<0.2	<0.02
Hanover Park Digester Drawoff (20)	<0.2	<0.2	<0.02-0.05	0.03-0.60	<0.02-0.17	<0.2-2.7	<0.2	<0.02
John Egan Digester Drawoff (30)	<0.2	<0.2	<0.02-0.10	0.02-0.11	<0.02-0.13	<0.2-2.5	<0.2	<0.02
WSW[c] Lagoon Sludge (13)	<0.2	<0.2	<0.20-0.03	0.11-0.27	<0.02-0.12	<0.2-7.8	<0.2	<0.02
Hanover Park Lagoon Sludge (2)	<0.2	<0.2	<0.02	0.06	<0.02	1.0-2.6	<0.2	<0.02
North Side Sludge (30)	<0.2	<0.2	<0.02-0.05	<0.04-0.25	<0.04-0.16	<0.2-3.8	<0.2	<0.02
Lemont Sludge (7)	<0.2	<0.2	<0.02	<0.04	<0.04-0.07	0.2-3.7	<0.2	<0.02

[a]Extractions performed according to the procedure on pp. 33127-33128 of the *Fed. Regist.* May 19, 1980.

[b]Results in milligrams per liter, except for Hg, which is given in micrograms per liter.

[c]WSW = West-Southwest Sewage Treatment Plant.

Table V. Ranges of Pesticide Concentrations in Extracts from Metropolitan Sanitary District of Greater Chicago Sludges[a]

Sludge Source	No. of Samples	Ranges of Pesticide Concentration (mg/L)					
		Toxa-phene	Endrin[b]	Lindane	Methoxy-chlor	2,4-D	Silvex
Heat Dried Sludge	2	<0.2	<1.0	<0.1	<0.1	<0.01	<0.01
Nu Earth	2	<0.2	<1.0	<0.1	<0.1	<0.01	<0.01
WSW[c] Digester Drawoff	2	<0.2	<1.0	<0.1	<0.1	<0.01	<0.01
Calumet Digester Drawoff	2	<0.2	<1.0	<0.1	<0.1	<0.01	<0.01
Hanover Park Digester Drawoff	1	<0.2	<1.0	<0.1	<0.1	<0.01	<0.01
John Egan Digester Drawoff	1	<0.2	<1.0	<0.1	<0.1	<0.01	<0.01
North Side Sludge	1	<0.2	<1.0	<0.1	<0.1	<0.01	<0.01
Lemont Sludge	1	<0.2	<1.0	<0.1	<0.1	<0.01	<0.01

[a]Extraction performed according to procedure on pp. 33127-33128 of the *Fed. Regist.* May 19, 1980.

[b]Results in milligrams per liter, except for Endrin, which is given in micrograms per liter.

[c]WSW = West-Southwest Sewage Treatment Plant.

Land Spreading

Land spreading of sludge has become increasingly popular in the United States. It offers the advantage of recycling nutrients back to the land and can be used to reclaim lands spoiled from strip mining operations. The sludge is usually first stabilized by anaerobic digestion or other suitable means before application to the land.

Table VI gives data on the plant nutrient content of sludge from 24 cities in Illinois (Zenz et al. 1975) and seven states in the United States (Sommers et al. 1976). These data clearly show that sewage sludge is a good source of nitrogen and phosphorus and is high in calcium and magnesium.

Municipal sludge also contains large amounts of organic matter, which, when applied to land, increases soil humus content. This organic matter increases the water holding capacity of sandy soils while, on the other hand, it can be used to break up or increase the water infiltration of clay soils.

Table VI. Major Plant Nutrients Present in Various Sewage Sludge Sources (% dry weight)

Constituent	Illinois—24 Cities[a]		7 States in the United States[b]	
	Range	Mean	Range	Mean
Total N	2.6-9.8	5.4	0.03-17.6	3.2
NH_3-H	0.1-6.1	1.8	0.0005-6.7	0.7
P	0.7-4.9	2.4	0.04-6.1	1.8
K			0.008-1.9	0.3
Ca			0.1-25	5.1
Mg			0.03-2.0	0.5

[a]Zenz, D. R., B. T. Lynam, C. Lue-Hing, R. R. Rimkus, and T. D. Hinesly. 1975. "USEPA Guidelines on Sludge Utilization and Disposal—A Review of Its Impact upon Municipal Wastewater Treatment Agencies," paper presented at the 48th Annual Water Pollution Control Federation.

[b]Sommers, L. E., O. W. Nelson, and K. J. Yost. 1976. "Variable Nature of the Chemical Composition of Sewage Sludges," *J. Environ. Qual.* Vol. 5.

The process of land spreading allows urban areas to recycle the fertilizer value of sludge to the farming areas where the nutrients originated. With the current shortage of fuels and the high energy requirements for inorganic fertilizer production, land spreading offers a ready opportunity to reduce the energy requirements of the agricultural sector.

CURRENT U.S. PRACTICES IN SLUDGE UTILIZATION AND DISPOSAL

Municipal sludge management agencies use various sludge management options. The choice of a particular option will depend on factors, many of which relate to unique local situations. It is interesting to see what sludge management options are utilized by municipalities today.

The EPA (1977) conducted a survey of the municipal sludge handling practices of members of the Association of Metropolitan Sewerage Agencies. This survey included 75 individual treatment plants handling over 2700 dry tons/d of sludge. Data from this survey are in Table VII.

The largest amount of sludge was being used in sale and giveaway programs. In these programs, sludge is sold or given away to private parties, usually for use as a soil amendment (topsoil or mulch) and for its fertilizer value. This type of operation is attractive because the municipality need not be involved in the actual final disposal of the

Table VII. Disposal Methods Used by Municipal Agencies[a]

	Plants	Dry Tons	Wt %
Sale and Giveaway	8	953	34.5
Ocean Disposal	19	833	30.2
Landfill	27	461	16.7
Land Application	14	402	14.6
Lagoons	7	109	4.0
Total	75	2758	100

[a]U.S. Environmental Protection Agency. 1977. "Sludge Handling and Disposal Practices at Selected Municipal Wastewater Treatment Plants," Report 430-0-77-077.

sludge. Application costs and the need for land space are nonexistent.

Ocean disposal of municipal sludge was used for 30% of the sludge being disposed of in 1977. As noted previously, the use of this option by municipalities remains uncertain.

Landfilling was used by 17% of the municipalities studied, whereas about 15% used land application. Less than 4% of the municipalities used lagoons.

The future of various options available to sludge management agencies will be influenced greatly by future regulations and public attitudes. In the past, there has been a concerted effort on the part of EPA to be conservative in environmental regulations. Often the agency has sought zero risk rather than achieving a realistic level of risk based on the relative risk of alternative sludge management options. The general populace has, in the past, been reluctant to accept municipal sludge management operations, especially when they are located near residential areas. Governmental agencies face a large challenge in trying to educate citizens about the aesthetics of well-managed sludge management operations.

COSTS

The cost of sludge management is an important consideration, especially since the general public has become increasingly concerned about the level of governmental expenditures. Table VIII presents costs for four sludge management operations conducted by the District.

The Fulton County land application program involves the transportation of about 200 dry tons of digested sludge from the 850-mgd West-Southwest Sewage Treatment Plant to a 15,000-acre site located 200 miles southwest from the plant. The sludge is applied mainly to

Table VIII. Metropolitan Sanitary District of Greater Chicago Solids Management Costs (dollars)

Item	Fulton County Land Application	Heat Drying and Sale	Landfill	Giveaway Program
Capital Costs	25.72			
Application Costs	94.88			
Land Costs	4.99			
Taxes	4.78			
Sludge Processing	21.74	242.16	58.76	94.87
Transportation	133.00	24.92	15.83	8.79
Landfill			8.66	
Return	−2.57	−23.0		
Total	282.54	244.08	83.25	103.66

corn at 25 to 30 dry tons/acre. Transportation costs ($132/dry ton) are high because of the large travel distance involved. Return on crops from similar sites where sludge is used cannot offset the costs for sludge transportation and application.

The District also produces a dry sludge product by vacuum filtering and heat drying its sludge. This sludge is sold to a broker for sale, in turn, to private parties who mainly use it for fertilizing citrus crops in Florida. The sale of the product ($23/dry ton) does not even begin to offset the cost for production, despite the fact that no application or transportation costs are incurred. The large amount of energy to dry the sludge makes the cost of processing the sludge quite high.

Some of the sludge from the District is digested, lagooned, and then landfilled. Costs are low because the sludge processes used have low energy demands and because landfill space is now available in Chicago.

Sludge, which has been digested, sand-bed dewatered, and then air dried, is given away in truckload quantities to private parties by the District. The costs are low because the energy requirements of the processes used are low and because application costs are not incurred.

REFERENCES

Council for Agricultural Science and Technology. 1976. "Application of Sewage Sludge to Cropland: Appraisal of Potential Hazards of the Heavy Metals to Plants and Animals," Report No. 64.

Dean, R. B. 1973. "Disposal and Reuse of Sludge and Sewage: What Are the Options?" In *Proceedings of the Conference on Land Disposal of Municipal Effluents and Sludges,* Rutgers University, NJ.

Olexsey, R. A., and J. B. Farrell. 1974. "Sludge Incineration and Fuel Conservation," *News of Environmental Research,* U.S. Environmental Protection Agency.

Patterson, J. W. 1979. "Hazardous Waste Management in Illinois," Illinois Institute of Natural Resources Document No. 79/32.

Sommers, L. E., O. W. Nelson, and K. J. Yost. 1976. "Variable Nature of the Chemical Composition of Sewage Sludges," *J. Environ. Qual.* Vol. 5.

U.S. Environmental Protection Agency. 1977. "Sludge Handling and Disposal Practices at Selected Municipal Wastewater Treatment Plants," USEPA 430-0-77-077.

Zenz, D. R., B. T. Lynam, C. Lue-Hing, R. R. Rimkus, and T. D. Hinesly. 1975. "USEPA Guidelines on Sludge Utilization and Disposal—A Review of Its Impact Upon Municipal Wastewater Treatment Agencies," presented at the 48th Annual Water Pollution Control Federation.

DISCUSSION

***Jens Hansen,* Aalborg University, Denmark:** I wonder why, under this heading of Sources and Management of Wastes, you present numbers for cadmium in sludge about two orders of magnitude higher than we find in Scandinavia? You say medians of 100 to 300 ppm, and we say medians of 4 to 7 ppm.

D. Zenz: If you have a nonindustrial type of wastewater, you're probably talking about around 10 ppm of cadmium. When you get into urban areas, you very often find that cadmium concentrations are much higher. This is a problem of industrial waste enforcement; obviously, by controlling the sources of cadmium to municipal source systems, you can reduce this amount. The problem, unfortunately, in many of our urban areas here in the United States where we do have a significant amount of combined sewers, is there are many nonpoint sources in urban areas for cadmium. For example, in automobile junk yards, we have cadmium in batteries and so forth, which is a problem.

***Terry Logan,* Ohio State University:** A couple of comments relative to cadmium concentrations—the data you showed was for mean cadmium, not median; in fact, if you look at a lot of data from the United States, as we did in the north-central part of the United States, the median is about 15; what you find is a lot of very high outliers, which really increases your mean value. I think, where you have a nonnormally distributed parameter like cadmium, you're much better off looking at the median rather than the mean.

D. Zenz: I tried to make an allusion to that. That's true. When you look at it from a median point of view, many of the small plants make up the majority of the municipal wastewater treatment plants as you might suspect, and they tend to generally be very low in cadmium content.

***David Friedman,* U.S. Environmental Protection Agency:** Have you noticed over the past 5 years, for example, any change in the heavy metal concentrations in the sludges? Do people seem to be cleaning up any better?

D. Zenz: I know that in terms of the metro-Chicago area, we did a survey of metal concentration in our sludge over the last 10 years, and we have seen a reduction in the cadmium content of our sludges; it's not as low as 10 ppm, as in some of these other nonindustrial sites, but yes, there has been a reduction.

***Ralph Franklin*, U.S. Department of Energy:** Would you estimate what fraction of sludges that are produced in the country would be unsuitable for land farming or disposal because of the metal content? And secondly, is it true that most of the metals that are in the sludge get there as a result of illegal introduction into the sanitary sewage system rather than from a failure on the part of the municipality to clean it up before it gets in?

D. Zenz: Well, municipalities really don't have a viable method for cleaning up their sludges, as was mentioned by Mr. Mayhew. There really isn't any practical way of removing metals from sludges once they're in the municipal sludge. It is just not a practical situation. You have to rely on an industrial waste standard. As far as which sludges are suitable for land application and which are not, it's really not in the present regulations. As I pointed out earlier, the regulations limit the amount of sludge that you can put on; they do not really discuss the cadmium content. Obviously, of course, the higher the cadmium content of the sludge, the lower the amount that can be put on land. However, there isn't a strict prohibition one way or the other.

***Chet Francis*, Oak Ridge National Laboratory:** One part of RCRA is the EP toxicity testing, and you have carried out your toxicity testing on the sludges. For example, most of the sludges that you have tested are nonhazardous; therefore, you have the freedom to put it in landfill. The other part of RCRA, as I understand it, is if you have a cadmium sludge or sludge that had high concentration of cadmium, but it does not exceed the cadmium concentrations to make it toxic, you have to live with a rather low standard in the groundwater. I think the concentration of cadmium should not exceed 0.01 ppm in groundwater at the landfill site.

D. Zenz: Correct. Part of the regulations are that you have to provide groundwater monitoring and check that for cadmium content. That is correct.

C. Francis: It seems to be a concern that if a sludge has 20 ppm of soluble cadmium, the concentration of the cadmium in the groundwater could very easily exceed 0.01 ppm if you don't have good soil attenuation and dilution there. Have you had any experience with monitoring the groundwater?

D. Zenz: I don't really think that soluble cadmium, unless it was a very unusual sludge, would approach that kind of amount. Most of the soluble cadmium is very, very low in sludges. Most of it is tied up in some kind of organic complexes, sulfide, or other things. You don't see those kinds of soluble amounts. You also have to realize that the amount of cadmium being applied in these situations is still relatively low given the soil capacity to absorb these metals. The studies that have been done show that the capacity of the soil to absorb these metals is extremely great and does not appear to be a major problem. I'm not saying that in the specific situation you couldn't have a problem. And again I didn't put this on the screen, but that is part of the regulations, as well as the amount to be applied. There is a strict regulation for groundwater monitoring, at least at the boundary of the land application site. I would seriously doubt, unless you had a very unusual situation, limestone formations, fissures, and so forth, that you would have a groundwater monitoring problem for cadmium. That does not appear to be a big concern at this point. I think more of the concern has been in the application of cadmium in the growing crops and the uptake of that cadmium into the crop and then into the food chain.

CHAPTER 4

COAL-FIRED POWER PLANT WASTE MANAGEMENT

John S. Maulbetsch and Ishwar P. Murarka
Electric Power Research Institute
3412 Hillview Avenue
Palo Alto, California 94304

ABSTRACT

Coal-fired power plants produced 45% of the 2.2 × 10^{12} kWh of electricity generated in 1978; this is expected to increase to 49% of a projected 3.5 × 10^{12} kWh by 1988, a 77% increase. Over 63 million tons of solid wastes were produced (primarily ash and scrubber sludge) in 1979. These wastes are disposed of on land; potential environmental effects are airborne (fugitive emissions) and waterborne (runoff to surface waters; leachate and permeate to groundwater).

This chapter summarizes briefly what is known and what needs to be known about the wastes and their characteristics; what constituents enter the environment from disposal facilities; where the wastes go and at what rate they arrive; whether potential harm exists from their eventual distribution; and what the methods and costs of treating, disposing of, and containing the wastes may be. A brief summary of utility disposal practices and current regulatory issues is given, and data and sources of data are provided.

INTRODUCTION

The generation of electricity in steam-electric power plants fueled by coal is an inherently dirty process, with the potential for environmental effects. The large quantities of coal, water, and air that must be processed through the plant result in large flows of stack gas and aqueous discharge and large amounts of solid residuals. The environmental effects and the applicable regulations have traditionally been considered in the separate categories of air, water, and solid waste. The air and water issues and the associated control technologies have been long established, but the solid waste area is the most recent subject of regulatory and scientific attention. This is due both to the recent attention paid to protecting the quality of the nation's groundwater and to the increasing awareness that solid residuals are the inevitable result of the cleanup process applied to the gaseous and aqueous discharges.

This chapter addresses the subject of managing the solid by-products produced by coal-fired power plants in the following sequence: A brief description of the coal-fired segment of the utility industry is provided to bracket projected coal utilization and quantities of waste that must be managed. The important physical and chemical characteristics of the waste are noted and references to data are cited. The regulatory and environmental effects issues are summarized. Finally, the present and projected waste management practices of the industry and the associated costs are presented.

UTILITY SOLID WASTES

The solid-waste management requirements of the utility industry are noteworthy because of both the sheer size of the industry and the large quantity of waste material produced. In 1980, the industry consisted of approximately 576,000 MW(e) generating capacity, of which 237,000 MW(e), or 41%, was coal-fired (EPRI, 1981c). The coal-fired units delivered over 50% of energy generated (1162×10^9 kWh out of 2286×10^9 kWh). Projections to 1988 anticipate an increase of over 50% in the total generation to approximately 3500×10^9 kWh (EPRI, 1982). The coal-fired capacity would again provide approximately one-half of the total. In terms of coal-consumption, the present utility coal usage of nearly 500 million tons per year would increase to 700 to 800 million tons per year, accounting for approximately two-thirds of the national coal consumption. The combustion of such quantities of coal produces similarly large quantities of solid residues.

The major solid waste by-products resulting from coal-fired power plants are ash from the burning of the coal and sludge from flue gas desulfurization. Ash production rates from a 10%-ash coal are approximately 1 ton/MW(e)·d of which approximately 75% is fly ash and

25% is bottom ash. In 1979, this amounted to a national total of over 60×10^6 tons (EPRI, 1982). Scrubber sludge is produced at approximately 1.5 tons of dry solid/MW·d [for 3.5% sulfur coal scrubbed to 1979 New Source Performance Standards (NSPS)]. With sludge moisture content ranging from 20 to 70% solids, the sludge volume can approach 5 acre·ft/d for a single 1000-MW(e) plant. In 1979, over 4×10^6 tons of sludge were produced from less than 30,000 MW(e) of scrubbed capacity (EPRI, 1982).

Physical and chemical characteristics of ash and sludge affect their potential environmental impacts and the methods suitable for their disposal. Fly ash is produced as a fine, noncombustible residue carried off in the flue gas with relatively uniform particle size distribution in the 1- to 100-μm range; bottom ash, on the other hand, leaves the furnace as a solid granular ash (or molten slag, which later solidifies) with a well-graded size distribution for 10-1000 μm. Fly ash is susceptible to fugitive emissions from dusting; the very high surface area-to-volume ratios of the fly ash particles make the trace elements susceptible to leaching. However, the permeability of a well-constructed fly ash landfill, compacted at a near-the-maximum density is low (10^{-4} to 10^{-7} cm/s) as compared to the more permeable but less leachable bottom ash (10^{-1} to 10^{-2} cm/s).

Trace element concentration in ash, and particularly in ash leachate, is of great interest because of its relationship to regulatory criteria under the Resource Conservation and Recovery Act (RCRA) regarding toxicity and hazardousness. Trace element concentrations are highly variable, depending on coal source and furnace type; moreover, measurements, particularly of leachate concentrations, are very uncertain (EPRI, 1979,1981b). However, data are widely available and are summarized in EPRI (1981a).

Scrubber sludge characteristics are a function of the scrubber type, scrubbing chemistry, and dewatering methods used. The physical nature can range from dry, stackable gypsum to thixotropic sulfite sludges. Again, extensive data and literature are available for both physical and chemical properties (EPRI, 1980).

Of equal or greater interest in disposal purposes are characteristics of ash-sludge mixtures. The addition of fly ash, particularly alkaline fly ash, normally results in increased strength, increased density, and reduced permeability. Data are summarized in EPRI (1981a,b), and a more detailed laboratory study is reported in EPRI (1980b).

In addition to the high-volume wastes, many waste streams from power plants are primarily aqueous in nature. These streams may be considered and regulated as solid waste if they are not discharged under a National Pollutant Discharge Elimination System (NPDES) permit or are to be treated prior to discharge, with the resultant production of a sludge which must be disposed of. Examples include water treatment sludges from clarifiers, softeners, ion exchangers, evaporators, or filters; cleaning wastes from boiler fire- and water-side wastes, air preheater wastes; cooling-tower basin sludges; and miscellaneous sanitary wastes, laboratory drains, sampling streams, screen backwashes, floor drains, and area runoffs.

Estimates of the amounts of these wastes are highly uncertain. A recent study (*Engineering Science*, 1979) places these at 0.2 to 2.5% of the combustion wastes in the range of 100,000 to 1,750,000 tons per year.

DISPOSAL ISSUES

The primary environmental regulations governing the disposal of solid wastes stem from RCRA. With respect to disposal, it classifies all waste into hazardous and nonhazardous categories, on the basis of specific criteria, and establishes disposal practices for each category, primarily motivated by the aim of protecting groundwater. These practices involve the location, isolation, monitoring, and closure of disposal sites, and for hazardous wastes, they require elaborate manifesting and record-keeping procedures. Large-volume utility wastes (ash and sludge) are temporarily exempt from the regulation under hazardous disposal guidelines, pending full-scale environmental studies of existing disposal practices and determination of the costs of more stringent control. This study is mandated by the 1980 Congressional Amendment to RCRA, and the study has already been contracted by EPA for completion by the end of 1982.

Some specific low-volume wastes, such as PCBs or PCB solids are separately regulated under Toxic Substance Control Act and are subject to particularly stringent disposal requirements. The legislation and resulting regulations are the result of concerns for protection of the environment and human health. Literature and data documenting any such demonstrated effects are sparse.

Environmental Assessment

Ecological and human health effects of land disposal are caused by the types and quantities of substances that are released from disposal sites. Unfortunately, very little information is available to define what constitutents are released, how they are transformed in the environment, and where they accumlate in land, surface, or groundwaters.

Ecological Effect

In an accidental spill from a fly ash pond, fish populations and insect communities of a small river were drastically reduced (Cairns et al., 1972). The principal lethal agent was the high pH (pH > 12). Two years after the spill, the small river had recovered its biotic communities, apparently leaving no permanent damage. Another study indicated that runoff from alkaline ash-settling ponds was lethal to catfish because of high pH, which caused precipitation of ferritic hydroxide, which in turn resulted in clogged gill mucous of the fish (Wasserman et al., 1974).

Cumbie and Van Horn (1978) found that fish populations of Belews Lake, North Carolina, were reduced because of reproductive failures which were caused by unusual selenium accumulation (>3 mg/kg wet wt) in the fish. The selenium entered the lake in soluble form via the power plant fly-ash sluice water return. These studies suggest that although the fly ash could pose ecological problems, it has shown toxic effects on organisms. However, the overall effects on ecosystems are not well understood. To develop this understanding, research is being conducted at Electric Power Research Institute to evaluate the effects of fly ash on terrestrial agrosystems, using a laboratory microcosm and field plots. Similar research may be necessary to study the effects of fly ash on aquatic ecosystems.

Human Health Effects

A recent report stated that there are three different ways fly ash can be hazardous to humans: (1) by direct action on external surfaces (e.g., skin, eyes), (2) action in the respiratory tract, and (3) action in the alimentary canal (Roy et al., 1981). Undoubtedly, many inorganic elements (e.g., Se, As, Cr, Cd, Co) are extremely toxic to many organisms (Waters et al., 1975; Eichorn et al., 1973; Hartung, 1973), and their presence in the solid wastes from fossil fuels (Roy et al., 1981) is known. Therefore, adding the toxic elements to the environment through fly ash could present a potential for human health hazard. Again, most of the studies on effects of fly ash are carried out in the laboratories with animals.

Hogue et al. (1980) observed that selenium in the blood of sheep was increased by a factor of 3 when the sheep were fed rations containing 10% fly ash. However, no adverse effects on the animals were observed. Fisher et al. (1979) concluded that fine fly ash was more mutagenic than coarse fly ash and that the fly ash from an electrostatic precipitator was not mutagenic. These studies indicate that the effects are not well understood and that conflicting results can be found in the literature.

DISPOSAL PRACTICES

The selection of a waste management strategy requires a series of decisions. The first of these is a choice between sale for reuse and simple disposal. The choice may be made separately for bottom ash, fly ash, and scrubber sludge. Although no large-scale utilization options are currently practiced for sludge, the decision to sell the ash may affect the sludge disposal method, because ash would no longer be available to mix with the sludge as a stabilizing agent.

If it is decided to dispose of the material as a waste product, the choice of an overall disposal strategy includes decisions on codisposal vs separate disposal of ash and sludge, and wet ponding vs dry landfill. These decisions are made on the basis of land availability, waste properties, local hydrology and geology, and local environmental regulations.

Current Practice

On a national average, the split between wet and dry ash disposal is almost even on a weight basis. Of 58.4 million tons produced in 1979, 9.7 million tons were utilized, whereas 25 million tons were placed in a wet disposal and 24 million tons were placed dry in landfill or mine disposal sites. There are substantial regional differences, however, with the Southeast using primarily wet disposal (81%) and the West and Southwest using primarily dry (70%) (EPRI, 1982). A relatively small (14%), but growing percentage of ash is codisposed with sludge, mostly in dry disposal, for stabilization purposes.

The trend appears to be toward dry disposal. This results from increasing emphasis on groundwater protection, potential imposition of zero discharge limits on certain metals from ash transport water, and a desire to preserve the ash for future reuse.

As of late 1979, 62 units representing 20,630 MW were scrubbing sulfur dioxide and producing sludge. Of the 62 units, 37 were using wet disposal, and 25 were using landfills (EPRI, 1982).

Many of these units have early generation scrubbers and, in the absence of other experience, the sludge disposal systems were chosen on the basis of how ash disposal had been handled. In contrast, of 51 units with scrubbers under construction and under contract or letter of intent as of late 1979, 33 will use dry disposal; 17 will use wet; and 1 was undecided.

Costs of Disposal

The effect of increasingly strict disposal regulations and increasingly complex disposal practices has been estimated (EPRI, 1982). First, a survey of the disposal practices of most, large [>200 MW(e)] coal-fired plants categorized the industry's disposal practices into 30 separate categories covering three site types (pond, landfill, and mine) and ten cases of separate or codisposal [bottom ash, fly ash, and scrubber sludge (fixed, unfixed)—separately and in all combinations]; results of the survey are summarized in Table I. Eight case studies were performed from which costs in each of the 30 categories were interpolated. The case studies included a reestimation of costs of current disposal

Table I. Current Solid-Waste Disposal Unit Costs (Annual, levelized 1980 dollars per ton)

Waste Stream	Disposal Method		
	Pond	Landfill	Mine
Bottom Ash	28.10[a]	28.10	No[b]
Fly Ash	28.10	28.10	5.82
Sludge, fixed	6.90	No	No
Sludge, unfixed	6.90	10.12	No
Bottom Ash and Fly Ash	14.24	7.82	5.82
Bottom Ash and Sludge	No	No	No
Fly Ash and Sludge, fixed	12.28	6.93	No
Fly Ash and Sludge, stabilized and unfixed	12.28	5.10	No
Bottom Ash, Fly Ash, and Sludge, fixed	12.28	5.10	4.88
Bottom Ash, Fly Ash, and Sludge, stabilized and unfixed	12.28	5.10	No

[a]Apparently, the unit costs for separate disposal of bottom ash and fly ash in ponds or landfills are somewhat higher than would normally be anticipated, largely due to typical site-specific characteristics of the case study (No. 6). However, they were the most representative costs available within the scope of this study (EPRI, 1982).

[b]No = no occurrence.

practices in 1980 dollars, as well as an estimate of the cost of facilities sited, built, and operated in accordance with both nonhazardous guidelines and hazardous regulations. Nonhazardous costs were estimated both with and without site lining for each case. The costs expressed in annual, levelized 1980 dollars per ton are shown in Tables I and III.

Table II. Solid Waste Disposal Unit Costs, Nonhazardous Scenario (annual, levelized 1980 dollars per ton)

	Disposal Method[a]					
	Pond		Landfill		Mine	
Waste Stream	Lined Site	Unlined Site	Lined Site	Unlined Site	Lined Site	Unlined Site
Separate Disposal						
Bottom Ash	115.89	80.25	110.62	74.87	No	No
Fly Ash	115.89	80.25	110.62	74.87	7.00	6.42
Sludge, fixed	25.10	9.99	No	No	No	No
Sludge, unfixed	25.10	9.99	12.34	11.93	No	No
Codisposal						
Bottom Ash and Fly Ash	53.29	41.08	8.68	8.45	7.00	6.42
Bottom Ash and Sludge	No	No	No	No	No	No
Fly Ash and Sludge, fixed	40.64	24.22	19.57	11.37	No	No
Fly Ash and Sludge, stabilized, unfixed	40.64	24.22	14.35	6.37	No	No
Bottom Ash, Fly Ash and Sludge, fixed	40.64	24.22	14.35	6.37	5.85	5.35
Bottom Ash, Fly Ash and Sludge, stabilized and unfixed	40.64	24.22	14.35	6.37	No	No

[a]No = no occurrence.

On the basis of these figures, current total industry costs for the disposal of 63 million tons of coal combustion wastes in 1980 are $0.8 billion. Applications of the full complement of hazardous waste rules would increase these costs to approximately $3.4 billion. Nonhazardous guidelines would result in intermediate costs, ranging from $1.9 to $2.7 billion, strongly dependent on the fraction of cases for which liners are assumed to be required.

CONCLUSIONS

The choice of the appropriate waste management techniques and the associated cost depend on the degree of containment deemed necessary

Table III. Solid-Waste Disposal Unit Costs, Hazardous Scenario (Annual, levelized 1980 dollars per ton)

Waste Stream	Disposal Method		
	Pond	Landfill	Mine
Bottom Ash	149.79	132.24	No
Fly Ash	149.79	132.24	9.53
Sludge, fixed	38.23	No	No
Sludge, unfixed	38.23	12.46	No
Bottom Ash and Fly Ash	67.54	8.81	9.53
Fly Ash and Sludge, fixed	57.99	21.50	No
Fly Ash and Sludge, stabilized and unfixed	57.99	16.52	No
Bottom Ash, Fly Ash, and Sludge, fixed	57.99	16.52	7.93
Bottom Ash, Fly Ash, and Sludge, stabilized and unfixed	57.99	16.52	No

[a]No = No occurrence.

to provide adequate protection for the environment, primarily groundwater aquifers. The techniques for any specified level of containment are available at a price. The greatest uncertainty lies in our ability to predict which contaminants leave a disposal facility, at what rate they enter the groundwater, and where they go. Development of validated predictive models or conducting a sufficient number of monitoring programs to reduce that uncertainty is a formidable task.

REFERENCES

Cairns, J., J. S. Crossman, and K. L. Dickson. 1972. "The Biological Recovery of the Clinch River Following a Fly Ash Pond Spill," in *Proceedings of the 25th Industrial Waste Conference*, pp. 182-192.

Cumbie, P. M., and S. L. Van Horn. 1978. "Selenium Accumulation Associated with Fish Mortality and Reproductive Failure," *Proceedings of the Annual Conference of the Southeast Association of Fish and Wildlife Agencies, Vol. 32*, pp. 612-624.

Eichorn, G. L., et al. 1973. "Some Effects of Metal Ions in the Structure and Function of Nucleic Acids," in *Metal Ions in Biological Systems: Advances in Experimental Medicine and Biology, Vol. 40*, S. K. Dahr (ed.), pp. 43-66.

Engineering Science. 1979. "Evaluation of the Impacts of Proposed RCRA Regulations and Other Related Federal/Environmental Regulations on Utility and Industrial Sector Fossil Fuel–Fired Facilities, Phase I—Utility Sector," Interim Report, prepared for the U.S. Department of Energy.

EPRI. 1979. "ASTM Leachate Test Evaluation Program," EPRI FP-1183.

EPRI. 1980a. "FGD Sludge Disposal Manual–2nd ed.," EPRI CS-1515.

EPRI. 1980b. "Studies of Long-Term Chemical and Physical Properties of Mixtures of Flue Gas Cleaning Wastes," EPRI CS-1533.

EPRI. 1981a. "Coal Ash Disposal Manual–2nd ed.," EPRI CS-2049.

EPRI. 1981b. "Extraction Procedure and Utility Industry Solid Waste," EPRI EA-1667.

EPRI. 1981c. "EPRI 1982-1986 Research and Development Program Plan: Overview and Strategy," EPRI P-2156-SR.

EPRI. 1982. "Engineering Evaluation of Projected Solid Waste Disposal Practices," EPRI CS-2627.

Fisher, G. L., C. E. Chrisp, and O. G. Raabe. 1979. "Physical Factors Affecting the Mutagenicity of Fly Ash from a Coal-Fired Power Plant," *Science* 204:879-881.

Hartung, R. 1973. "Biological Effects of Heavy Metal Pollutants in Water," in *Metal Ions in Biological Systems: Advances in Experimental Medicine and Biology, Vol. 40,* S. K. Dahr (ed.), pp. 161-172.

Hogue, D. E., et al. 1980. "Soft Coal Fly Ash as a Source of Selenium in Unpelleted Sheep Rations," *Cornell Vet.* 70:67-71

Roy, W. R., R. G. Thiery, R. M. Schuller, and J. L. Suloway. 1981. "Coal Fly Ash: A Review of the Literature and Proposed Classification System with Emphasis on Environmental Impacts," Environ. Geol. Notes 96, Ill. State Geol. Surv.

Wasserman, C. S., M. S. Chung, and D. B. Rubin. 1974. "Environmental Responses of the Secondary Discharges from Fly Ash Ponds," NTIS FE-1536-30, Vol. 2, p. 45.

Waters, M. D., et al. 1975. "Metal Toxicity for Rabbit Alveolar Macrophages in Vitro," *Environ. Res.* 9:32-47.

DISCUSSION

***Robert Griffin,* Illinois State Geological Survey:** During this past year our laboratory has sampled fly ash from all the major power plants in Illinois, and we found that about 10% of the fly ashes tested out as a hazardous waste according to the EPA extraction procedure. Could you comment on the impact to the utility industry if that trend were to hold up nationwide?

J. Maulbetsch: I am not familiar with your data. You say 10% tested out as a hazardous waste; is that on the basis of 100 times primary drinking water? (Yes.) I am surprised; I have not seen your data, and I have not seen data from other sources that support that. I cannot give you a real dollar figure relative to the disposal costs if 10% of the ash disposal capacity of the utility industry had to be treated as hazardous waste. However, for those 10% the economic costs would be enormous. In the study that we recently completed, the cost was estimated to be about $800 million for the 60 million tons under current practices. If liners were required, we are talking about perhaps a tripling of that cost.

***John Brown*, Nottingham, England:** This is a comment, not a question. I don't think you should get hung up on acetic acid tests. When the EPA program was started, we were invited to comment on the actual nature of the program, and we pointed out that the mechanics of the percolation of water through ash determines what comes out at the other end. To do a simple acetic acid test and result in the classification of large tonnage of hazardous waste is in my view misguided, either nationally or from the point of view of the industry. Also, where are you going to put it all?

CHAPTER 5

CEGB EXPERIENCE IN FUEL ASH DISPOSAL

John Brown

Central Electricity Generating Board
Midlands Region
Scientific Services Department
Ratcliffe-on-Soar
Nottingham, England

ABSTRACT

The Central Electricity Generating Board (CEGB) coal-fired power stations produce about 12×10^6 t of ash per year. After sales for commercial use, about 7×10^6 t are left for disposal to landfill sites. The environmental aspects of CEGB experience in this area are described.

In the handling of ash in wet disposal systems, the discharge of water after use to the appropriate water course is governed by consents set by water authorities. With good design and operation of lagoons, the limits are achieved with reasonable margins. The potential problem of dust nuisance as the ash dries out can likewise be adequately controlled by good practice.

The possible impact of ash disposal on groundwaters is governed by several factors—the ash itself, the size of the area to be reclaimed, the rainfall pattern, and the local hydrogeology, including the size and rate of recharge of the aquifer itself. Each potential site needs to be considered on its merits. The basic CEGB approach is discussed; it requires the combination of laboratory studies of percolate quality with hydrogeological assessments to evaluate potential risk, if any.

The successful restoration of various derelict or low-grade sites to agricultural use has been achieved with obvious beneficial effects on visual amenity. Reclamation with pulverized fuel ash (PFA) can usually be achieved with minimal application of topsoil, which has obvious economic benefits. In the early stages, attention to the nutrient status of the ash is required, and appropriate selection of crops is practised until the site is weathered and established. In 25 years of experience with such reclamation practice, no adverse effects have been observed in man or animals from trace element uptake in growing crops on PFA, and this is consistent with the associated scientific studies in this area.

INTRODUCTION

In the United Kingdom, the Central Electricity Generating Board (CEGB) is the authority responsible for the generation and primary distribution of electricity for public supply in England and Wales, which include a land area of about 58,000 square miles. The annual demand for electricity is now about 2.1×10^{14} Wh, of which more than 80% is now met by coal-fired plants. This represents a coal burn of 80×10^6 t, and the ash contents of the available varieties are such that about 12×10^6 t of ash is produced in the process. As is typical for conventional pulverized-fuel firing, about 75% of the material is fine dust (PFA), and the rest is furnace bottom ash.

Depending on the level of activity in the construction industry, about 40% of the ash is sold for commercial use. The applications include building block manufacture, partial replacement of cement in concrete, production of lightweight aggregates, use as a grouting material, and road building. Promoting such use for power station ash is doubly important because, in addition to alleviating a disposal problem, the ash replaces materials that would otherwise have to be excavated, thus producing attendant effects on the environment.

The 7×10^6 t or so of ash that is not used commercially must be disposed of in some other way. Fortunately, the properties of PFA make it particularly attractive for land reclamation, and most of the surplus ash is used to reinstate areas such as old gravel workings and clay pits so that the land can be returned to agricultural or amenity use. In the United Kingdom several hundred sites of varying sizes, covering a land area of about 15 km^2, have already been reclaimed in this way.

This chapter describes CEGB experience relating to the environmental aspects of ash disposal and discusses both the disposal process itself and the subsequent use of the reclaimed site.

ASH HANDLING AND DISPOSAL

Because ash is required in the dry condition for most commercial purposes, special arrangements are made at the power stations for its transfer to sealed road or rail wagons for transport to the point of use. However, ash disposal systems are highly site-specific, depending in part on the distance to the point of disposal. Most schemes involve hydraulic methods at some stage, and the first environmental consideration is to procure a suitable supply of water for handling large volumes of material as slurry and to return the process water to some water course in an acceptable condition after use.

The CEGB practices three basic variations in hydraulic disposal (Wright and Brown, 1979). If lagoons are available indefinitely on or near the generation site, the PFA is extracted from storage silos hydraulically and, after being mixed to give a handleable slurry, it is pumped through relatively short pipelines to the local lagoon. When such a lagoon is full, it is allowed to drain for a period, after which it is emptied by excavator. The contents are then transported by road to fill and reclaim some appropriate area. Where permanent ash disposal grounds are available at intermediate distances from the station site (approximately 10 km), the slurried ash is pumped directly to the disposal area by using long pipelines with booster stages if necessary. In this case the water course used as the source of supply may differ from that used to receive the decanted water (lagoon effluent). When the disposal area is distant from the power stations, as at the Peterborough site used by the CEGB, the ash is transported dry in sealed rail wagons for distances of up to 130 km; it is then slurried directly from the rail wagons on arrival before disposal to the lagoon being worked. A peculiarity of the Peterborough site is that, because makeup water is in short supply, the process water is stored in a central reservoir and is continuously recycled. In all hydraulic handling schemes, the second environmental consideration is the potential dust nuisance that may arise as a lagoon is filled and dries out, especially the 1% of the ash that forms hollow spheres and floats on the surface of the water.

At smaller power stations, a more economic method of disposal might be to condition the ash with 10 to 20% water to improve handling and prevent wind-blown dust nuisance, and then to transport it by road to the disposal site. Ash is also handled in the conditioned state at the relatively new Drax Power Station, where wasteland adjacent to the station is being reclaimed in a landscaped mound. In this case, the conditioned ash is transported by conveyor belt and is graded and consolidated immediately by conventional vehicles such as bulldozers and scrapers.

ENVIRONMENTAL CONSIDERATIONS

Characteristics of PFA

The PFA produced by the burning of finely pulverized coal is composed of fine particles, which are largely spherical and glassy in nature. The particle size distribution of the PFA is similar to that of a light, silty soil. About 60% of the particles are within the fine sand category (0.2 to 0.02 mm), and about 40% are within the silt category (0.02 to 0.002 mm). Moreover, although coal ash contains much of the inorganic material present in the plants from which the coal was formed, the bulk of the material is extraneous mineral matter mined with the coal and resembling the soil in which the coal-forming plants grew. Consequently, PFA is similar in general composition to many present-day mineral soils. These properties make the material particularly attractive as a basis for land reclamation for agricultural use.

Chemically, the material is largely composed of insoluble combined alumina, silica, and other oxides. It contains only a small amount of readily soluble material (about 2% w/w) and lesser amounts of material that must diffuse to the surface before it can be dissolved at all. In addition, the dry ash particles are not uniform in composition. After formation in the main furnace chamber, the particles pass through a series of convective heat exchangers in contact with the combustion gases until they are collected in the electrostatic precipitators. The dry ash collected, therefore, has a thin outer layer, which is effectively an acid sulfate, surrounding an alkaline core that contains some free lime. Consequently, although most of the soluble matter would finally appear in water as sodium, potassium, and calcium sulfates, when water is first added to the dry ash, the more-acid layer dissolves first and the water is temporarily on the acid side of neutral. The alkaline core then reacts with the outer material in solution for typically about 1 h until the water becomes alkaline. This behavior is illustrated in Figure 1 for the main parameters. The exact shape of the curve, the time taken to reach neutrality, and the final pH depend on the source and particle size of the ash and the water/ash ratio. This behavior may influence not only the quality of the discharge from an ash-handling scheme to the environment but also the technical problems that must be considered in the design of such systems.

In addition to the major elements already mentioned, PFA contains significant amounts of trace elements, of which only a few milligrams per kilogram are soluble. Although the amounts of such material are

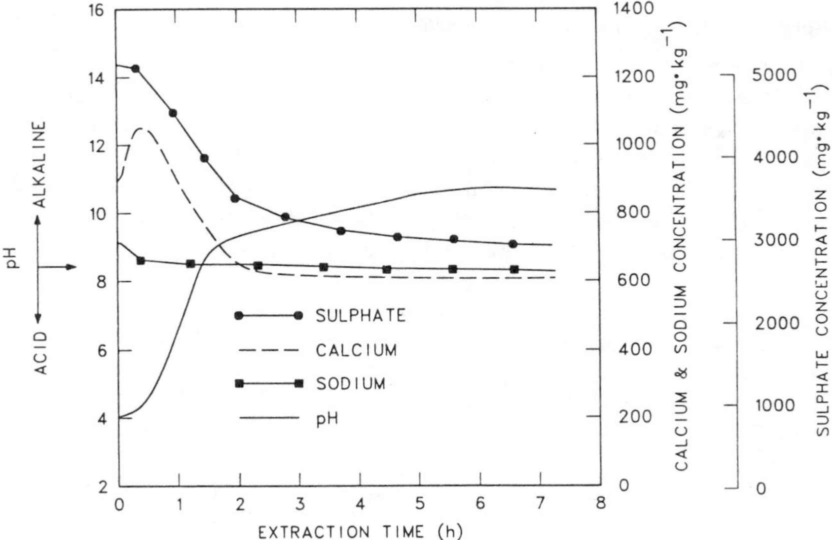

Figure 1. Extraction of 40 parts (by wt) of West Burton PFA with 60 parts water.

very small in relative terms, they are important in the context of current legislation regarding discharges to surface waters and groundwaters. The chemical compositions of three fairly typical British ashes studied in this context are shown in Table I (Brown et al., 1976).

Ash Lagoons

Lagoon Effluents

Ash can be readily handled as a slurry at concentrations up to 40% by weight, but in most hydraulic disposal schemes the average burden is probably nearer to 20%. Depending on whether the scheme is local to or distant from the power station, the slurry may be in the pipeline for times ranging from a few minutes to a few hours. These two factors and the composition of the ash determine the makeup of the slurry discharged to the lagoon and thus the quality of the water decanted after the ash has settled.

In most hydraulic handling systems the process water is taken from a nearby water course, usually a river, and is used only once before being returned to a water course as lagoon effluent. The conditions for the return of this effluent are set out in consents granted by the responsible water authority, and for PFA these include pH, suspended solids, and total trace metals (see Table II). Despite the range of ashes encountered and the differences already noted in the ash-handling systems, there is little difficulty meeting these consent conditions. A typical ash lagoon effluent composition is shown in Table III along with the original river water composition (Ray and Jenkins, 1979). The main difficulty encountered is meeting the stringent levels for

Table I. Ash Compositions from Different Coal Sources

Element or Radical	Units of Measure	Source of Ash		
		1	2	3
Aluminum	ppm Al	145,000	140,000	120,000
	% Al_2O_3	27	27	23
Arsenic	ppm As	110	80	70
Cadmium	ppm Cd	5	6	8
Calcium	ppm Ca	22,000	13,000	28,000
	% CaO	3.1	1.8	3.9
Chromium	ppm Cr	120	120	100
Copper	ppm Cu	210	160	120
Iron	ppm Fe	58,000	56,000	45,000
	% Fe_2O_3	8.2	8.0	6.5
Lead	ppm Pb	110	90	200
Magnesium	ppm Mg	10,000	9,600	16,000
	% MgO	1.7	1.6	2.7
Manganese	ppm Mn	600	300	1,300
	% Mn_3O_4	0.08	0.04	0.18
Molybdenum	ppm Mo	23	19	8
Potassium	ppm K	23,000	31,000	21,000
	% K_2O	2.8	3.8	2.6
Selenium	ppm Se	5	3	12
Sodium	ppm Na	12,000	10,000	6,700
	% Na_2O	1.6	1.4	0.9
Titanium	ppm Ti	5,800	5,800	5,300
	% TiO_2	1.0	0.8	0.9
Zinc	ppm Zn	130	90	300
Boron	ppm B	210	170	150
Chloride	ppm Cl	200	200	300
Phosphate	ppm PO_4	3,500	1,800	2,900
	% P_2O_5	0.26	0.13	0.22
Silicate	ppm SiO_2	510,000	480,000	500,000
	% SiO_2	51	48	50
Sulfate	ppm SO_4	19,000	14,000	6,300
	% SO_3	1.6	1.1	0.5

Table II. Typical Consent Conditions for Discharges of Trade Waste into Rivers

Parameter	Limit
Biochemical Oxygen Demand	20 mg/L
Arsenic, Cadmium, Chromium, Copper, Lead, Nickel, and Zinc	1.0 mg/L (as total or individually)
Cyanide	0.1 mg/L
Nonvolatile Matter Extractable by Carbon Tetrachloride (e.g., oil)	5 mg/L
pH	5 to 9
Temperature	30°C
Suspended Solids	30-100 mg/L (depending on type of discharge and flow rate)

suspended solids at the outfall from the lagoon when the lagoon is approaching the full condition. Designing the lagoon system in cascade and placing an additional settlement lagoon between the main area and the discharge point ensures the high quality of the decanted water.

Prevention of Dust Nuisance

About 1% of the ash particles float on the surface of the lagoon and may cause dust nuisance if picked up by the wind as they dry out.

Table III. Comparison of a Typical PFA Lagoon Effluent with River Water

Element	Ash Lagoon Discharge to River (mg/L in solution)	River at Station Inlet (mg/L in solution)
Arsenic	0.07	<0.05
Boron	2	0.3
Cadmium	<0.01	<0.01
Chromium	<0.05	<0.05
Copper	<0.1	<0.1
Cyanide	<0.1	<0.1
Fluoride	1.3	0.3
Iron	<0.05	<0.05
Lead	<0.05	<0.05
Manganese	<0.1	<0.1
Molybdenum	0.3	<0.05
Selenium	<0.05	<0.05
Sulfate	530	230
Zinc	<0.1	<0.1
pH	8.0	7.7
Suspended Solids	5-20	10-100

Because this material, consisting entirely of hollow spheres, is a particularly valuable source of extremely lightweight insulation, CEGB attempts to skim it from the lagoons and stockpile it for sale either directly or through subcontractors. When a lagoon is full and the main body of the ash begins to dry out, there is a further possibility of dust nuisance before final reclamation. The permanent solution is to reclaim the lagoon for its final purpose as soon as possible, but until this stage is reached, several mechanical palliatives are available for temporarily sealing the surface of the ash. One involves spraying the surface with an organic material in solution, which forms a film on drying out. For this purpose the solution is colored so that the progress of the spraying can be clearly seen. In another technique known as hydroseeding, a light coating material mixed with grass seed is sprayed on the surface, and the coating holds the surface together until the grass can germinate and stabilize the surface as it grows.

Disposal of PFA in Water Supply Catchment Areas

For many ash-disposal schemes, the only interaction with the aqueous environment that must be considered is the surface water aspect. (Landscaped mounds built up on waste ground are a special case and will be considered later.) However, where ash disposal is to occur over or near potable groundwaters, the considerations are very different.

If rainfall percolates through a bed of ash, it will displace any water already held there, which is in equilibrium with the ash, and will itself be able to pick up any remaining soluble matter. As would be expected, the initial composition of the percolate from an ash bed is considerably more concentrated than an ash lagoon effluent and will generally exceed the permitted composition for potable waters in several respects. Therefore, the local authority that issues the licences for ash disposal and the appropriate water authority must be assured that a particular scheme will not constitute a risk to nearby groundwater. When queries began to arise in the early 1960s, virtually no experience was available on which to base such applications. The CEGB therefore conducted a substantial program of investigation into the mechanics and chemistry of percolation in PFA beds to provide such a basis (Brown et al., 1976).

Research Program

The basic approach in the studies was to determine the composition of percolate vs time from beds of various ashes subjected to (1) continuous slow irrigation in columns in the laboratory and (2) natural rainfall in the field, by using purpose-built tanks from which all the percolate could be collected, measured, and analyzed. The results were expected to reflect accurately the composition of percolates likely to be realized in actual practice and used to estimate the dilution needed to render the percolate acceptable.

In applying these results to actual schemes, the volume of percolate expected from a proposed ash disposal scheme would first be estimated by making the worst assumptions that all rainfall would appear as percolate and pass directly to the groundwater supply. The volume and probable composition could then be related to the known hydrogeology of the site to ascertain whether the scheme would be acceptable based on these assumptions. In actual practice, one would expect that (1) immobilization and dispersion of trace elements in the ground between the ash and the water source would significantly reduce the levels of material reaching the groundwater; and (2) management of the ashed area could reduce the net amount of percolate relative to the theoretical (e.g., by doming the ground to encourage runoff). Nevertheless, these factors would not be incorporated in the initial assessment and they constitute an unquantified safety margin.

Because the results of the investigations described have already been reported in detail in the literature (Brown et al. 1976), I note here only those main findings that were both surprising and encouraging from the point of view of ash disposal practice:

1. For a given ash the initial percolate quality was virtually independent of bed depth, as shown by column studies over the range 1.8 to 14 m.
2. The levels of dissolved species fell dramatically after about 1 bed volume of water had been displaced. (A bed volume is the amount of water required to completely fill the voids in an ash bed and is about 50% of the superficial volume.)
3. The percolate concentrations were reduced by dispersion if an ash bed was followed by an equivalent depth of gravelly subsoil.
4. Dilutions of only about 100 would reduce the worst concentration of a trace element to the levels considered suitable for potable supplies.
5. In the field tanks, if the bed was not initially saturated, percolation was delayed until enough voids were filled to satisfy the hydraulic requirement at the base.

Practical Application

The research program indicated that the pattern of leaching from an ash disposal area can be predicted from laboratory data if the geometry of the ash bed and the rainfall are considered. If available hydrogeological data showed that the prevailing conditions would remove percolate from potable sources, disposal of the ash without further exploration should be acceptable. However, if the documented hydrogeology were inadequate to establish the point with certainty, a borehole survey could provide the required information, but at some cost.

Where the known hydrogeology or the results of a survey suggest that mixing of ash percolate with potable groundwater is likely to occur, the situation needs to be defined more precisely. As already noted, the simplest assumption is that no attenuation of the percolate occurs and that its volume and composition need to be related to the rate of use and recharge of any affected aquifer. If such an appraisal indicates that the dilution criterion is met with a good safety margin, there is no reason, in principle, why the scheme should not go ahead. If there is any doubt, a borehole survey could be done before the event or could be installed as part of the scheme (and charged against it), with some provision for limiting the total amount of ash deposited if the monitoring reveals any unforeseen factors. Such monitoring would also indicate any beneficial effects of dispersion and fixing of particular elements underground before the percolate reaches the aquifer.

Where there is some choice over the actual geometry of a site, the following factors should be considered.

1. For a given tonnage of ash to be laid down, a deep bed with a small surface area would give an initial percolate no worse than a shallow bed, and the absolute rate of percolation (volume/time), which is proportional to the surface area, would be low, which is favorable from the point of view of dilution underground. It would, however, take longer to yield all its soluble material, the time being directly proportional to depth.
2. For the same amount of ash, a shallow bed of correspondingly larger area would yield its first percolate more quickly and in higher volume, but the soluble material would be leached from it in a much shorter time.
3. As the initial percolate is expected to be the worst and the amount of material leached diminishes rapidly after the first bed volume, the risk assessment should be based on both long- and short-term considerations.

Field Experience

Despite the amount of research that has been done on PFA and the amount of ash that has been disposed of since the advent of pulverized-fuel-fired boilers, few field studies have been necessary. However, with increasing stringency of environmental regulations and controls, such studies may become more common.

Recently, the CEGB collaborated with the Severn Trent Water Authority in examining the effects of ash already deposited in old gravel pits over an aquifer in an attempt to assure the authorities that an extension of infilling would be justified and acceptable. The study included boreholes placed directly in the ash to sample the interstitial water and immediately outside it to check drift of percolate (Table IV). The results showed the following:

1. The interstitial water was of better quality than expected, reflecting the benefits of wet disposal methods.
2. The relative permeabilities of the aquifer and ash deposit were such that movement of water out of the ash was small.
3. The variability of the groundwater itself was such that key ash-characteristic elements were difficult to find for monitoring purposes. Boron and possibly potassium were the most likely indicators.

In two schemes involving landscaped mounds, monitoring boreholes have been sunk around the ash disposal area. These boreholes have indicated no adverse effects on local water supplies after several years of operation (Owens). Mounds present the further possibility of runoff

Table IV. Field Study—Comparison of Interstitial Water from Ash with Water from Boreholes in the Vicinity of the Ash

	Interstitial Water from Ash	Borehole Just Below Ash Site	Borehole Adjacent to Ash Site (~10 m)	Borehole More Remote (~400 m) from Ash Site
pH	11.4	7.4	6.7	6.6
Ca, mg/L	88	167	370	560
Mg, mg/L	0.11	40	68	167
Na, mg/L	170	358	325	530
K, mg/L	121	19	17.5	4.1
SO_4, mg/L	208	635	990	1740
Cl, mg/L	157	430	450	720
B, mg/L	4.0	1.2	1.5	0.25
As, mg/L	<0.01	<0.002	<0.002	<0.002
Se, mg/L	0.11	<0.002	0.017	<0.002
Cr, mg/L	0.39	0.02	0.01	0.03
Mo, mg/L	4.4	0.1	<0.05	<0.05

into low-flow surface water courses. If monitoring reveals problems, they may be avoided by collecting the relatively small quantities of runoff and returning it to the main process water.

Reclamation of Ash Disposal Sites

The ultimate objective in any ash-disposal scheme is to restore the land for further use. To date, more than 4000 ha of land have been used for ash disposal, and the end use has generally been for agriculture or amenity; only small amounts have been restored to residential or industrial use. About 60% of the total area referred to is still being filled or restored. The diversity of reclamation schemes, which has been described in the literature (Snell and Brown, 1979), includes the following:

1. infilling of worked-out sand and gravel quarries, usually lying along river valleys;
2. infilling of old clay workings associated with the brick and earthenware products industries;
3. the reclamation of salt marsh;
4. construction of carefully contoured mounds of PFA above original ground level; and
5. infilling of a geological cleft.

The subsequent uses of such sites have included dairy and arable farming, public parks and playing fields, and light industrial estates. For major schemes, landscape consultants are appointed to prepare detailed plans for surface restoration aimed at producing a final appearance that is both useful and esthetically pleasing.

Restoration for Agricultural Use

Although PFA has many of the characteristics of a mineral soil, it is completely sterile and virtually devoid of nitrogen as a result of the combustion process by which it is formed. Other plant nutrients are present in acceptable quantities, but initially, small amounts of elements to which some plants are sensitive (principally boron) are also present. Initial use of restored ash sites for agriculture requires careful fertilizer management until soil structure is established, and intelligent planning of the sequence of crops until phytotoxic elements have been removed from the upper layers by weathering processes (Snell and Brown,

1979; Townsend and Gillham, 1973). The available boron is reduced to acceptable levels after about 2 years, and until that stage is reached, the site should be used for cultivating boron-tolerant plants (e.g., grass or clover ley). If the ash is laid down from a water-based slurry, which is usually the case, then by the time the material is firm enough to work, it is capable of growing a good grass sward.

A common feature of restored sites is that topsoil is usually in short supply. Consequently, the CEGB has done a substantial amount of work to establish the topsoil requirements for good restoration (Townsend and Gillham, 1973). Although crops can be grown on bare ash, addition of topsoil generally improves yields and limits wind erosion, which affects bare ash in much the same way that it would affect a sandy soil. However, the amount of topsoil required for successful restoration is quite small. Ash has good water retention properties and, in dry areas, mixing only small amounts of topsoil into the ash can actually yield better results than are achieved on normal soils. Townsend and Gillham (1973) have reported experiments at Besthorpe at a site with a high water table where, in a dry year, the best yields of winter wheat were obtained with only 8 cm of topsoil mixed into the ash. By contrast, 30 cm of unmixed topsoil gave results that were no better than those obtained on bare ash. This is an important finding because the cost of importing unnecessarily large quantities of topsoil to reclaimed sites could make the difference between a scheme being economically viable or not. However, the *optimum* amount of topsoil may vary with site conditions. At another site in England, at Peterborough, a ready supply of topsoil, which is available in the form of sugar beet processing waste, is used at a depth of 15 cm or above. It has the advantage of allowing earlier use of the infilled areas at Peterborough than would otherwise be possible. (The scheme involves the infilling of old brick clay workings and, because the site is wet, it dries out very slowly after ash disposal has been completed.)

Agriculture: Trace Elements in PFA

PFA contains many of the trace elements that were present in the original coal or in the mineral matter mined with it. Some of these elements are phytotoxic, whereas others would be regarded as inimical to animals or man if they were absorbed in plants on restored sites and thus entered the food chain. Consequently, in addition to carrying out work into the mechanics and husbandry of restoring sites for agriculture, the CEGB has also looked at the uptake of elements from PFA by plants and animals (Townsend and Gillham, 1973). In addition, the available literature on the compositions of soils and ashes was reviewed comparatively to identify differences and similarities (Gillham, 1980). Many of the elements found in British coal ashes are found in quantities comparable to those in normal soils. In addition, uptake of PFA by plants is not great because many of the elements in PFA, which is an alkaline material, are not very mobile. These general observations are reflected in the findings of Gillham and co-workers (Townsend and Gillham, 1973) that the composition of grass and arable crops grown on PFA does not appear to present problems. An obvious exception is that the molybdenum content of the ash may lead to symptoms of copper deficiency in grazing animals, but this phenomenon is also found on normal soils and can be easily corrected by adding copper to the diet of the animals.

These findings are based on a much smaller data base for ashes than is available for soil, but work is continuing within the CEGB to increase the number of data available. In particular, since the work of Gillham and Townsend was initially published, concern about the environment has increased, and specific trace elements, such as cadmium, lead, mercury, and selenium, have been emphasized. In addition, the analytical methods available for the determination of trace elements have improved, and many analyses that have previously been considered impractical can now be performed routinely. The results of this additional work will be reported in the future. Notwithstanding this caution, it is reassuring that in more than 25 years of experience with restored ash sites, on scales ranging from small experimental plots to full-sized fields in normal farming use, there have been no instances of toxic effects on humans or animals directly attributable to trace elements from coal ash.

Site Registers

CEGB decided to prepare a National Register of Sites to be used for the disposal of power station ash, because the knowledge of such sites, after they had ceased to be operational, depended heavily on the individual records of scattered personnel. To this end, a survey was done, and the material was coded in such a way that it could be presented in a document no bigger than an average research investigation report. The sites were numbered, and each was provided with certain basic information: the grid reference at the site center, its area in hectares, the date on which it was restored, the nature of the site before restoration, the method of filling, the type and depth of ash, the type and depth of soil, the initial seed mixtures used, the subsequent crops with dates (if available), and the landowner and farmer. A final column was added to allow incidental comments to be included where required.

CONCLUSION

Despite the rapid development of nuclear power, coal will continue to provide a major part of Europe's primary fuel needs, and ash disposal is likely to grow as countries that have traditionally burned oil switch to coal. Clearly, every effort will be made to find commercial uses for ash, but the use of the material for reclaiming derelict and marginal land areas is likely to continue.

The CEGB has played a major part in establishing ground rules for ash restoration and, as we have seen, it has done a great deal of associated environmental work to provide reassurance. With careful planning and good initial management, ash-disposal schemes can provide a net environmental gain at low risk. Nevertheless, as environmental legislation grows and, if anything, becomes more stringent, new work or reassessment of existing data and modification of disposal and reclamation practices may be necessary.

REFERENCES

Brown, J., N. J. Ray, and M. Ball. 1976. "The Disposal of Pulverised Fuel Ash in Water Supply Catchment Areas," *Water Res.* 10:1115–1121.

Gillham, E. W. F. 1980. "Trace Elements in Soils and Ashes—A Comparative Review," CEGB Report No. MID/SSD/80/0002/R.

Owens, P. M. Private communication.

Ray, N. J., and M. A. Jenkins. 1979. "Ash Lagoon Effluents in the Midlands Region," CEGB Report No. SSD/MID/R36/79.

Snell, P. A., and J. Brown. 1979. Re-development of Derelict Land as Part of the Operation of the Central Electricity Generating Board," presented at the Reclamation of Contaminated Land Conference, Eastbourne, England, October 22–25.

Townsend, W. N., and E. W. F. Gillham. 1973. "Pulverised Fuel Ash as a Medium for Plant Growth," presented at the 15th Symposium of British Ecological Society, London, July 10–12.

Wright, D., and J. Brown. 1979. "Hydraulic Disposal of P.F. Ash from CEGB Midlands Region Power Stations," Hydrotransport 6, Sixth International Conference on the Hydralic Transport of Solids in Pipes, Sept. 26–28.

DISCUSSION

***Ed Protner*, Johns Hopkins University Applied Physics Laboratory:** Could you elaborate a little bit more on the test that you did with percolate concentrations as a function of bed depth; in particular, whether it was only TDS that you looked at or whether you were also looking at trace metals?

Also, I've seen a lot of data for experiments of this type where the core was only 10 cm thick. Do you have any feel for whether you were working with very short cores, or whether the percolate concentration would also be independent as you went from centimeters to meters?

J. Brown: Let me say that there are two categories here. The first thing is, we worked in columns up to 50 ft (or 14.5 m) which we felt was typical of the kind of pits we were likely to fill in, and we found no effect over the whole of that range. Let me qualify what I mean by no effect. If you dispose of ash that is wet, the void volume in the ash is already full of material which is equilibrated, and if you think about it a few moments, there is no reason why there should be any depth effect at all. You may then ask why doesn't the water get more saturated as it passes deeper and deeper in ash columns? The answer is that there is so little soluble material in the ash in the first place that the interstitial water, the first bed volume, is capable of mopping up a good deal of it. What comes out depends on the ash—don't deny that—and we can show you variations in ashes; but for that given ash, it doesn't matter how deep the bed is—you'll get the same results.

***Terry Logan,* Ohio State University:** In the first couple of years when you are trying to establish a vegetation pond, do you ever encounter salt problems? An excess of salt?

J. Brown: The only site on which I'm conscious of a salt problem is the Peterborough site, because there is no purge from the site at all. The water is in very short supply, which is why it is reused. The reservoir is 500-million gallons in capacity. You have to have some makeup at the site, which you would get from flash streams, but we don't put the effluent anywhere. That means the concentration of water in that particular ash field is growing as the scheme continues to be used. It's very high in dissolved solids until the salt is washed out of the surface layers by rainfall. That's why you need the topsoil at Peterborough; it's not just the mechanical question. I'm glad you raised it.

***David C. Wilson,* Harwell Laboratory, England:** I would like to make a comment on leachate generation rates from fly ash. We have been looking for some years at a landfill site that has been receiving 1000 tons a day of domestic refuse using fly ash in approximately equivalent quantities as the intermediate cover material. This is part of our much larger research program, which is aimed at looking at differences in leachate and gas generation rates for municipal refuse landfills. Our interest in this particular site is that, thanks to the ash, it is an extremely dry site. After 4 years we have obtained virtually no leachate at all; on gas production, because of the dryness, we have obtained our peak hydrogen concentration of around 20%.

J. Brown: What I've tried not to mention in my paper at all are some of the beneficial uses of PFA. I've given you a paper that regards it as a waste that we have to account for. What David is talking about is the actual use of ash to beneficiate other more toxic materials. And I think in that case, you have to look very carefully at the mechanical properties of the material. One thing that can clearly happen with fly ash as an overblanket is that it can suck water out of what is underneath it because of the fine structure. On the other hand, it has a very high alkaline buffer capacity. And even if you have gas works liquor underneath it, as long as you equated the quantities of gas liquor with your actual bed thickness, you know that you save. It's an extensive problem, not an intensive one.

CHAPTER 6

WASTES FROM EMERGING ENERGY TECHNOLOGIES

Lawrence Kronenberger
Synthetic Fuels Department
Exxon Company
Houston, Texas 77001

ABSTRACT

When synthetic fuels are produced by oil shale retorting and coal gasification or liquefaction, large quantities of solid waste materials must be handled and disposed of in an environmentally acceptable manner. Although solid waste management plans have been prepared and, in some cases, permits have been obtained, no commercial-scale synthetic fuels facilities are operating in this country today. However, many major components of these facilities are similar to other commercial operations. For example, a coal conversion project would include mining, steam generation, and gas and liquid processing steps not unlike those in petroleum refining and chemicals manufacture.

A commercial 50,000-bbl/d shale oil project involves mining and retorting of about 70,000 tons/d oil shale. After recovery of the kerogen, the spent shale would be conveyed to a suitable surface disposal site and compacted in place. Over a 20- to 30-year project life, the accumulation would cover hundreds of acres to depths of several hundred feet or more. Important environmental concerns include pile stability, runoff water quality, and degree of leaching before and after revegetation has been completed.

Lurgi coal gasification technology is being studied for possible application to East Texas lignite. Large quantities of solids to be handled include boiler fly ash, boiler bottom ash, and gasifier ash. The project would be supplied by a lignite surface mine. Some wastes may have characteristics that would allow disposal in the mine before reclamation.

INTRODUCTION

When synthetic fuels are produced by oil shale retorting, coal gasification, or coal liquefaction, large quantities of solid waste materials must be handled and disposed of in an environmentally acceptable manner. Although solid waste management plans have been prepared and, in some cases, permits have been obtained, no commercial-scale synthetic fuels facilities are operating in the United States today. However, several commercial-scale projects are in the construction stage. This chapter presents solid waste management plans for an oil shale project and for a coal gasification project. The solid waste management problems associated with coal liquefaction may be considered similar to those associated with coal gasification, particularly with gasification technologies that produce by-product liquids.

REGULATORY SITUATION

The federal statute covering solid waste disposal is the Resource Conservation and Recovery Act (RCRA) of 1976. The two parts of the act primarily affect solid waste disposal practices by synfuels facilities, or by industry in general: Subtitle C, which covers the practices associated with the generation, transport, storage, and disposal of hazardous wastes; and Subtitle D, which covers disposal activities for nonhazardous waste. Solid wastes in these definitions encompass a variety of materials, many of which are aqueous sludges. Wastes are hazardous if they are listed under Subtitle C or if they fail the hazardous waste classification criteria. The November 1980 amendment to RCRA, referred to as the mining exemption, excluded all waste associated with mining and upgrading of coal and minerals from both parts of RCRA for a period of 3 years. The U.S. Environmental Protection Agency (EPA) subsequently interpreted this exemption to apply to wastes generated through direct upgrading, that is, retorting or gasification. Wastes from hydrotreating, sulfur recovery, or wastewater treatment, for example, would still be regulated by the applicable RCRA subtitle. Hence, the exact coverage by RCRA is not clear for the long term. In addition to the federal statutes, many state laws and agencies administer programs that regulate solid waste disposal. For example, the Texas program was in place before the federal RCRA program. Coverage by other state agencies include the surface mining control and reclamation program administration by the Texas Railroad Commission and the Mined Land Reclamation Board permit system in Colorado.

SHALE OIL PRODUCTION

Exxon and Tosco are jointly constructing a commercial oil shale retorting facility in northwest Colorado at a site about 15 miles north of the Colorado River on Parachute Creek. The Colony Project concept includes the new town of Battlement Mesa just south of the Colorado River, in addition to the extensive mining and processing facilities.

Commercial-scale facilities will typically produce 50,000 bbl/d and will cost $2 to $3 billion. Operations include (1) an underground mine of the room and pillar type, which would be several hundred feet below the surface; (2) primary oil shale crushing; (3) transport of crushed shale in conveyors; (4) stockpiling of crushed shale followed by final crushing; (5) retorting to separate the kerogen or oil from the rock; (6) upgrading of the oil in facilities analogous to those in a petroleum refinery; and (7) disposal of the spent shale on the surface followed by reclamation.

The setting for most of the shale oil operations, as well as for many other synthetic fuel projects such as coal liquefaction and gasification, will be the West, typically in areas where

- the terrain is rugged;
- scenic quality is a prime consideration;
- water supplies are limited, and surface waters tend to have strict in-stream water quality requirements; and
- population is sparse, and infrastructure is relatively undeveloped.

After mining, the three basic processing steps in the Colony Project (Figure 1) are shale crushing, retorting operations, and product upgrading. Retorting is a unique feature of the project. Raw shale, which has been crushed to 0.5 in. topsize, is preheated in a dilute-phase fluid-bed preheat system by hot flue gas from a downstream vessel. The flue gas is scrubbed before it is released to the atmosphere. Purge water containing raw shale particulates is one of the wastes from the oil shale retorting.

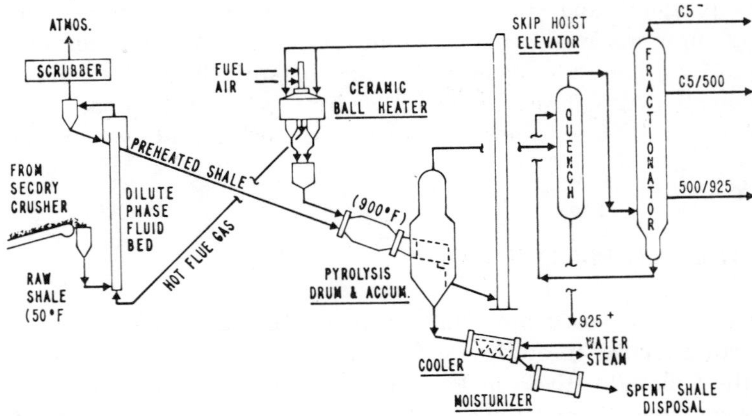

Figure 1. Tosco II oil shale retort, Colony Project.

The preheated shale is then fed to the pyrolysis drum. Retorting of the oil shale is achieved by solid-to-solid heat transfer between the shale and hot ceramic balls, flowing together through the rotating pyrolysis drum. Complete retorting of shale is achieved at about 900°F (480°C) during a short residence time.

The shale oil vapors, spent shale, and ceramic balls exit together and are separated in the accumulator. The balls are lifted by a skip hoist and reheated in a ball heater. The ball heater flue gases are recycled through the lift pipe shale preheat system to recover the waste heat.

Spent shale (Figure 2) exits from the accumulator vessel at about the retorting temperature [900°F (480°C)] and passes through a special heat exchanger designed to cool the spent shale and generate steam for plant use. The spent shale is then cooled further by direct contact with water and is moisturized to about 14%, which has been determined to be the proper level for subsequent handling.

The solid waste produced by the oil shale operations includes (1) raw shale dust rejected in the crushing step; (2) spent shale and

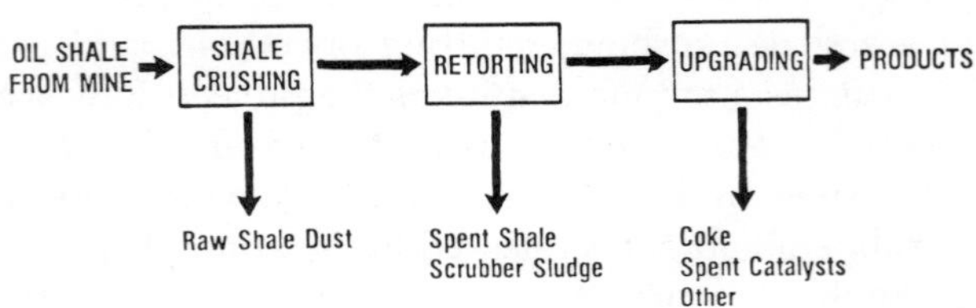

Figure 2. Solid waste production, shale oil project.

MAJOR WASTES	T/D	PLANNED DISPOSITION
Raw Shale Dust	425	Non-hazardous Landfill
Spent Shale	53,300	Non-hazardous Landfill
Scrubber Sludge	860	Non-hazardous Landfill
Coke	800	Sale or disposal in Non-hazardous Landfill
Catalysts	700 T/YR	Hazardous Waste Facility
	550 T/YR	Metals Reclamation
Other	8,000 T/YR	Non-hazardous Landfill
	3,000 T/YR	Hazardous Waste Facility
Total	20.2×10^6 T/YR	

Figure 3. Solid waste management, shale oil project.

scrubber sludge, which leave the retorting step; and (3) coke, spent catalyst, and miscellaneous wastes from the upgrading processing.

The solid waste management plan for the shale oil project must address each of these waste streams (Figure 3). Raw shale dust, produced at the rate of 425 tons/d, will be transported in a covered conveyor to a nonhazardous landfill. Spent shale, produced at a rate of about 53,000 tons/d, will be moistened and transported to the nonhazardous reclamation area, where it will be spread, compacted, contoured, and revegetated. Scrubber sludge purged from the air pollution control devices associated with the retorting includes some 860 tons/d dry solids. This stream will be mixed with the spent shale upstream of the moisturizer. The water associated with these collected particulates provides part of the moisturization water for the spent shale.

At full-scale operation, a coke by-product of 800 tons/d is also produced. If this coke cannot be sold, it would probably be stored for future use or disposed of in a nonhazardous landfill. Catalysts from the upgrading facilities will need to be replaced periodically. Quantities of these catalysts could amount to 700 tons per year, some or all of which may have to be disposed of in a hazardous waste facility, and about 550 tons per year that appear to be amenable to metals reclamation techniques. Other wastes include about 8000 tons per year of construction debris and raw-water treating sludges, which could be included in a nonhazardous landfill. Another 3000 tons per year of aqueous materials, such as storage tank bottoms and API separator sludge, would probably need to be disposed of in a hazardous waste facility. The notable solid waste feature of a commercial oil shale project is the 20 million tons per year that must be handled and disposed of in an environmentally acceptable manner. By far the largest single component in this total is the spent shale, which accounts for about 96% of the solids on a water-free weight basis.

A considerable amount of research has already been conducted, and more is underway or planned to gain a better understanding of leachate qualities, potential water migration in spent shale piles, and reclamation requirements. For the Colony Project, the permit issued by the Mined Land Reclamation Board established requirements for revegetation, site maintenance, monitoring, and reporting.

If a major oil shale industry develops, resource recovery considerations point to large-scale surface mining of the oil shale, especially in the western and central Piceance Creek Basin of Colorado. Some developers are already looking at surface mining to supply shale to retorts, which would recover 50,000 to 100,000 bbl/d. Surface mining presents the opportunity to return the processed shale and perhaps other solids to the mine. This process would minimize topographical effects and revegetation problems, but would need to be accomplished without contaminating potentially usable groundwaters.

COAL GASIFICATION

Exxon is studying a coal gasification project that would involve mining and processing East Texas lignite. The lignite resource is located southeast of Dallas and north of Houston. The coal is mineable with current technology, and several utility mines are already operating in the area. At full capacity, the lignite would be surface-mined at a rate of about 15 million tons/year.

The design of the plant is based on the Lurgi gasification process, a German technology that has been proved commercially in South Africa since 1955. The plant under consideration would convert the lignite into intermediate-Btu gas. The gas can be sold as a chemical feedstock or a premium industrial fuel gas, or it can be converted to coal liquids such as fuel-grade methanol. The plant would have a capacity to produce fuel oil product equivalent to 60,000 bbl/d.

The Lurgi coal gasification project (Figure 4), as planned for the East Texas site, includes coal preparation, gasification, and gas and gas liquor processing as the main processing operations. Also, support facilities include oxygen plants, boilers with attendant flue gas desulfurization, and wastewater treating facilities.

In the coal preparation step, lignite from the mine is crushed and sized before being separated in a dry/wet system. Coarse lignite, which is about three-fourths of the total, will be transported by covered conveyors to the gasifier feed system, which will distribute it to the individual gasifiers. Fine lignite down to a 100 mesh will be dewatered to

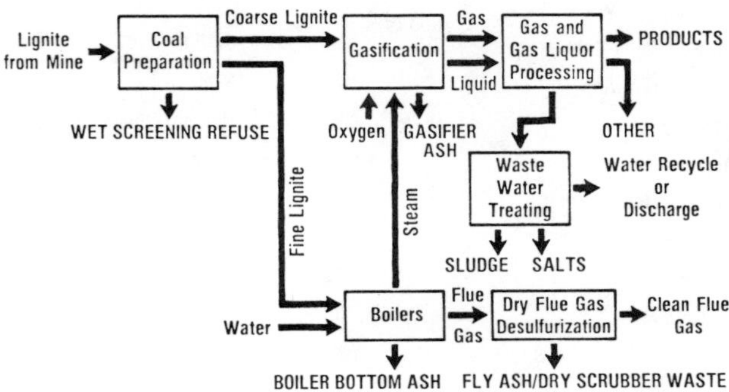

Figure 4. Solid waste production, Lurgi coal gasification.

the extent necessary and placed in silos for use as boiler fuel. The gasification facilities would consist of about 30 Lurgi gasifiers. Lignite enters each gasifier through a lock hopper system, which is intermittently pressured to over 400 psig to allow the lignite to flow by gravity into the gasifier. As it travels from the top to the bottom of the gasifier, the lignite will be dried, devolatilized, and gasified. A mixture of oxygen and superheated steam is used in the gasification. Oxygen is consumed in the partial combustion of the lignite to supply heat for the gasification reactions. These reactions involve the steam, which is a hydrogen source for the process products. Heat is recovered from the system by generation of steam. The raw gas product is sent to the gas cooling system, where heavier hydrocarbons and unreacted steam are condensed before the gas is purified. The gas purification unit utilizes the Rectisol process, in which the raw gas is contacted with cold methanol to recover naphtha and to remove carbon dioxide, sulfur compounds, and other impurities. Purified intermediate-Btu product gas contains less than 0.1 ppm sulfur compounds.

The solid waste production from this processing includes wet screening refuse, gasifier ash, boiler bottom ash, fly ash/dry scrubber waste, sludge and perhaps salts from wastewater treating, and other wastes from the gas and gas liquor processing operations.

For this particular project the solid waste management plan has not been finalized through the permitting stage. Hence, the dispositions for solid wastes are still tentative (Figure 5). Wet screening refuse, consisting of 200 tons/d (300 ton/d on a wet basis), would be directed to a settling pond. Water decanted from the pond would likely be recycled to the coal preparation operations. Boiler bottom ash amounting to 700 tons/d would be returned to the surface mine if this process could be demonstrated to be environmentally acceptable. In general,

MAJOR WASTES	T/D	TENTATIVE DISPOSAL
Wet Screening Refuse	200	Settling Pond
Boiler Bottom Ash	700	Mine
Boiler Fly Ash	2,800	Non-hazardous Landfill
Gasifier Ash	4,800	Mine
Waste Water Treating		
Sludge	30	Landfarm
Salts	150	Hazardous Landfill
Total	8,700	
	2.9×10^6 T/YR	

Figure 5. Solid waste management, Lurgi coal gasification.

the primary emphasis will be on returning as much of the material to the mine as is environmentally, economically, and operationally feasible. All the streams are being evaluated for this alternative. Those determined to be suitable on the basis of regulatory requirements may be returned to the mine. Placement of the waste in the mine spoils "vees" before final reclamation is one alternative being evaluated. Even though wastes returned to the mine will have been determined to be nonhazardous by EPA classification procedures, consideration will be given to locating these wastes in those pits having the maximum depth to groundwater and the most suitable overburden from the perspective of soil types. Boiler fly ash represents a significant quantity of 2800 tons/d, but would be essentially the same as the corresponding material from nearby power plants. At this time disposal in a nonhazardous landfill is planned.

The major solid waste component, gasifier ash, amounts to 4800 tons/d. This material is a leading candidate for in-mine disposal. Wastewater treating solids include sludge from oil-water separation and biological treating operations. These sludges amount to 30 tons/d (270 ton/d on a wet basis) and are being considered for handling by way of landfarming. One wastewater treating option involves evaporation, which would produce significant quantities of soluble salts. These salts, which would have to be placed in a secure location to avoid subsequent re-solution into rainwater or other waters, are shown as requiring disposal in a hazardous landfill. Evaporation facilities, if installed, would be used because of site-specific considerations, because the only effluent-discharge receiving waters available are the small freshwater streams of the area; the imposition of a significant salt burden on these streams would not be acceptable.

Thus, a commercial coal synthetics project would also involve management of a significant quantity of solid waste material, about 8700 tons/d, or almost 3 million tons per year on a dry basis. Adding the water from sluicing of gasifier ash and other sources would raise the total to 3.7 million tons per year. The major constituents of this large total are the various ashes from lignite gasification and from combustion of lignite in the boilers.

COAL LIQUEFACTION

Solid wastes from coal liquefaction will be related to the coal source and processing used. Coal liquefaction, by both the direct and indirect routes, is even closer to refinery or chemical plant operations than Lurgi coal gasification. Compared with the Lurgi gasification, steam generation requirements would likely be less, resulting in less solid waste from boilers. Also, wastewater volumes would probably be lower, with correspondingly smaller volumes of wastewater treating sludges to be handled. However, several types of spent catalyst would need to be disposed of periodically.

SUMMARY

When we consider wastes from emerging synthetic fuels technologies, the following points stand out. First, the solid waste management plans are dominated by the need to handle large volumes of nonhazardous or low-hazard potential wastes. Total project plans often include mining operations, power generation, and product upgrading, in addition to the conversion step itself. The wastes from mining, power generation, and the hydrocarbon processing associated with product upgrading are not unique to synfuels. Before entering the construction phase, the first U.S. commercial-scale synthetic fuels projects typically must obtain more than 100 permits. Comprehensive permits cover the proposed waste disposal practices and land reclamation procedures. These permits require record keeping and monitoring systems to be put in place, and in some cases they specify additional study programs. As stated at the outset, the plans developed so far are for projects in the planning or construction stages; commercial projects are not yet in operation. Research is continuing. One aim should be to maximize the potential for in-mine disposal of solid waste. This option could be economically attractive in large-scale oil shale or coal surface-mining projects.

DISCUSSION

***Rogene F. Henderson*, Lovelace Inhalation Toxicology Research Institute:** It was curious that you didn't mention in the gasification process having any tars or oils in your scrubber water. I wonder if salt is the only thing that is in there or are any tars left over from the process? I am familiar with other gasification processes where that is a problem, and I was curious whether you anticipate any?

L. Kronenberger: In this particular case, basically the operation of the gasifier would be such that we would recycle the tars to extinction, although there would be some heavy fuel oil and there would be some of these aqueous sludges that I mentioned. But we are not anticipating a tar product.

***K. E. Cowser*, Oak Ridge National Laboratory:** Do you have any active work going on now at Exxon, or do you plan any in terms of perhaps bench scale or process-development unit-size investigations of the wastewater treatment systems for both your shale as well as your gasification concepts?

L. Kronenberger: Yes, but not a whole lot. There are some plans on the oil shale, and we have done some tests on the gasification. For example, we ran a demonstration test on the full-scale gasifier in South Africa several years ago. We brought back water from that facility, but before bringing it back, we had to do some bench-scale treating in South Africa to remove phenols and to do some stream stripping, so it was several months old by the time we got it. In the U.S., we ran various tests to get an idea of how it would respond to biological oxidation. Right now we are not doing any more simply because we don't have representative water to work with.

SECTION 2
PROBLEMS IN CHARACTERIZATION

CHAPTER 7

OVERVIEW

Robert A. Griffin

Illinois State Geological Survey
615 E. Peabody Drive
Champaign, Illinois 61820

Data acquisition for a complete environmental assessment of solid wastes includes physical, chemical, and biological analyses. As the level of sophistication increases relative to environmentally acceptable disposal options, so do the needs for more complete characterization of the wastes. The goals for this section were to bring together some of the different approaches to addressing the problem of characterizing solid wastes and to provide meaningful environmental assessments for wastes in a timely and economical manner.

In the chapter by Simmons and Stephens, the classical approach to chemical characterization of wastes as used in the California hazardous waste management program is outlined. The characterization process is illustrated by use of an example case history dealing with an abandoned acid petroleum sludge site. They point out that although the general approach to characterization is the same for all wastes they encounter, each waste has particular problems that may have to be addressed on an individual basis. Major waste characterization tasks have evolved into an iterative scheme. The five steps in the scheme are (1) background information search, including a review of the appropriate industrial processes and the pertinent physicochemical reactions that may have altered the composition of the wastes; (2) initial sampling and analyses, which includes screening tests for hazard assessment; (3) review of the initial data and design of a comprehensive characterization plan, including a detailed safety plan; (4) detailed sampling and analyses; and (5) interpretation of results for legal purposes, health risk assessment and environmental impact assessment. The characterization scheme outlined by Simmons and Stephens is designed for unknown wastes but incorporates the major elements of the waste characterization process. As such, it provides an overview of the general procedures used to characterize wastes and also provides an overview for the remaining chapters, which tend to deal with specific aspects of the generalized waste characterization process.

One approach to development of screening tests for hazard assessment has been leaching tests of solid wastes. Tests are under development to both define if a waste has characteristics that would classify it as a hazardous waste and to make predictions relative to the leaching characteristics of the waste. Major research efforts have been sponsored by the U.S. Environmental Protection Agency, the American Society for Testing and Materials, and the U.K. Department of the Environment, to name a few. The chapter by Wilson and Young presents the approach used by the Harwell Laboratory in England, and some of their results from development of leaching test methods for assessment of hazardous wastes prior to landfill disposal.

The Harwell group has taken a pragmatic approach to assessing the suitability of a specific hazardous waste for disposal at a particular landfill site. The approach is based on measuring the reduction in leachate composition due to specific mechanisms. A general framework for decision-making, an attenuation map, has been developed, within which laboratory tests were designed. The solid-waste leaching test is simple and based on repetitive extraction using high solid-to-liquid ratios. Wilson and Young point out that laboratory tests should be designed to allow correlation with expected field behavior over time. Their results demonstrate the necessity to validate small-scale tests with field experiments.

The chapter by Kingsbury, Mack, and Harkins presents an approach to interpretation of results with respect to the relative environmental hazard of pollutants in solid wastes and leachates. Multimedia environmental goals (MEGs) are chemical-specific goals expressed as concentrations in air, water, and soil (or solid waste). Quantitative values for over 600 chemical pollutants have been developed and shown to have great value as criteria for interpreting data to support environmental assessments. Assessments using the MEG methodology can include both potential human health effects and ecological effects, thus increasing their value for characterization of solid wastes relative to health risk assessment and environmental impact assessment. The authors discuss applications and limitations of the MEGs; revisions currently under consideration are presented.

Complete characterization of a solid waste includes a biological assessment. The chapter by Kenaga addresses problems associated with biological toxicity characterizations of solid wastes. Because these characterizations are expensive and time consuming, Kenaga presents a framework for screening the chemicals contained in wastes so that only those tests and bioassays are conducted that are necessary for a sound hazard evaluation under the expected use and distribution conditions of the specific waste.

Kenaga presents a number of useful early-stage evaluation and predictive regression equations that are based on physical, chemical, and structural properties of chemicals; the growing body of comparative toxicology data; and information on the fate of chemicals in the environment. He points out that we may never be able to predict the exact fate of a chemical in the environment or its toxicity to all life stages and species of organisms because of the huge number of permutations and combinations of factors that are possible. We must nevertheless attempt to predict the effects of chemicals on organisms so that we do not exhaust our resources and energy testing chemicals unnecessarily.

The complete characterization of solid wastes can be an expensive and time-consuming process subject to numerous problems. A general framework for characterizing wastes and various approaches to solving the problems have been presented by the authors of this section. More research is needed to expand the work presented here and to provide improved methods for characterization of solid wastes.

CHAPTER 8

TESTING METHODS FOR HAZARDOUS WASTES PRIOR TO LANDFILL DISPOSAL

David C. Wilson and P. J. Young

Waste Research Unit
Harwell Laboratory
Didcot
Oxfordshire, England OX11 ORA

ABSTRACT

For several years, the Waste Research Unit of the Harwell Laboratory has been developing testing methods on behalf of the United Kingdom's Department of the Environment. The goal was to assess the suitability of a specific hazardous waste for disposal at a particular landfill site. The approach is based on measuring the reduction of contaminant concentrations in leachate caused by one or more specific mechanisms.

The leaching test we have devised is simple and is based on repetitive extraction using a very high solid-to-liquid ratio, defined in such a way as to allow correlation with expected field behavior over time. Validation experiments using large laboratory columns to simulate leachate generation in a landfill have yielded surprising results. These experiments demonstrate the need to verify any small-scale test by comparison with field, or simulated field, experiments.

In addition to specific measurements of one attenuation mechanism in isolation, the unit is currently developing test methods to simulate codisposal of hazardous with municipal wastes on a laboratory scale. To illustrate the techniques, preliminary results will be presented from a series of large-column experiments containing a "sandwich filling" of four different pesticide wastes.

INTRODUCTION

In determining the suitability of a hazardous waste for landfill disposal, the following three general criteria have to be considered:

1. the health and safety of site personnel and of local residents;
2. the long-term persistence of the waste, which could adversely affect future reuse of the land; and
3. the potential for pollution of surface or groundwater resources.

For the last eight years, the Waste Research Unit (WRU) of the Harwell Laboratory has been developing simple testing methods to measure performance against various aspects of these criteria. Examples include a flammability test suitable for use in the field (Waring and Hudson, 1979) and procedures to measure the rate of vaporization of solvents absorbed on solid waste (Jones and McGugan, 1978). However, much of our effort has focused on the third criterion, that of preventing water pollution.

All of the test methods that we develop are aimed at pragmatic application, to answer the question: "Can this waste safely be disposed of at this particular landfill site?" To achieve this, we have developed a general framework for decision-making within which the tests are designed. This framework may be illustrated schematically as an "attenuation map" (Figure 1). Many mechanisms can reduce contaminant concentrations in leachate from those present in the original deposit of hazardous waste (Department of the Environment, 1978). For example, Figure 1 depicts codisposal of hazardous waste with municipal waste at an unsealed site underlain by an unsaturated zone. Attenuation at such a site may be due to

1. the initial leaching of the contaminant from the deposit of hazardous waste;
2. processes within the municipal waste, including biodegradation, sorption, chemical interactions such as neutralization, oxidation-reduction, and precipitation;
3. processes at the base of the landfill, notably dilution with leachate from other areas of the landfill and coprecipitation of metal ions with ferric hydroxide;
4. processes occurring in the unsaturated zone, including physical filtration, dispersion, dilution, biodegradation, sorption, and ion exchange; and

5. processes within the saturated zone (groundwater aquifer), notably dispersion and dilution.

These mechanisms, either singly or in combination, reduce the concentration of contaminants by many orders of magnitude at a groundwater abstraction point. If the criterion assumes that the drinking water standard for contaminant X [i.e., concentration (8) in Figure 1] must not be exceeded at the abstraction point, then it is necessary to show in any particular case that attenuation of X from concentration (2) to concentration (8) will occur. Our testing methods are designed to measure attenuation caused by initial leaching or by specific mechanisms within the landfill. When these are combined with information on rainfall infiltration, on the quantities of waste deposited (and thus on dilution by leachate from the rest of the site), and on the hydrogeology of the site (yielding an estimate of attenuation beneath the site), the likely concentration of X at the water abstraction point can be estimated. If this is less than the drinking water standard, disposal of that waste, in that quantity and at that site, will not pollute the water supply. Because measurements are usually restricted to a few of the most important attenuation mechanisms, which leaves attenuation from many other mechanisms unquantified, this decision-making procedure incorporates a built-in safety factor.

The map in Figure 1 was deliberately drawn to include a wide range of attenuation mechanisms. Similar diagrams can be constructed for different modes of landfill: for example, at a sealed site, the map may be terminated at the landfill boundary, and the decision would be based on comparing concentrations in the collected leachate (5) with some standard such as that for discharge of trade effluents to sewer. For a chemical landfill site, a similar approach can be used, except that the appropriate attenuation mechanisms within the landfill will vary from case to case. Two types of tests, primary leaching and total attenuation in a codisposal landfill site, are illustrated.

LABORATORY LEACHING TEST

Objectives

The initial leaching of the toxic component from the waste as deposited will often be a significant factor in attenuating its concentration in the leachate below that in the waste. There has been much international interest in small-scale leaching tests (e.g., Lowenbach, 1978;

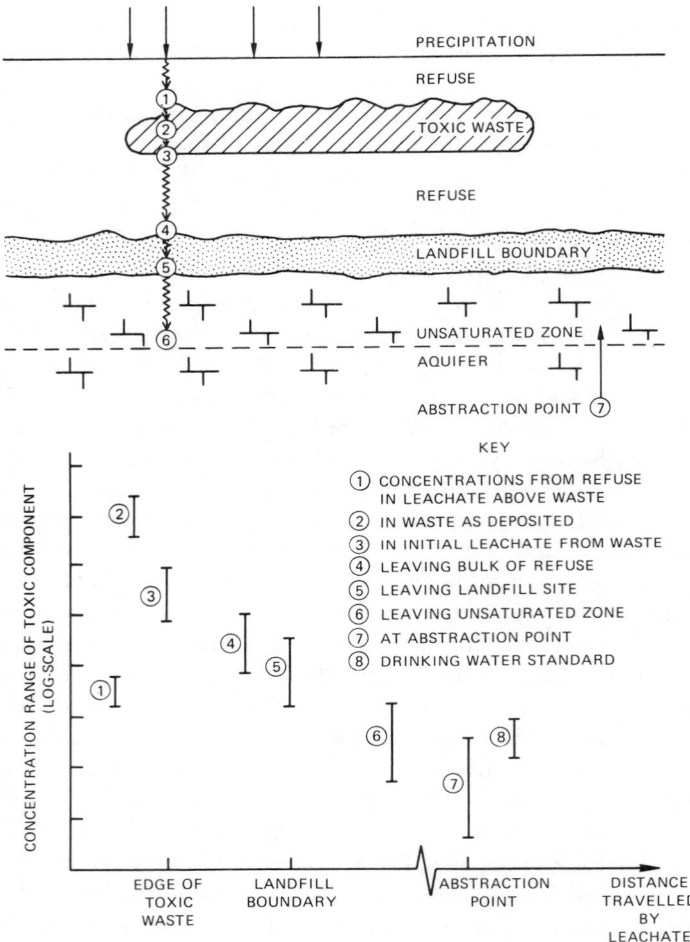

Figure 1. Attenuation map showing relative concentrations of a contaminant as leachate moves through and away from a deposit of toxic waste. Illustrates the case of codisposal with municipal waste at an unsealed site with a significant unsaturated zone.

Anderson et al., 1979; Perket and Webster, 1981), but much of this has been directed primarily at legal definition of what constitutes a hazardous or toxic waste. From a review of the state-of-the-art methodology, the main objectives for the WRU leaching test were defined as follows:

1. easy interpretation of results for use in decision-making,

2. accurate prediction of the pattern of leaching behavior over time, and
3. validation of the results against field behavior.

Outline of the Test Procedure

The WRU leaching test is basically very simple. It is a repetitive batchwise shaking test, using either distilled water or dilute, buffered acetic acid as the leaching fluid. The duration of each extraction is determined from a preliminary equilibrium test. The solid-to-liquid ratio is approximately 1:1 based on the concept of bed volume. The extraction is repeated five times with fresh fluid to provide leachate samples corresponding to successive bed volumes, and the remaining waste (or a portion of it) is then depleted using a solid-to-liquid ratio of 1:10 (indicating an average concentration corresponding to about ten bed volumes). Depending on the analytical results, the test may be extended to study either very rapid leaching, by replacing the waste rather than the leaching fluid, or slow leaching, by carrying out further extractions with a solid-to-liquid ratio of 1:10.

Solid-to-Liquid Ratio

The novel feature of this test procedure is the initially high solid-to-liquid ratio of 1:1 based on the concept of bed volume. The bed volume of a waste is a measure of its void space or the volume of liquid required to just saturate it. It is defined in the laboratory as the volume of leaching fluid required to just cover and render mobile a sample of waste of known weight.

For any given field situation, it is possible to calculate the time required for one bed volume of leachate to pass through a deposit of the waste, and the laboratory results may thus be scaled up effectively to yield information on the time dependence of leaching. As an example, consider a waste with a void space of 50%, giving a bed volume about 1 L/kg waste. If the waste is deposited in a layer 1 m thick, with a rainfall infiltration rate of 25 cm per year, then one bed volume of leachate corresponds to about 4 years in the field. The more normal solid-to-liquid ratios of 1:5 or 1:10 usually quoted for laboratory leaching tests (e.g., Anderson et al., 1979) thus yield results corresponding to the average concentration of a contaminant in leachate over many years (Anderson, 1979).

From applications of the test procedure to many wastes, three different patterns of leaching behavior with time may be distinguished:

1. The most common situation occurs when the repetitive test does indeed locate the maximum concentration within the first five bed volumes (Figure 2a). Note that a single shaker test with a solid-to-liquid ratio of 1:10 would grossly underestimate such a maximum.
2. The maximum concentration sometimes corresponds to the first bed volume of leaching fluid. In such a case, the true maximum in the field could be much higher, because all measurements were an

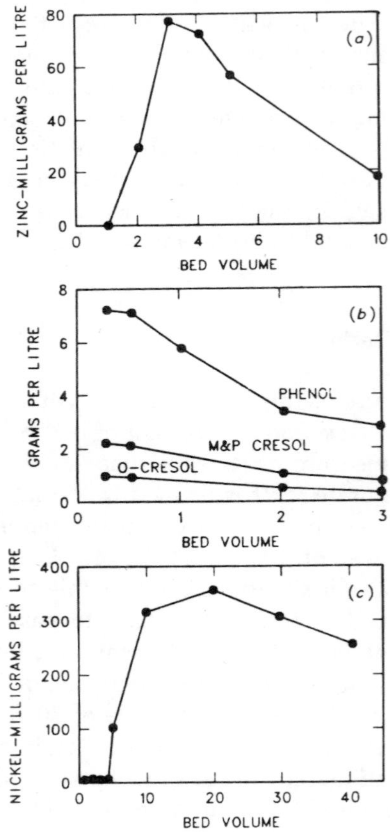

Figure 2. Typical leaching profiles. (a) Normal case, illustrated by zinc leaching from a tin-arsenic slag. (b) Early leaching, illustrated by leaching from a phenol-lime sludge. (c) Delayed leaching, illustrated by leaching of nickel from a synthetic metal hydroxide sludge.

average of those in the first bed volume. In this case, the test should be repeated, extrapolating to even higher solid-to-liquid ratios by replacing the waste rather than the leaching fluid at each stage. The example of leaching from a phenol-lime sludge is shown in Figure 2b.

3. After the normal procedure of extractions over ten bed volumes, concentrations may still be increasing, or no maximum may be indicated. In such cases, the test should be extended by depleting the waste with further samples of ten bed volumes of leaching fluid (Figure 2c). Such cases of delayed leaching are the only ones where use of a low solid-to-liquid ratio will yield a good estimate of the maximum concentration.

Choice of Leaching Fluid

The leaching fluid is chosen according to the landfill situation that is being modeled. Distilled water is useful both for comparative purposes and for the case of monodisposal of a bulk waste. For codisposal of hazardous waste with municipal waste, a simulant for municipal refuse leachate is required. In a very extensive series of trials, we tested ten leaching fluids by carrying out the leaching test on a synthetic metal hydroxide sludge and comparing the leaching profiles for nine metal species. The leaching fluids comprised the following:

1. distilled water;
2. 5000-ppm acetic acid, buffered to pH 3, 4, 5, 6, and 7 with sodium hydroxide;
3. an anaerobic model leachate developed by Ham's group at Madison, comprising buffered acetic acid (pH 4.5) with added glycine, pyrogallol, and ferrous sulfate to model both the redox potential and complexing effects of real leachate (Stanforth et al., 1979);
4. two aerobic model leachates, one the same as above but with the ferrous sulfate removed and the pyrogallol replaced by salicylic acid. The second was a complex mixture of organic acids and inorganic salts, buffered to pH 5.5, developed by the Dutch Institute of Waste Disposal (SVA, 1979); and
5. a sample of real leachate collected from a landfill cell containing 800 t of 2-year-old refuse (total fatty acid concentration about 2.5%).

The results for each metal generally followed a similar pattern (Figure 3). The three model leachates and acetic acid at pH 3 or 4 invariably gave vigorous leaching, yielding a massive early peak and, in one

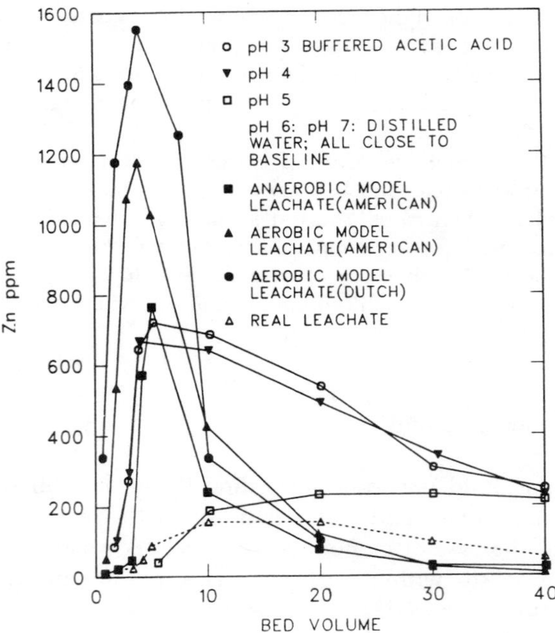

Figure 3. Typical results from comparative leaching tests on a synthetic metal hydroxide sludge, using ten different leaching fluids.

case, completely dissolving the sludge by the end of the experiment. Both acetic acid at pH 5 and real leachate showed less vigorous behavior, generally yielding similar, much lower peaks over a longer time. Acetic acid at pH 6 or 7 and distilled water were nonaggressive, often showing negligible leaching. Based on these results, a solution of 5000-ppm acetic acid buffered to pH 5 was chosen as a simple but effective model of municipal refuse leachate.

Test Duration

The duration of each stage in the repetitive extraction is determined by a preliminary equilibrium test. A sample of waste is shaken with one bed volume of fluid, and small samples are taken (and replaced by fresh fluid) at intervals of 1, 2, 4, and 8 h, etc. In many cases, pH and conductivity measurements will give an adequate guide as to whether equilibrium is being approached, and other analytical measurements at this stage are confined either to easily determined or to fundamental toxic species. For many wastes, a shaking period of 4 h is adequate to attain equilibrium, but in other cases, equilibrium has not been attained even after 80 h. A very long "equilibrium time" may indicate kinetic control of leaching; in such cases, the time chosen should be based on a balance between the observed behavior and the degree of contact between the waste and percolating leachate expected in practice.

Validation Experiments

At the development stage, it is necessary to validate any small-scale leaching test against results from real life to ensure that the maximum concentration has been estimated reliably. Too often, leaching tests appear to have been developed in isolation (Anderson et al., 1979), with the justification that the conditions of the test are so strenuous that only an overestimate of the problem could result. It is indeed difficult to obtain reliable field results, because the leachate from a landfill is subject to many attenuation mechanisms other than initial leaching, and observed concentrations are thus almost always lower than a small-scale leaching test would suggest. Our solution has been to use large-column experiments to simulate only the primary leaching process in the field situation.

A series of 15-cm-diam columns were packed with 5 cm of toxic waste followed by 1 m of pulverized domestic refuse. The columns were irrigated with water at a natural rainfall rate and allowed to follow their natural degradation cycle. The progress of the columns was followed by monitoring gas evolution: although some air ingress did occur, equilibrium levels of 35–45% were attained for both carbon dioxide and methane after about 150–200 d, showing a satisfactory transition to anaerobic conditions. The construction of the columns was such that one bed volume of leachate passed through the waste about every 10 d. Leachate was collected both at the base of the column and above the layer of toxic waste.

The comparison of results from these column experiments with those from the shaking tests proved to be most instructive, and it was this comparison that led to an extension of the leaching test procedure to cover the cases of both early and delayed leaching. Some of the early laboratory tests underestimated observed leaching in the columns by a factor of up to 100, but improvements in the protocol have eliminated such gross discrepancies. A compilation of leaching test vs column results is shown in Figure 4. The general correlation is good, to within

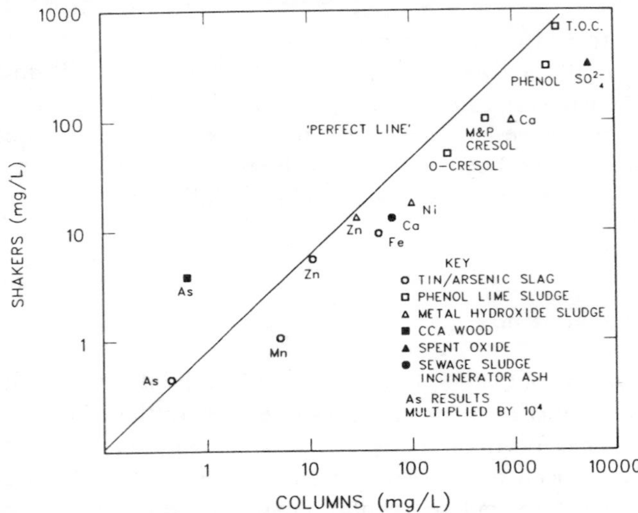

Figure 4. Comparison of results from laboratory leaching tests, "shakers," and column validation experiments.

a factor of 3. (It should be remembered that attenuation factors are generally measured as powers of 10.) Remaining underestimates by the laboratory test can generally be explained either by very early leaching or by experimental error (the tests were carried out over a 4-year period, and conditions did not always correspond exactly).

The major conclusion from these validation experiments is simple: any small-scale laboratory test must be verified by comparison with field, or simulated field, behavior.

Interpretation of Results

The results of the leaching test may be expressed either as a concentration plot (as in Figure 2) if the maximum concentration is of primary concern, or as a histogram of absolute releases (milligram per kilogram per bed volume) if one is interested in the total release. Attenuation factors may be calculated as the ratio of the maximum concontration in leachate to that in the waste as deposited. Typical values range from 10^{-5} for arsenic leaching from a blast furnace slag, to 10^{-3} for various metals from a metal hydroxide sludge, to 1 for phenol from a lime sludge. These measured attenuation factors are then used for decision-making within the framework outlined in Figure 1.

SIMULATED CODISPOSAL TEST

The WRU leaching test is one example of a test method designed to measure attenuation due to a single mechanism. We have also been working for several years with various experimental systems designed to measure attenuation of components of hazardous waste within a municipal waste landfill site as a whole. These systems are all based on laboratory columns ranging in size from 60 by 10 cm to 150 by 15 cm i.d.

One example of this work is a series of six large-column experiments, each containing a "sandwich" filling of pesticide waste between layers of pulverized domestic waste, the columns being irrigated at a natural rainfall rate. Leachate is collected both at the column base and about 1 m above, just below the pesticide layer. Three of the pesticides studied (atrazine, gamma-HCH or lindane, and pirimiphos-methyl) effectively did not leach, whereas relatively high levels of the supposedly nonpersistent 4-chloro-2-methylphenoxyacetic acid (MCPA) were still being leached after two years. Thus, solubility of the pesticide seems to be the dominant mechanism affecting leachate concentrations. These experiments have also yielded information on any effect of the pesticides on the microbiology within the refuse, and destructive sampling to examine persistence is currently in progress. A report on this work is in preparation (Stevens and Wilson, 1982).

CONCLUSIONS

The WRU at Harwell has developed a variety of simple tests for hazardous wastes prior to landfill disposal. The tests have been developed within a framework for practical decision-making, and attention has been paid to verifying the results against field behavior. The two major tasks for the future are to encourage the use of the methods by both landfill operators and controlling authorities, and to extend and develop the tests to cover a wider range of important attenuation mechanisms.

ACKNOWLEDGMENTS

This work has been sponsored by the United Kingdom Department of the Environment. The authors are grateful to their many colleagues who have contributed to the program, notably Dr. J. Bromley, Dr. S. Waring, Dr. C. J. Jones, Dr. C. Stevens, Mr. B. C. Hudson, and Mr. G. Baldwin. The paper is published by permission of the Director of the Harwell Laboratory.

REFERENCES

Anderson, M. A., R. K. Ham, R. Stegmann, and R. Stanforth. 1979. "Test Factors Affecting the Release of Materials from Industrial Wastes in Leaching Tests," in *Toxic and Hazardous Waste Disposal, Vol. 2*, R. B. Pojasek, Ed. (Ann Arbor, MI: Ann Arbor Science Publishers, Inc.).

Department of the Environment. 1978. *Cooperative Programme of Research on the Behaviour of Hazardous Wastes in Landfill Sites* (London: HMSO).

Jones, C. J., and P. J. McGugan. 1978. "An Investigation of the Evaporation of Some Volatile Solvents from Domestic Waste," *J. Hazardous Materials* 2:235-251.

Lowenbach, W. 1978. "Compilation and Evaluation of Leaching Test Methods," EPA-600/2-78-095, U.S. Environmental Protection Agency.

Perket, C. L., and W. C. Webster. 1981. "Literature Review of Batch Laboratory Leaching and Extraction Procedures," ASTM Symposium on Wastes, Florida.

Stanforth, R., R. Ham, M. Anderson, and R. Stegmann. 1979. "Development of a Synthetic Municipal Landfill Leachate," *J. Water Poll. Control Fed.* 51(7):1965-1975.

Stevens, C., and D. C. Wilson. 1982. Paper to be presented at ASTM Symposium on Wastes.

Stichting Verwijdering Afvalstoffen (SVA). 1979. "Laboratory Leaching Test," Report No. SVA/3014.

Waring, S., and B. C. Hudson. 1979. "A Simple Flash-Point Measurement Apparatus," *Lab. Pract.* 28(5):512-513.

DISCUSSION

***Milton Schloss*, Bureau of Land Management:** I'm interested in knowing if you've done any research on landfill liners regarding their durability, reliability, and if you've done any monitoring and for what length of time?

D. Wilson: We have done some small-scale work on the compatibility of landfill liners with municipal refuse leachate. We have designed an accelerated procedure in which we put liners under tension in a small pressure vessel, which contained a strong municipal refuse leachate held at 40°C under nitrogen pressure. We believe that 3 months of testing in this environment gives a reasonable estimate of long-term field behavior. We have not done any testing on the compatibility of landfill liners with hazardous wastes.

***Ishwar P. Murarka,* Electric Power Research Institute:** You made comments about the attenuation factors occurring when you are running the leaching experiments and tests. Could you talk about what the properties of the soil are that would contribute to the attenuation. Obviously, it would be different depending on the particular species coming out and the particular soil medium that you may have there.

D. Wilson: We began in the landfill business about 9 years ago with a fairly large research program on behalf of the U.K. Department of the Environment. This program identified the sites in the country which were suspected from the hydrogeology of having the potential for groundwater pollution. There were some 50 sites from a total of 2500. We then looked at 20 of these in detail, carrying out detailed field investigations, looking at migration of pollution plumes, and so on. From that we got a good idea of the sorts of geological strata which are likely to give you attenuation, at least in my country. We also carried out large-scale lysimeter experiments in which we isolated 3-m^3 blocks of undisturbed sandstone. We irrigated them with synthetic leachate solution, for example, a heavy metal solution, and we monitored the migration of the plume through this. Our first suction probe was 40 cm below the surface, and none of the heavy metals ever reached that probe. We got our information by drilling from the top, and migration of all the metals, even nickel (which was the most mobile), was extremely slow; so we believe that saturated sandstone (and we have information on other materials) will give you significant attenuation. If you have a substantial unsaturated soil, then the attitude in my country is fairly relaxed about the philosophy of what we call dilute and disperse.

***David Friedman,* EPA:** As you might surmise, I was very interested in what you said. I was curious about any work on the less polar organics, the mobility of those, and in fact, are they mobilized by the landfill leachate?

D. Wilson: I think I probably require notice of that question. As usual in our development program, and this may surprise some of you, but I guess it won't surprise many of you, one of our main problems was identifying wastes that we could test. There just don't seem to be many problem wastes about, so in our leachate-testing program we have not looked at many nonpolar organic compounds. Of the field situations that we have investigated, we have tried to look at a range of things, and I guess we have some information. We certainly have information, for example, on mineral oils. That partially answers your question.

CHAPTER 9

DEVELOPMENT OF MULTIMEDIA ENVIRONMENTAL GOALS FOR POLLUTANTS IN SOLID WASTES AND LEACHATES

Garrie L. Kingsbury, Karen L. Mack, and Scott M. Harkins

Research Triangle Institute
P.O. Box 12194
Research Triangle Park, North Carolina 27709

ABSTRACT

Multimedia environmental goals (MEGs) are concentration goals for specific pollutants in air, water, and soil. They have been developed for use in environmental assessments to identify and rank chemicals of concern in discharge streams. MEGs are being used extensively by the Environmental Protection Agency and others to aid in data interpretation.

A methodology to generate MEGs from available data was introduced and demonstrated in 1977. To date, goals for more than 600 chemical pollutants have been specified. Compounds addressed by MEGs are assigned to categories on the basis of their chemical functional groups.

MEGs for solid waste contaminants are derived from water goals based on the assumption that the significant exposure routes will involve leaching of pollutants into ground or surface waters. Water MEGs reflect existing federal guidelines, potential acute and chronic effects (indicated by human and/or animal data), genotoxic potential, and aquatic toxicity.

Applications and limitations of the MEGs are discussed, and revisions to the methodology now under consideration are presented.

INTRODUCTION

Multimedia environmental goals are chemical-specific goals expressed as concentrations in air, water, or soil (or solid waste). Separate sets of values reflect potential human health and ecological effects. The quantitative goals serve as criteria for interpreting data to support environmental assessments. MEGs may be used to rank a large number and variety of chemicals including many nonregulated pollutants. The goals are not intended to be used as standards; instead, they are comparative indexes used to guide decision-making related to desirable control levels and to focus additional data-gathering efforts on the pollutants and streams of *potentially* greatest environmental impacts.

Development of the MEGs began in 1976 as part of the environmental assessment methodology for synthetic fuels sponsored by the U.S. Environmental Protection Agency (EPA), Industrial Environmental Research Laboratory at Research Triangle Park, North Carolina. Environmental assessment requires that pollutant levels in waste streams be related to their environmental impacts. The need for MEGs arises in this context. The MEG program sets forth a procedural approach to characterize pollutants in terms of their potential effects for the purpose of environmental assessment.

Since the initial methodology was introduced in 1977, MEGs have been used extensively by the EPA and others in assessments ranging from coal conversion to textile plants. The methodology, background information summaries for each chemical, and the resulting MEG values are presented in three separate EPA reports (Cleland and Kingsbury, 1977; Kingsbury et al., 1979,1980). More than 600 chemicals have been addressed. Updating of the goals to reflect new information continues, along with evolution of the methodology.

Practical Considerations

The MEG methodology has been developed with several important practical considerations in mind. These are summarized as follows:

1. Goals must not conflict with existing federal guidelines.
2. Worst-case assumptions should be applied in employing available data (i.e., most-conservative values should be used when discrepancies or uncertainties arise).
3. Pollutants should be classified in a manner that permits comprehensive coverage without having to consider thousands of com-

pounds individually. Organization of the chemical data and goals should allow its systematic utilization.

4. A procedure is needed for quantifying relative toxicity of classes of compounds as well as individual compounds.
5. A uniform approach in applying existing health and ecological effects data is needed as well as a means for directly delineating health/ecological effects data gaps.
6. A wide variety of potentially adverse effects, including oncogenic and teratogenic effects, must be reflected in the goals.
7. Sufficient information should be provided to indicate the type and extent of information supporting each goal.
8. Goals for continuous ambient exposure [ambient MEGs (AMEGs)] and goals for discharges [discharge MEGs (DMEGs)] are needed. The DMEGs are to reflect potential impacts of short-term exposure.
9. Previously established methods to set forth goals should be adopted where practical. New methods should be developed to supplement existing methodology.
10. Provision is needed for evaluating goals to determine if they are reasonable in light of all available information.

MEG Categories

Organic compounds addressed by MEGs are grouped in categories on the basis of their chemical functional groups (e.g., aliphatic or aromatic hydrocarbons, alcohols, amines, and polycyclics). Inorganic species are categorized according to the principal or most toxic element. The chemical categories serve to combine pollutants that behave in toxicologically similar fashion. Presently, 26 organic categories and 73 inorganic categories are specified. Six-digit identification numbers indicating the category and chemical complexity are assigned to each MEG compound to facilitate the association of structurally similar compounds. These unique codes also help to alleviate many nomenclature problems.

Background Information Summaries and Goals

A collection of background information is assembled for each compound addressed by MEGs. In addition to the quantitative data used to compute goals, each background information summary describes physical properties, occurrence, associated compounds, toxic and oncogenic effects, and existing standards and guidelines for the subject chemical. Information is compiled largely from collective-type references.

Numerical goals for each chemical reflect existing federal guidelines or are computed from available toxicity data via equations designed to consider the severity of the reported effect for which dose/response data are reported. Types of data used in the equations include occupational exposure guidelines, acute animal toxicity data such as lethal dose or lethal concentration, dosages causing a significant increase in tumor incidence, concentrations affecting vegetation, and aquatic toxicity data such as 96-h threshold limit median (TLm-96) and bioaccumulation potential. Safety factors are incorporated to offset uncertainties inherent in extrapolating adverse effects levels to goals. A hierarchy among the various equations is established to designate the preferred goal when several types of information are available. The preferred value reflects the data type most related to a goal definition. Existing guidelines are the first preferred basis for MEGs.

Background information compiled for each chemical and the corresponding MEG values are grouped for presentation by chemical category. By combining information for the several chemicals in a given category, it is frequently plausible to provide MEG values by analogy for compounds for which little or no toxicity data are available.

DISCHARGE MEGs FOR ASSESSMENT OF SOLID WASTES

Although the goals have been applied primarily in the assessment of atmospheric discharges and aqueous effluents, the DMEGs are also useful for interpreting chemical characterization data for solid waste. To identify constituents that may pose environmental problems, chemical concentrations analyzed in the subject solid waste may be compared with corresponding $DMEG_{solid\ waste}$ values. This approach incorporates an *assumed* worst-case leaching scenario to permit ranking of the contaminants. Alternatively, chemical concentrations that may be anticipated to occur in leachate from the subject solid waste may be compared to $DMEG_{water}$ values. The latter approach is preferred because it results in a more realistic appraisal of the potential hazard. The methodology used to establish $DMEG_{water}$ and $DMEG_{solid\ waste}$ is discussed briefly in the next two sections. Further details are contained in the MEG reports.

Basis for $DMEG_{water}$

The preferred $DMEG_{water}$ may reflect primary or secondary drinking water standards or water quality criteria for protection of human health or aquatic life. In the absence of existing guidelines, the basis for $DMEG_{water}$ may lie in the results of laboratory bioassays or industrial hygiene experience. Equations used to generate discharge goals for water are presented in Table I.

Table I. Equations to Generate $DMEG_{water}$ (mg/L)

Basis	Equation
Health Effects	DMEG[a] = 5 × Drinking water standard or water quality criterion for protection of human health (mg/L)
	DMEG = 15 × Occupational exposure limit or $DMEG_{air}$ (mg/m^3)[b]
	DMEG = 0.3[c] × Lowest effective dosage to elicit genotoxic response in animals (mg/kg)[c]
	DMEG[a] = Level in drinking water (mg/L) associated with a cancer risk of 10^{-3}[a, d]
Ecological Effects	DMEG[a] = 5 × Water quality criterion for protection of aquatic life (mg/L)
	DMEG = 0.1 × TLm-96 (mg/L) for most sensitive species

[a]Indicates preferred value.
[b]$DMEG_{air}$ may be based on lethal dose data such as oral rat LD_{50} (mg/kg) or on lethal or effective concentrations (mg/m^3) reported from inhalation studies.
[c]The multiplier varies between 0.12 and 0.5 depending on extent of positive test results reported.
[d]This equation pertains to EPA 1980 Water Quality Criteria and is not included in the 1977 MEG methodology.

The equations used to generate MEGs involve multipliers based on a variety of assumptions. The multiplier, 5, applied to water quality criteria for aquatic life is suggested by the relationship between minimal risk levels and hazard levels for aquatic life as recommended by the NAS/NAE (1973). Aqueous discharge concentrations below the $DMEG_{water}$ derived by this equation, upon dilution by a factor of 5, will not exceed the ambient water criteria. Short-term exposures of aquatic life to the undiluted aqueous discharge would probably not result in *irreversible* effects. Similarly, the multiplier, 0.1, applied to the TLM may be interpreted as a modified application factor (i.e., the ratio of safe-to-lethal concentration). To predict safe ambient levels for toxic chemicals, application factors of 0.05 and 0.01 for nonpersistent and persistent toxicants, respectively, have been recommended (NAS/NAE, 1973). Discharge concentrations at the DMEG level will, upon dilution by a factor of 5, not exceed the predicted safe continuous exposure limits for nonpersistent toxicants. Dilution by a factor of 10 is required for persistent contaminants.

The multiplier, 15, is applied to acceptable contaminant levels in workplace air or to $DMEG_{air,\ health}$ to translate a permissible air exposure level to a comparable water exposure level. This approach is adapted from a model introduced by Stokinger and Woodward (1958).

The assumptions incorporated in this model are as follows:

1. Total contaminant loading from 30 m^3/d of air is absorbed.
2. Permissible daily dosage from air is equal to the permissible air concentration (mg/m^3) multiplied by 30 m^3/d.
3. Permissible daily dosage from 2 L of water is equal to permissible daily dosage from air.
4. Unity absorption factors are assumed.

MEGs for Solid Wastes

Concentration goals for contaminants in solid waste are based on the assumption that the principal route of exposure to contaminants in solid waste is through leaching into drinking water sources or surface waters supporting aquatic life. A worst-case scenario involving leachability of toxic contaminants assumes that (1) all contaminants in 1 kg of waste would be leached by 2 L of water, and (2) the resulting concentration of the contaminant in the leachate after dilution by 100 should not exceed the value of the $DMEG_{water}$ for the subject pollutant. The hundred-fold dilution might occur through a combination of site- and chemical-specific variables including chemical degradation, evaporation, adsorption, and attenuation through soil.

The resulting equations are given as follows:

$$DMEG_{solid,\ health}\ (\mu g/g) = 0.2 \times DMEG_{water,\ health}\ (\mu g/L)$$
$$DMEG_{solid,\ ecology}\ (\mu g/g) = 0.2 \times DMEG_{water,\ ecology}\ (\mu g/L)$$

APPLICATION OF THE GOALS

Two solid wastes from coal gasification, Lurgi gasifier ash, and heavy tar, have been selected to illustrate the use of DMEGs in assessing the "cause for concern" in solid discharges. Potential contaminants in the gasifier ash are almost exclusively inorganic, whereas tar contains both organic and inorganic species. The following analyses are based on the Source Analysis Model (SAM/IA) (Schalit and Wolfe, 1978), which was developed for use with the MEGs. The key to the MEG/SAM approach is the quantity called discharge severity (i.e., the ratio of the chemical concentration in the discharge stream to the DMEG value for that chemical). An inherent assumption is that the toxic potential of a discharge stream is an additive function of the hazards associated with the individual components present in the stream.

Gasifier Ash Analysis

Contaminant levels of selected inorganics in gasifier ash from Illinois No. 6 coal are presented in Table II along with current solid waste DMEGs. Only the reported ash constituents with discharge severities exceeding unity are listed; 53 elements were actually analyzed in the ash. Slurry supernate concentrations from the gasifier ash are compared with the constituent water DMEGs in Table III. Characterization data in these tables are reported by Griffin et al. (1980).

Comparison of the ash concentrations for the various contaminants with their appropriate DMEGs indicates that 20 species (of 53 ana-

Table II. Gasifier Ash Constituent Levels Compared to Solid Waste DMEGs

Constituent	Concentration in Lurgi Gasifier Ash[a] (mg/kg)	$DMEG_{solid}$ (mg/kg)		Discharge Severity (Concentration/DMEG)	
		Health	Ecology	Health	Ecology
Aluminum	108,121	4,000	200	27	540
Arsenic	3	0.44	40	7	<1
Barium	950	1,000	200	<1	5
Beryllium	12	1.4	5.4	9	2
Boron	355	200	220	1.8	1.6
Cadmium	<1.6	1.0	0.012	<1.6	<133
Calcium	16,652	48,000	3,200	<1	5
Chromium	212	50	0.3	4	7.1
Copper	57	1,000	2.4	<1	24
Iron	143,780	300	50	479	2,876
Lead	45	50	0.76	<1	59
Manganese	1,859	50	20	37	93
Molybdenum	30	15,000	10	<1	3
Nickel	89	640	56	<1	2
Silver	<0.4	50	0.12	<1	<3
Strontium	370	168,000	2	<1	185
Sulfide	1,500	960	0.4	1.6	3,750
Titanium	6,295	18,000	160	<1	39
Vanadium	184	150	2	1.2	9
Zinc	400	5,000	36	<1	11

[a]Data are from Illinois No. 6 coal as reported by Griffin et al. 1980.

Table III. Gasifier Ash 10% Slurry Supernate Concentrations Compared to Water DMEGs

Constituent	Concentration in Supernate, pH 5.10 (mg/L)[a]	$DMEG_{water}$ (mg/L)		Discharge Severity (Concentration/DMEG)	
		Health	Ecology	Health	Ecology
Aluminum	2	20	1.0	<1	2
Ammonium	8	2.5	0.6	3.2	13
Antimony	<1.0	0.73	3.1	<1.4	<1
Arsenic	<1.0[b]	0.002	0.2	500	5
Barium	<0.1	5	1.0	<1	<1
Beryllium	<0.02[b]	0.007	0.027	3	<1
Boron	4.5	47	1.1	4.5	4.1
Cadmium	0.03	0.05	0.00006	<1	500
Calcium	480	240	16	2	30
Chromium	0.02	0.25	0.0015	<1	1.3
Copper	0.02	5.0	0.012	<1	1.7
Iron	0.19	1.5	0.25	<1	<1
Lead	0.1	0.25	0.004	<1	25
Manganese	1.94	0.25	0.10	8	19
Nickel	0.13	3.2	0.28	<1	<1
Selenium	<0.1	0.05	0.01	<2	<10
Sulfate	943	1300	43	<1	22
Sulfite	<0.2	4.8	0.002	<1	<10
Titanium	<0.5[b]	90	0.82	<1	<1
Zinc	5.5	25	0.18	<1	30

[a]Source: Griffin et al., 1980.
[b]Detection limit of the method of analysis. Concentration could not be determined.

lyzed) exceed the $DMEG_{solid\ waste}$. Table III also indicated that sulfide and iron may pose significant health problems if they are leached at a significant rate; aluminum, strontium, cadmium, manganese, and lead are also flagged as potential problems, although less severe. In the 10% slurry supernate, arsenic and cadmium are indicated to be of notable concern. Dilution of the supernate by a factor of 20 is necessary to achieve a final dilution analogous to that assumed in the solid waste leaching scenario (i.e., 1:200). Therefore, only discharge severities greater than 20 should be considered significant for the 10% supernate. Calcium, lead, and sulfate are flagged as potential problems. Although the gasifier ash is a large-quantity waste, most of the trace elements that are present are unlikely to pose a significant environmental hazard.

The arsenic level in the supernate is not determined, the level reported merely indicating the lowest detectable level. Additional testing would be necessary to determine whether the leaching of arsenic from the ash poses a potential environmental problem. The health-based $DMEG_{water}$ for arsenic is the level associated with a cancer risk of 10^{-3}

in water used for continuous human consumption (EPA, 1980). The ecology-based DMG_{water} reflects the 1980 water quality criterion for protection of freshwater aquatic life. Examination of background data for calcium is needed to assess the significance of the ecology-based discharge severity. Calcium contributes to the level of dissolved solids in water and influences the water hardness, but it is essentially nontoxic to aquatic life. There is evidence that increasing hardness decreases the effective aquatic toxicity of certain elements (i.e., copper and zinc). Total dissolved salt content of the leachate from gasifier ash may warrant further consideration, however.

Coal Tar Analysis

Most tar from coal gasification is likely to be used as boiler fuel on-site, or it can be refined into a variety of products. A portion of the heavy tar, however, may be disposed as solid waste. In the operating coal gasification facility in Kosovo, Yugoslavia, approximately 15% of the tar is landfilled (Bombaugh et al., 1980). An analysis to determine the potential severity of impacts from disposal of such tars is of interest.

Organic constituents that have been identified in coal tar are listed in Table IV by MEG category (Cleland, 1981). The compounds of concern, based on discharge severity using $DMEG_{solid\ waste}$, are in italics. Although specific goals have not been published for all of these constituents, surrogate goals may be assigned based on values for structurally similar compounds. A conservative approach is to assign the lowest DMEG of the category to compounds without specific values. This is the approach that is customarily used in performing a MEG/SAM analysis. An alternative method (and the one used in this analysis) is to determine the MEG compound most closely related structurally to the new species of interest and to assume that goals for the two should be equivalent.

The magnitude of the discharge severity is significant for several compounds. Organic constituents that exceed their solid waste DMEGs by more than a factor of 10 are phenolic compounds, benzidine, benzo(a) anthracene, dibenzo(a,h) anthracene, and benzo(a) pyrene. The latter four compounds are recognized as potent carcinogens. In addition, arsenic and chromium levels are high. Although leachate data are unavailable for the tar sample, it is likely that several of the components could be extracted in concentrations well in excess of water goals.

The organic chemicals of concern in tar, as indicated by the MEG/SAM analysis, are not addressed by drinking water standards. Application of the Resource Conservation Recovery Act extraction procedure

Table IV. Organic Constituents in Coal Gasification Tar[a]

MEG Category	Constituents[a]
Aliphatic Hydrocarbons	Alkanes (C_6+)
Ethers	Anisoles
Alcohols	Alkyl alcohols (C_6+)
Carboxylic Acid Derivatives	*Phthalates,*[b] adipates, aliphatic esters (C_9+)
Amines	Aniline, *alkylanilines,* benzofluorenamines, methylamino-acenaphthylene, benzidine, aminonaphthalenes, aminotetralin
Aromatic Hydrocarbons	*Biphenyl,* indene
Phenols	*Phenol,* alkyl phenols, indanol, naphthols, alkyl naphthols, acenaphthols, alkyl acenaphthols, hydroxy alkyl anthracene, hydroxy alkyl pyrenes, hydroxy benzofluorene
Fused Polycyclic Hydrocarbons	Naphthalene, alkyl naphthalenes, anthracene, alkyl anthracenes, *phenanthrene,* acenaphthene, acenaphthylene, alkyl acenaphthylenes, *benzo(a)anthracene,* triphenylene, chrysene, pyrene, *dibenzo(a,h)anthracene, benzo(a)pyrene, benzo(e)pyrene,* perylene, benzo-(e)perylene, benzo(g,h,i)perylene
Fused Nonalternant Hydrocarbons	Methyl fluorene, benzofluorenes, benzofluoranthrenes, indeno (1,2,3-cd) pyrene
Heterocyclic Nitrogen Compounds	Alkyl pyridines, *quinolines,* acridine, alkyl acridines, phenanthridine, alkyl benzoquinolines, indole, carbazole, methyl carbazole
Heterocyclic Oxygen Compounds	Dibenzofuran
Heterocyclic Sulfur Compounds	Benzothiophene

[a]Composite list of constituents. Source: Cleland, 1981.
[b]Italics indicate discharge severity exceeds unity.

toxicity criteria would not flag this material as a hazardous waste. Nonetheless, there is a sound basis for treating this material as hazardous.

PHASE II MEGs

Over the last four years, several issues regarding the MEG methodology have been raised, and weaknesses in the approach in deriving the MEGs are widely recognized. Presently, development of a second generation of MEGs is under way. The Phase II methodology will retain many of the successful features of the original approach, such as the category organization, and it will incorporate refinements that will lead to a more defendable set of goals.

Phase II MEG values for water will be derived through models analogous to those used to derive the EPA 1980 water quality criteria. One method will involve isolating the threshold level through lowest observed adverse effects levels and no response levels. Safety factors or uncertainty factors will be determined according to guidelines given by the National Academy of Sciences (1977). Goals for carcinogens will be associated with specified risk levels (10^{-5} for AMEGs, 10^{-3} for DMEGs). Phase II MEGs for solid waste will continue to be based on a leaching scenario because this appears to be the most realistic and sensitive exposure route for solid waste contaminants.

Data that are required to derive MEGs are not always available for the particular compound of interest. Often, however, a pattern of biological activity may be identified for many of the compounds within a specific chemical class. In these instances, it may be possible to estimate missing data by comparing structural features of the untested compound to others within its chemical class. Structure-activity relationships will be used to determine Phase II MEGs for certain substances that are not individually characterized by toxicological testing.

For example, the biological activity levels of chloro- and alkyl-phenols are generally reflected by the hydrophobic and electronic characteristics of the ring substituents. If specific quantitative biological data can be obtained for a sufficient number of phenols, the data may be statistically correlated to values that represent the substituent characteristics. A biological data point can then be estimated, with confidence limits, for any chloro- or alkyl-phenol based only on its substituent characteristics.

In some cases, a particular type of biological data is available for only a few compounds of a chemical class, disallowing the statistical correlation described above. However, based on correlations involving other toxicity endpoints, if it is likely that certain chemical or structural characteristics are good indicators of a specific kind of biological activity of that class, then the nearest-neighbor approach may be used to estimate a missing value. Several types of molecular descriptors are useful for determining the best surrogate.

CONCLUSION

The current MEG methodology and the resulting set of numerical concentration goals provide valuable guidance for the initial step in data interpretation. They provide a reasonable basis for comparison to assess potential discharge hazard as a result of one or more toxic components.

Limitations in the current data pertinent to environmental effects preclude establishing acceptable exposure levels for every chemical toxicant. There are many patterns of biological activity, however, among certain chemical groups that can be used to estimate goals and to guide decision-making.

Thoughtful use of the collective set of goals can focus attention on the chemical species of major concern in discharge streams and eliminate from further consideration those species unlikely to result in adverse environmental impacts.

REFERENCES

Bombaugh, W. E. C., K. W. Lee, and W. S. Seames. 1980. "An Environmentally Based Evaluation of the Multimedia Discharges from the Lurgi Coal Gasification System at Kosovo," presented at the Symposium on Environmental Aspects of Fuel Conversion Technology, V, St. Louis, MO.

Cleland, J. G. 1981. "Environmental Hazard Rankings of Pollutants Generated in Coal Gasification Processes," EPA-600/57-81-101, Research Triangle Institute, Research Triangle Park, NC (NTIS PB81-231 698).

Cleland, J. G., and G. L. Kingsbury. 1977. "Multimedia Environmental Goals for Environmental Assessment, Vol. I," EPA-600/7-77-136a, Research Triangle Institute, Research Triangle Park, N.C. (NTIS PB 276919); "MEG Charts and Background Information, Vol. II," EPA-600/7-77-136b (NTIS PB 276920).

Griffin, R. A., R. M. Schuller, J. J. Suloway, N. F. Shimp, W. F. Childers, and R. H. Shiley. 1980. "Chemical and Biological Characterization of Leachates from Coal Solid Wastes," Illinois Institute of Natural Resources, State Geological Survey Division, Champaign, IL.

Kingsbury, G. L., R. C. Sims, and J. B. White. 1979. "Multimedia Environmental Goals for Environmental Assessment: Vol. III," EPA-600/779-176a; "MEG Charts and Background Information Summaries" (categories 1-12), Research Triangle Institute, Research Triangle Park, NC, EPA-600/779-176a (NTIS PB 80-115108); "Vol. IV. MEG Charts and Background Information Summaries" (categories 13-26), EPA-600/7-79-176b (NTIS PB 80-115116).

Kingsbury, G. L., J. B. White, and J. S. Watson. 1980. "Multimedia Environmental Goals for Environmental Assessment, Vol. I (Supplement A)," EPA-600/7-80-041, Research Triangle Institute, Research Triangle Park, NC (NTIS PB 80-19619).

National Academy of Sciences, National Academy of Engineering (NAS/NAE). 1973. "Water Quality Criteria 1972," EPA-R3-73-933, National Academy of Sciences, Washington, DC.

National Academy of Sciences (NAS). 1977. "Drinking Water and Health," National Academy of Sciences, Washington, DC.

Schalit, L. M., and K. J. Wolfe. 1978. "A Rapid Screening Method for Environmental Assessment of Fossil Energy Process Effluents," EPA-600/7-78-015, Acurex Corporation, Mountain View, CA, (NTIS PB 276088).

Stokinger, E. H., and R. L. Woodward. 1958. "Toxicologic Methods for Establishing Drinking Water Standards," *J. Am. Water Works Assoc.* p. 515.

U.S. Environmental Protection Agency (EPA). 1980. "Water Quality Criteria Documents," *Fed. Regist.* 45(231):79316-79357.

DISCUSSION

***John Brown,* CEGB:** We've gone down a similar pathway in England. We don't actually call them by the same name, but there are attempts to set criteria for disposals for particular kinds of land reclamation. One of the points that we've put to the department that is making these kinds of arrangements is this: in your scheme, if I understand it properly, if you do your DMEG calculation and the result is a good one, then because of the conservative calculation you're "in the clear," so to speak. If the result isn't a good one, then all of a sudden you're in the spot light, and you have to be further evaluated to see what the answer will be. The point I would like to make to you, and which you might like to comment on, is that in practice these decisions are not made by people who are scientifically trained. They have people in local authorities who have license arrangements at their disposal, and we are afraid that the conservatism you've built in will, in some cases, be used unimaginatively and turned into rigid standards.

G. Kingsbury: That's a very good comment and a good question, and I would like to respond to it. Whenever you write down a name or a number you're sticking your neck out because it can be guaranteed that number is going to be misused by somebody, sometime, probably within the hour. This is true with the MEGs and, at the risk of having these numbers misused, we've done it anyway. It does put one's reputation at stake to go ahead with specifying these numbers. Frequently, there is, as you know, an urgency in decision-making relative to environmental impacts. There is a need to make sense out of chemical analyses that will link the presence of one or more specific chemicals to potential problems. If used properly, the MEG/SAM approach will raise questions in most of the right places and will flag a few false positives. In my opinion, a method to guide the data interpretation on the basis of conservative estimates reflecting the available data base on effects is far better than no guidance at all.

MEGs should be updated continually to take into account the state of the art in predictive toxicology and risk assessment as well as to reflect the most up-to-date acute and chronic toxicity determinations.

***David C. Wilson,* Harwell:** I would like to congratulate you on what seems to be an excellent approach, and I have no argument with the basic structure of what you are doing as long as it is contained within the guidelines you set down. I would like to ask you a question about something that is slightly different, which I guess is what Mr. Brown was asking about. One thing our department is trying to do is set guidelines for contaminated soil. This is at a site used in the past by industry which is now being proposed for the redevelopment. They are trying to set guidelines which will, if you like, raise a flag as to whether or not that site needs to be considered in depth before you build houses or whatever else you are going to do on the site. They have published some discussion values on that one, I have had the misfortune to have the job of trying to work out similar values for contaminants from old coal gasification works trying to extend metals research to organics, which is not such an easy problem, and I was wondering if you have had any experience in this field?

G. Kingsbury: Not with the real problem, only conceptually and that is basically the methods that I have talked about. The difference is that where I talked about discharge MEGs you would now want to consider AMEGs, for continuous exposure. You are still talking about the exposure method or the exposure route that is most significant for that waste, which is still likely to be through contamination of water. Presumably, your task involves specific sites allowing much more sophisticated methods to predict the extent to which materials might be leached.

David Wilson: Certainly water is a problem. Water is the problem that you seem to think about first. What we are concerned about is mainly just contact. Children living in a house with a garden for 20 years—is that going to be a problem to them? And I do not see anything in the MEGs that you have described which apply to that particular problem.

G. Kingsbury: We have done some thinking about it. We do not have good answers. Much research is needed in the area of uptake by plants. I believe it is reasonable to assume that any uptake is via water, however. Certainly, absorption through skin contact, particularly for children, could be of concern. Skin painting studies in mice, as well as much industrial experience, demonstrates the importance of this route of exposure for many coal-derived materials. I believe that dermal exposure to soils containing high concentrations of heavy metals (e.g., lead) could also result in significant effects. I could not estimate the levels to trigger concern based on skin contact.

***Arthur Hounslow*, Oklahoma State University:** I was very interested in your classification of organics. This has been a major problem. How do you get over the problem of multifunctional groups and, secondly, have you tried a classification based on absorption techniques of any kind?

G. Kingsbury: We have not based anything on absorption techniques. How do we get around the multifunctional group? We attempt to identify the functional group that will result in the most significant effect and categorize a pollutant on that basis. Another approach would be to identify the functional group that will react first, because in many cases you do not know which group will contribute most to the toxic potential. Chances are that the most reactive group is the most important. There are many compounds that fall into gray areas somewhere in between these chemical categories that I showed. In general, a reasonable assignment can be made. We have categorized perhaps 1000 compounds, and the system seems to be effective. It is plausible to assign compounds to at least one of these 26 categories, but one should write down the reason for exceptional substances.

***R. A. Griffin*, Illinois State Geological Survey:** In the original four-volume report, discharge MEGs were called MATE values, and the concept of discharge severity was not mentioned. My question is where can the audience get a copy of the revised version, or is there a publication available?

G. Kingsbury: Yes. Five volumes on MEGs have been published to date. They are available through NTIS; they are all referenced in this chapter. The latest one was December 1980. The concept of discharge severity is set forth in another document that is being revised; it is also referenced in this chapter.

CHAPTER 10

CASE STUDIES IN INDUSTRIAL WASTE CHARACTERIZATION

Barton P. Simmons and Robert D. Stephens

Hazardous Materials Laboratory Section
California Department of Health Services
2151 Berkeley Way
Berkeley, California 94704

ABSTRACT

The Hazardous Materials Laboratory Section and the Hazardous Materials Management Section of the California Department of Health Services, as part of the California hazardous waste management program, have developed a scheme for characterizing industrial wastes: (1) background information search, including a review of the appropriate industrial processes and a review of pertinent physical, chemical, and biochemical reactions that may have affected the composition of the waste; (2) initial sampling and analysis, using such techniques as gas chromatography/mass spectrometry, X-ray fluorescence, in addition to screening tests for hazard assessment; (3) review of initial data and design of the comprehensive characterization; (4) detailed sampling and analysis, including a detailed safety plan; and (5) interpretation of results for legal purposes, health risk assessment, and environmental impact assessment.

A sample case history of an abandoned acid petroleum sludge site is given to illustrate this characterization process. Although the general approach to characterization is the same, each waste presents particular problems. These problems are discussed, with emphasis on the chemical and physical reactions that complicate the problem of adequate characterization. The goal in any problem is the same: to provide a characterization that is adequate for health assessment, legal interpretation of environmental assessment, or mitigation design.

CASE STUDIES IN INDUSTRIAL WASTE CHARACTERIZATION

The California Department of Health Services (DHS), as the lead agency in the state for hazardous waste control, has dealt with a wide variety of problems in waste stream characterization since inception of the state program in 1972. The characterization work done by DHS is for assessment of existing and potential human and environmental hazards. An iterative scheme has evolved for major characterization tasks. The effectiveness of this scheme is demonstrated by a case study of the abandoned McColl disposal site.

CHARACTERIZATION SCHEME

Background Information Search

The importance of background information cannot be overemphasized, particularly when dealing with unknown waste streams. The information typically uncovered includes the industrial processes likely to have generated the wastes in question, usual composition and properties of those waste streams, documented chemical and physical reactions that may have affected the waste since its generation, and any peculiar conditions of storage of the waste.

For onsite disposal investigations of a pesticide formulation facility, for example, the best background information typically comes from an audit of the historical formulation processes and waste disposal practices. Some offsite problems may also be tied to particular industries through haulers' manifests, drum labels and markings, interviews with haulers, historical aerial photos, or other information.

Initial Sampling and Analysis

In this stage of characterization, composite samples of the waste are taken and analyzed for the range of parameters indicated by the background information in addition to those parameters of general concern. Samples are usually composited to minimize the number of samples at this stage. Table I indicates some methods and parameters that may be used.

Table I. Typical Parameters for Initial Analysis

Hazard Assessment	
pH	
Flash Point	
Reactivity	
Spot Tests	Gas Chromatography
Cyanides	PCBs
Sulfides	Pesticide Scans
Other Selected Inorganics	Selected Organics
Gas Chromatography/Mass Spectrometry	HPLC
Volatiles	Herbicides, PAHs
Extractables	
X-Ray Fluorescence or Inductive Coupled Plasma AES	
Metals Scan	

Review of Initial Data and Design of a Comprehensive Characterization Plan

As will be illustrated in the case study, the initial results provide valuable guidance when designing a detailed sampling and analysis scheme. Expensive techniques can often be replaced by less expensive, more selective techniques. For example, if a few elements of concern have been identified by X-ray fluorescence or inductively coupled plasma atomic emission spectroscopy, they may be quantitated by using atomic absorption spectroscopy in a more cost-effective manner.

The results of the initial sampling and analysis are also used when designing the safety plan for field and laboratory personnel.

Detailed Sampling and Analysis

This stage is aimed at producing data that are completely adequate for the problem at hand, be it health assessment, environmental assessment, legal interpretation, or other purposes. For in situ characterization, the sampling often involves drilling to determine the vertical and areal disposition of waste.

Interpretation of Results

This stage again requires an assessment of pertinent physical and chemical reactions, including modeling, where necessary, to discuss the properties, distribution, and movement of materials.

ABANDONED SITE WASTE CHARACTERIZATION

Acid Petroleum Sludge Site

The McColl disposal site is located in Orange County, California, adjacent to residential developments. In response to complaints from local residents, a task group representing federal, state, and local agencies was formed in October 1980 to assess the nature and extent of impacts from the site. The discussion presented here deals primarily with waste characterization as the primary step toward environmental assessment, health impact assessment, and mitigation design.

Background Information

A background information search coordinated by Ecology and Environment, Inc., a U.S. Environmental Protection Agency (EPA) contractor, revealed that the site had received two major waste streams: refinery acid sludge and rotary drilling mud waste. Because the site had had free access for many years, the possibility of deposition of other waste streams could not be ruled out. The refinery acid sludge is believed to have resulted from the production of high-octane fuel during World War II. The waste was probably generated by alkylation of secondary alkanes with alkenes to form tertiary alkanes (e.g., the reaction of isobutane with isobutylene to form isooctane). Sulfuric acid sludge was created by many other processes used during World War II, including the removal of oxygenated compounds, nitrogen compounds, and sulfur compounds from various petroleum products (Kalichevsky and Stagner, 1942).

Initial Sampling and Analysis

As the first stage in characterization, shallow subsurface (0–0.5 m) composite sludge/soil samples were taken from each area that had been identified as a separate sump from historical aerial photos or visual observation. Samples were taken with hand augers by personnel wearing chemical cartridge respirators for organic and acid vapors plus other recommended clothing for sampling acid waste.

The samples were split in the field and sent to three laboratories for analysis: the EPA National Enforcement Investigations Center (NEIC) in Denver; and the Air and Industrial Hygiene Laboratory (AIHL) and the Hazardous Materials Laboratory (HML), both of the California Department of Health Services in Berkeley.

Review of Initial Data

Tables II and III contain results for analysis of the composite samples. The analysis results are generally consistent with the composition of sulfuric acid petroleum sludge, which is known to generate significant quantities of sulfur dioxide with time (Kalichevsky and Stagner, 1942). Although large amounts (up to 38,000 ppm) of surfur dioxide were found in the shallow subsurface samples, ambient air monitoring performed in October 1980 did not show high ambient sulfur dioxide concentrations (Air Resources Board and South Coast Air Quality Management District, 1980).

Table II. Results of Initial Sludge Analysis[a]

Characteristic	Sumps	Background
Inorganics		
Arsenic, mg/kg	<5 to 190	12
Lead, mg/kg	<20 to 40	<20
Be, Ba, Ca, Mg, Mn, Mo, Ni, Si	Not significantly above background concentrations	
Sulfate, %	7.2 to 14	0.015
pH	1.2 to 1.6	7.4
Volatile Organics		
Benzene, mg/kg	<90 to 880	
Toluene, mg/kg	130 to 810	
Tetrahydrothiophene, mg/kg	<90 to 140	
Neutral Fraction Organics Identified		
o,m,p-Xylenes	Trimethylbenzene	
Quinoline	Methyl Naphthalene	
Naphthalene	Trimethylpyridine	

[a]Scans for chlorinated pesticides and PCBs were negative in all samples.

Table III. Results of Sludge Headspace Analysis

Major Components
Sulfur Dioxide
C_4-C_{14} Alkanes and Cycloalkanes (or Alkenes)
Benzene
Toluene
Ethyl Benzene
Xylenes
Minor Components
C_3-C_4 Alkyl Benzenes
Naphthalene
Tetrahydrothiophene and C_1-C_4 Alkyl-Substituted Tetrahydrothiophenes
C_2-C_5 Alkyl Sulfides

The presence of large amounts of sulfur dioxide indicated that the detailed sampling includes continuous sulfur dioxide monitoring and a respirator program for all personnel in the work area.

Detailed Sampling and Analysis

Drilling was done to collect subsurface samples at 5- to 10-ft intervals from 12 holes in and around the apparent sump boundaries. Drilling mud was used in most of the sump areas to reduce air emissions. Temporary wood platforms were needed to support the drilling equipment in some of the low-viscosity sumps.

A composite sample from each hole was prepared and sent to the National Earthquake Information Center (NEIC) for analysis. Results are shown in Table IV. On the basis of the initial results, selected samples were analyzed for metals, pH, and sulfurous acid.

The initial results indicated abnormally high arsenic concentrations at the surface of one set of sumps. Additional sampling and analysis indicated that arsenic was distributed inhomogeneously, with concentrations ranging from less than 1 to 10,100 mg/kg. The source of the arsenic is still uncertain, although it appears to be confined to the surface of sumps exposed to the air. The pH values were consistently low within the apparent sump boundaries. Figure 1 shows the variation of pH with depth inside and outside one of the sumps. The lower pH value outside the sump was associated with a small quantity of perched water.

Table IV. Organic Results from Core Samples (National Earthquake Information Center)

Organic	Range[a] (mg/kg)	Median (mg/kg)
Volatile Organics		
Benzene	5.6-220	72
Ethylbenzene	5.8-41	14
Toluene	PBL-150	42
2-Methylbutane	ND-1.9	<0.1
Cyclohexane	ND-1.7	<0.1
Methylcyclopentane	PBL-2.4	0.43
Tetrahydrothiophene	PBL-160	34
Extractable Organics		
Xylenes	ND-660	135
1,2,4-Trimethylbenzene	ND-360	99
Naphthalene	46-350	145
2-Methylnaphthalene	21-160	43
Quinoline	ND-38	<5

[a]ND = not detected; PBL = present below lower limit of detection for reliable quantitation.

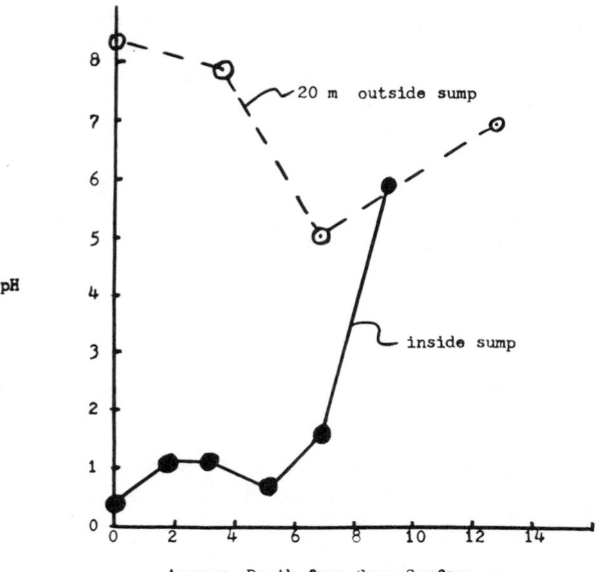

Figure 1. Variation of pH with depth.

Sulfurous acid was inhomogeneously distributed throughout the waste; the concentration varied from less than 240 to 12,000 mg/kg.

INTERPRETATION OF RESULTS

The results are generally consistent with the identification of the waste as a refinery sulfuric acid sludge. Large amounts of sulfur dioxide and volatile hydrocarbons are contained in the sludge. Initial indications are that the material has not spread laterally to a great extent. Arsenic in high concentrations is distributed on the surface of one sump. Tetrahydrothiophene (THT) and substituted tetrahydrothiophenes have been found in significant levels. Because THT has an odor threshold of 0.5 ppb (Whisman et al., 1978), it presents potential nuisance and health effects considerations. Work is now under way to determine the existing and potential vectors for movement of the identified waste components and to quantitate the existing and potential environmental and health effects.

REFERENCES

Kalichevsky, V. A., and B. A. Stagner. 1942. *Chemical Refining of Petroleum*, American Chemical Society (New York: Reinhold Publishing).

State of California Air Resources Board and South Coast Air Quality Management District. 1980. "Determination of the Air Quality of the Abandoned Waste Disposal Site in Fullerton, California," Report No. LE-80-008.

Whisman, M. L. et al. 1978. "Odorant Evaluation: A Study of Ethanethiol and Tetrahydrothiophene as Warning Agents in Propane," *Environ. Sci. Technol.* 12(12):1285-1288.

DISCUSSION

***Betty Willis*, EPA, Region IV:** In cleaning up sites, have you run into storage facilities where there have been a lot of drums of lab packs and, if so, how did you handle that?

B. Simmons: We have encountered some, but nothing on the order of Chemical Control, where I understand they encountered considerable amounts of that. It is not a problem that we have seen so far.

CHAPTER 11

CORRELATIONS OF PHYSICAL AND CHEMICAL PROPERTIES OF COMPOUNDS WITH THEIR TOXICITIES AND DISTRIBUTIONS IN THE ENVIRONMENT

Eugene E. Kenaga

The Dow Chemical Company
Health and Environmental Sciences
Midland, Michigan 48640

ABSTRACT

A detailed study of the environmental fate and toxicity of a chemical is necessary for persistent, toxic chemicals that bioconcentrate and that are widely distributed in the environment in significant volume. Few chemicals meet all these qualifications. Since such studies are expensive and time-consuming, only those tests that are necessary for good hazard evaluation under the expected use and distribution conditions of the specific chemical being evaluated should be conducted. From a growing background of information on comparative toxicology and fate of chemicals in the environment, we now have many useful early stage evaluation and predictive techniques. Predictive regression equations for physical and chemical properties and toxicology of chemicals allow calculation or estimation of distribution coefficients in air, water, soil, and organisms. Other correlations include comparative species toxicity and bioconcentration factors. Chemical structure and chemical and physical properties are useful for limited prediction of chronic-acute toxicity ratios (application factors).

INTRODUCTION

Predicting the exact fate of a chemical in the environment or its toxicity to all life stages and species of organisms is impossible because of the immense permutations and combinations of factors affecting results. We can accurately predict orbital events, such as the rise and set of the sun and moon, ebb and flow of the tide, beginning and ending of a day, of the four seasons, and of a year. Then things become more complicated. Such cyclic events directly affect temperature, wind, precipitation, growth of organisms, and distribution of soil, water, air, organisms, and associated chemicals. These events, although not predictable by the minute, can often be predicted on the average or within specified limits on the basis of past events that have been recorded by meteorologists, biologists, and phenologists. Critical temporal events for organisms occur daily, usually as gradients, such as changes in quantity and quality of light, humidity, oxygen availability in water, temperature, growth, energy, and speed of movement and metabolism of organisms. Each organism has survived because it can live within the range and interaction of certain critical biotic and abiotic events occurring during its daily existence and life cycle. The need for useful prediction of the effects of chemicals on organisms is nonetheless important because we must conduct representative testing without exhausting our resources and energy needlessly.

ACCURACY OF STANDARD TEST METHOD RESULTS AND ACCURACY NEEDED FOR HAZARD EVALUATION

In an attempt to learn the effects of chemicals on organisms, many laboratory tests are devised to isolate variables. Standardization of variables, such as temperature, exposure period, species, life stage and strain of organism, and food and water intake, is valuable because it isolates these variables that most strongly influence changes in the toxicity of chemicals to organisms.

Trying to obtain consensus standardization of a test method for minimum variation of results between and within laboratories is extremely difficult. One housefly toxicity test method, which began its period of standardization in about 1938, is still not finished. Results of "round robin" tests, conducted by the Organization for Economic Cooperative Development in 1979, are interesting. Simple physical property determinations such as for water solubility or *n*-octanol–water partition coefficients were conducted internationally at four to nine laboratories. Variation within the range of values was from 1.1 to as much as 17.9-fold. The results obtained by different laboratories from the standard acute oral toxicity LD_{50} test on rats, when using the same chemical, vary from 1.1 to 25-fold (Kenaga, 1979b). Thus, results of some simple physical and toxicological test methods, where the variables are or should be minimal, show variations of over tenfold. Fewer variations usually occur in replicated tests for the same chemical in the same laboratory. The 95% confidence limits for standard acute LC_{50} fish and bird tests commonly show a range of two- to four-fold (Hill et al., 1975; Johnson et al., 1980).

These observations indicate that experimental numbers from tests used for hazard evaluation must be viewed as representative of a large permissible range so that 1.5- to 2-fold differences are often not more useful for decision-making than order-of-magnitude, or "ball park," differences. This conclusion, along with the fact that laboratory tests do not cover the range of environmental variables and are not representative of every variable, leads to the potential for using scientifically estimated values if experimental data are unavailable. Estimated data often may be just as valuable as experimental data, depending on the accuracy needed for decision-making. With this assumption in mind, let us examine the minimum information necessary to establish a scientifically based use of prediction factors and equations, which are, in turn, useful for hazard evaluation.

Predictor terms are needed for converting known data to estimated values, based on correlation of both environmental fate and toxicity data.

NEED FOR PREDICTIVE CORRELATIVE COEFFICIENTS AND FACTORS

Initial physical factors needed for evaluating the potential hazards of a chemical are those related to distribution and persistence in various environmental media, particularly air, water, soil, and animal and plant organisms. Where experimental data are limited, predictive equations and factors are particularly useful. This is also true of toxicology. We need to estimate acute and chronic toxicity to a wide range of phylogenetically representative organisms early in the evaluation of a chemical. Because of this, we must use data that are easily available and inexpensive. The ability to predict the ratio of the acute-to-chronic toxicity [application factor (AF)] from other species or other chemicals is also important.

Physical distribution of a chemical depends on the use and disposal of the chemical, physical and chemical properties of the environment in which the chemical occurs, and properties of the chemical itself. The last factor comes first in the development of a chemical evaluation.

Identifying the medium preferred by the chemical is important because this tells one where to perform monitoring tests in the field for possible harmful concentrations. Minimal physical property data are needed to predict such distribution. Equilibrium distribution concentrations can be calculated from molecular weight, water solubility, vapor pressure, Henry's law, and correlations shown later in this chapter.

BASIC PROPERTIES OF CHEMICALS USED FOR CALCULATION OF DISTRIBUTION COEFFICIENTS

The basic data needed for calculating distribution coefficients for each chemical are (1) molecular structure, (2) molecular weight, (3) vapor pressure, and (4) melting point.

Molecular Weight

Molecular weight (MW) can be calculated with great accuracy from the atomic weight in the molecular formula of the chemical. It is used to the nearest whole number as sufficient accuracy needed for the purpose of these comparisons and calculations. Molecular weight is used in Henry's law [water-air partition coefficient (K_w)] calculations. MW is an overall indicator of the size of the molecule and its ability to penetrate cells, which is also related to biological activity.

Water Solubility

Water solubility (WS) data come from many sources and are not always verifiable; WS is measured here at 25°C, unless otherwise noted. The accuracy of experimentally determined water solubility is most difficult to obtain below 10 ppm and therefore more questionable. WS is predictable to sufficient accuracy through the use of melting point and entrophy of fusion data for nonelectrolytes and weak electrolytes (Yalkowsky, 1979; Banerjee et al., 1980). The applicability of the water solubility data of the chemicals that are gases (vapor pressures above 760 mm Hg at 25°C) is questionable. The water solubility of such chemicals should probably be given in parts per million at a partial pressure of the gas at 1 atm (760 mm Hg). For example, under these conditions the solubility of ethane would be 55 ppm. Water solubility is useful because it is inversely correlatable with the bioconcentration factor (BCF) and with soil-organic matter water partition coefficient (K_{oc}) as shown by Kenaga and Goring (1979) and others.

Vapor Pressure

Vapor pressures (VP) given here are measured at 25°C and 760 mm air pressure; VP data taken from the literature are calculated by several different techniques. A straight-line representation of VP at different temperatures can be determined from known values and plotted on "Cox charts" for VP at 25°C. Other values can be extrapolated with sufficient accuracy if used within the plotted ranges; VP for chemicals below 1 mm Hg are hardest to measure accurately.

BASIC PARTITION COEFFICIENTS NEEDED TO CALCULATE AIR-WATER-SOIL-ORGANISM DISTRIBUTION OF CHEMICALS IN THE ENVIRONMENT

Water-Air Partition Coefficient

By using MW, VP, WS, and Henry's law, the water-air distribution ratio of many chemicals can be calculated under equilibrium conditions. Organic chemicals of the medium range of volatility are most accurately determined. Unfortunately, Henry's law does not apply accurately to K_w values of chemicals having a high water solubility (>1200 ppm) *and* high VP (>760 mm). In general, the accuracy of K_w data is to within one or two orders of magnitude and serves the function of identifying whether air or water is the medium preferred by the chemical being tested and therefore the medium where the highest chemical concentrations are most likely to occur.

Octanol-Water Partition Coefficient

Octanol-water partition coefficient (K_{ow}) data illustrate the distribution of chemicals between two immiscible liquids, namely *n*-octanol and water, imitating in a general way the distribution of the chemical between water or blood and fat in animal organisms.

The use, determination, and calculation of K_{ow} from chemical structure is discussed by Kenaga and Goring (1980). By use of equations, under the stated conditions, K_{ow} can be used to calculate the WS, bioconcentration factor, or soil organic carbon-water coefficient of the same chemical.

These K_{ow} values are accurate to within several orders of magnitude. The Environmental Protection Agency (EPA) has used a K_{ow} of 1000 or more as an indicator of possible BCF problems worth further investigation. This seems low because the K_{ow}-BCF equation relates a K_{ow} of 1000 to a BCF of 20. However, the use of K_{ow} data to estimate BCF and K_{oc} values seems rather artificial and redundant unless WS or K_{oc} data are unavailable.

Soil Sorption Coefficient

The soil sorption coefficient is the concentration of chemical sorbed by the soil, expressed on a soil organic carbon basis, divided by the concentration of the chemical in the soil water. This coefficient represents the most useful and reproducible measure of comparative leachability in soil and establishes the medium preferred by the chemical being tested and therefore the medium where possible problems are most likely to occur. Kenaga and Goring (1980) give a thorough review of these relationships. Predicted K_{oc} values for chemicals with high potential for cation exchange capacity of the soil or subject with certain ionized chemicals are more susceptible to error. Chemicals having a K_{oc} around 1000 or more are quite tightly bound to organic matter in soil and considered immobile (Kenaga, 1980).

Bioconcentration Factor

BCF, as used here, is the concentration of a chemical in an organism, divided by the concentration in water. It is usually related most specifically to the fat content of the organism and, thus, varies somewhat within and among aquatic species of animals, depending on their fat content. Kenaga and Goring (1980) have reviewed the literature on BCF data. Equations were developed to predict BCF values from WS, K_{oc}, and K_{ow} values on the same chemical interchangeably. Predicted BCF values are usually higher than experimental values for chemicals that are rapidly biodegradable and may not be valid for metals and inorganic chemicals.

Compounds having a BCF of over 1000 are those which from experience are most likely to need further evaluation for hazard assessment. Chemicals such as DDT, PCBs, and dieldrin have BCFs well in excess of this value. BCFs vary by as much as one to two orders of magnitude between species and life stages, probably because of differences in fat content and metabolism, even with one compound (Kenaga and Goring, 1980). Thus, a given BCF value should not be used as a precise representation of bioconcentration under varying conditions. Nevertheless, these data suggest that compounds having predicted BCF values of 100 or less probably do not merit experimental confirmation. Calculated values from water solubility are good enough. A BCF value of 100 relates by equation to WS, K_{oc}, and K_{ow} values of about 5 ppm, 2000, and 8400, respectively. BCF values in fish and *Daphnia* are nearly equivalent, and one can be predicted from the other. BCF data from *Daphnia* require shorter exposure periods to reach equilibrium (Kenaga and Goring, 1980).

In terrestrial organisms the principal contact with chemicals appears to be via the diet, in contrast to fish, whose principal contact is via water. BCFs in fat tissues, derived from 23 chemicals in feeding studies with cattle, are often three to four orders of magnitude smaller than those reported for fish because different denominators are used for calculating BCFs. This is true even though the absolute residue values in cattle and fish are similar in magnitude. This difference in BCFs is further enlarged if the comparisons of BCFs in fish are made on a fat residue basis instead of on a whole body basis (contains 5 to 10% fat). However, BCFs in cattle, swine, and fish show significant correlations with each other and can be predicted one from the other via regression equations with useful accuracy for preliminary estimates. Bioconcentration factors from fish and from cattle feeding studies are also significantly correlated with K_{oc}, K_{ow}, and WS, and each can be predicted from the other values. (Kenaga, 1979c; Kenaga and Goring, 1978.) See Table I for examples of equations.

Another step in predicting the distribution of chemicals in the environment can be taken by combining the K_{oc}, K_{w}, K_{oc}, and BCF equations, which can be computerized for use with known proportions of media in the environment to be studied (water, soil, air, and organisms). The proportion and concentration of each chemical can be calculated in the different media at equilibrium concentrations (McCall et. al., 1981). These concentrations and properties, of course, are subject to correction because of climatic and degradative variables. Knowl-

Table I. Accuracy of Correlation Equations Suitable for Predictive Purposes[a]

Correlation	Regression Equation	Plus-Minus Order of Magnitude from Calculated Value (95% confidence limits)	Correlation Coefficient (r)[b]
WS-K_{oc}	$\log WS = 5.09 - 1.28(\log K_{oc})$	1.94	−0.84
WS-BCF	$\log WS = 2.531 - 0.916(\log BCF(f))$	2.54	−0.72
K_{oc}-WS	$\log K_{oc} = 3.64 - 0.55(\log WS)$	1.23	−0.84
K_{oc}-BCF	$\log K_{oc} = 1.963 + 0.681(\log BCF(f))$	1.52	+0.87
BCF-WS	$\log BCF(f) = 2.791 - 0.564(\log WS)$	1.99	−0.72
BCF-K_{oc}	$\log BCF(f) = -1.579 + 1.119(\log K_{oc})$	1.95	+0.87

[a]Source: Kenaga and Goring, 1980.
[b]Correction coefficients are all significant at less than the 1% level.

edge of the important climatic and geographic features and water and air flow patterns are important local variables to be factored into the specific estimated environmental concentrations. From these predictive mechanisms, one can determine the principal medium that should be investigated or monitored for environmental problems and the approximate concentrations to be expected.

CONCENTRATIONS OF PERSISTENT CHLORINATED HYDROCARBONS OCCURRING IN WATER

Monitoring tests on the concentrations of persistent chlorinated hydrocarbons such as DDT, DDE, dieldrin, and many other pesticides in water have been conducted for years (Manigold and Schulze, 1969; Schulze et al., 1973; Lichtenberg et al., 1970). The persistent chlorinated hydrocarbons, pesticides, and PCBs were found most frequently in the highest concentrations in rivers and surface waters. Concentrations were usually less than 1 ppm and more frequently, in the low parts per billion level. Other data suggest that water concentrations of chemicals are mostly well below 1 ppm. On the basis of this premise, if 1 ppm is a good estimation of the upper limits of expected water concentrations, what are the chances that a given chemical chosen at random will be toxic to various organisms at 1 ppm or lower?

RANGE OF CONCENTRATIONS OF CHEMICALS THAT KILL VARIOUS KINDS OF ORGANISMS

Over the years, the Dow Chemical Company has screened many chemicals on species diagnostic of pest problems. These tests represent a large sampling of organic and inorganic chemicals far greater than normally available in the literature. These data demonstrate that a certain percentage of these chemicals caused growth inhibition of one or more species of microorganisms over a range of concentrations differing by four orders of magnitude. When the percentage of chemicals causing 100% inhibition is plotted on log-log paper against the four concentration ranges (in parts per million), an excellent straight-line slope function is formed. The regression equation relating to this function is log (% of chemicals) = 0.216 + 0.533 (log concentration). The correlation coefficient (r) is 0.9992. This slope line serves as a guide to the probability that, from a random sample of chemicals, a certain number will cause 100% growth inhibition of one or more of the species of microorganisms tested when using a given concentration range (Kenaga and Chambers, 1979).

When eight species of closely related organisms (fungi and bacteria) were compared, the range of sensitivity to chemicals between species did not increase much over the most sensitive species in the range of one order of magnitude. The difference between concentrations necessary to kill approximately 100% vs approximately 0% is very much within this same tenfold range of concentrations. Differences in sensitivities of different species of fish or birds are usually within one or sometimes two orders of magnitude, rarely more (Kenaga, 1978).

Screening test results of the effects of a large number of chemicals at levels below 1 ppm on a variety of organisms are shown in Table II. Among the species tested (*Daphnia,* fish, alga, aquatic vascular plants, terrestrial plants, fungi, and bacteria), *Daphnia* and fish are sensitive to the largest percentage of chemicals at levels below 1 ppm (Kenaga and Moolenaar, 1979; Kenaga, 1981). For the other groups of organisms, less than 1% of the chemicals cause 100% lethality at levels below 1 ppm in water or soil. Thus, 1 ppm is the concentration of organic chemicals that is rarely exceeded in water and soil away from effluent sources and is the concentration below which most chemicals are not toxic to organisms.

Table II. Comparison of the Range of Lethal Concentrations of Chemicals to Various Microorganisms and Plant and Animal Organisms

Organism	Chemicals Tested (No.)	Chemicals Causing 100% Mortality by ppm Range (%)	
		0.01-0.90	0.1-0.99
Daphnia	33,909	0.6	2.4
Composite of Fish Species[a]	35,305	0.14	1.3
Composite of Aquatic Vascular Plant Species[a]	27,781	0	0.1
Alga (*Chlorella*)[a]	49,082	0.006	0.02
Composite of Terrestrial Plant Species[b]	131,596	0.006	0.17
Composite of Bacterial Species[c]	13,409		0.22
Composite of Fungal Species[c]	13,409		0.19

[a]Source: Kenaga and Moolenaar, 1979.
[b]Source: Kenaga, 1981.
[c]Source: Kenaga and Chambers, 1979.

SELECTION OF REPRESENTATIVE ORGANISMS FOR EARLY TOXICOLOGICAL EVALUATION

Selection of species acting as surrogates for good representation of all important taxonomic groups is difficult, mainly because there is not a large toxicological base of information on "benchmark" chemicals using comparative test methods. Kenaga (1978) selected 75 pesticides and eight organisms from which to make choices for animal species surrogates for acute toxicity testing. From this comparison, correlation coefficients showed the following highest correlations for acute toxicity: rat LD_{50} vs bobwhite LC_{50}; rat vs mallard LD_{50}; rainbow trout vs saltwater fish LC_{50}; saltwater fish vs shrimp LC_{50}. *Daphnia* was the most sensitive organism and the honey bee the least sensitive in range of concentration of chemicals toxic to these organisms.

The rat LD_{50}, *Daphnia* LC_{50}, and a fish LC_{50} were found to be the most useful screening organism test methods for acute toxicity to animals. Equations were derived to estimate LC_{50} data between rainbow trout, bluegill, and the fathead minnow from the LC_{50} data on one species.

Fish and *Daphnia* were found to be good surrogates for aquatic vascular plants and algae at concentrations below 1 ppm in water (Kenaga and Moolenaar, 1979).

ESTIMATION OF CHRONIC TOXICITY FROM ACUTE TOXICITY

Kenaga (1979a,b) reviewed the literature where comparable data on a number of chemicals were available and found no consistent relationship of acute toxicity to chronic toxicity of a chemical among mammals, fish, or *Daphnia*. Because of the lack of chronic toxicity data on aquatic organisms with most chemicals, a judgment must be made on the need for chronic toxicity studies. A common practice, useful for predicting chronic toxicity concentrations from acute toxicity data, especially among closely related species, or chemicals, is to calculate the ratio between the LC_{50} and chronic no-effect concentration (MATC = maximum acceptable toxicant concentration in a sensitive life stage or life cycle test). This is known as an application factor (AF). The AF (as used here) = LC_{50}/MATC. The AFs for 31 compounds of various chemical structures ranged from about 2 to 3725.

Only one species is necessary to establish the chronicity and sensitive life stage relationship between organisms that are as closely related as fish. The relationship between acute and chronic toxicity (AF) in one species of fish, determined by using the experimental test results, provides an AF for calculating chronic toxicity from the LC_{50}s of other species of fish with the same chemical (Kenaga, 1979a). Regression equations for predicting AFs from one species to another have been made. Using the AF established for sensitive life stages of one fish to determine the AF of various species to the same chemical usually results in no more than a twofold difference than full-life-cycle chronic tests, as shown in work done by EPA at Duluth (McKim, 1977). Tests on saltwater fish life stages are not needed if freshwater life-stage data are available or vice versa. Interspecies sensitivity relationships are already sufficiently established from acute tests (Kenaga, 1978; EPA, 1978).

Another method of estimating chronicity (AF) of a chemical and the no-effect level (MATC) for aquatic organisms is by comparing the AFs of other closely related chemicals.

For example, 1,1,1-trichloroethane (TCE) is a member of a rather large, structurally related group of commonly used chlorinated aliphatic chemicals, which have not been tested for chronic toxicity to aquatic organisms. The AF data are available on nine chlorinated aliphatic chemicals and one of three aquatic animal organisms, as shown in Table III. The AFs ranged from about 1 to 35. On the basis of the AF data for related chemicals, a conservative AF of 35, applied to the lowest

Table III. Application Factors for Aquatic Organisms with Chlorinated Aliphatics

Chemical	Organism[a]	MATC (ppm)	LC_{50} (ppm)	AF
Chloroform	DM	1.8-3.6[b]	29[c]	8-16
1,2-Dichloroethane	DM	10.6[d]	220,[c] 315[d]	21-30
1,1,2-Trichloroethane	DM	26.0[d]	18,[c] 174[d]	0.7-7
1,1,2,2-Tetrachloroethane	DM	3.45[d]	9.3,[c] 56.9[d]	2.7-16
Pentachloroethane	MS	0.53-0.63[b]	5.06[b]	8-9.5
Hexachloroethane	FHM	0.38-0.54[b]	1.53[b]	2.8-4
Trichloroethylene	MS	0.3-0.67[b]	10.2[b]	15-34
Tetrachloroethylene	DM	0.51[d]	18,[c] 9.1[d]	18-35
1,3-Dichloropropene	FHM	0.18-0.33[b]	4.1[e]	12-22

[a]Key: DM = *Daphnia magna;* MS = mysid shrimp; FHM = fathead minnow; AF = application factor = LC_{50}/MATC; MATC = maximum acceptable toxicant concentration (chronic toxicity).

[b]U.S. Environmental Protection Agency, 1978.

[c]LeBlanc, 1980.

[d]Call, et al., 1981.

[e]Johnson and Finley, 1980.

LC_{50} for aquatic organisms with TCE of 5 ppm, would result in a calculated MATC of 145 ppb in water. This MATC would give at least a 14-fold margin of safety over an average concentration of less than 10-ppb TCE occurring in the environment.

Thus, a small amount of data can be used to predict a great deal of environmentally useful data with a degree of accuracy sufficient for hazard evaluation of chemicals.

REFERENCES

Banerjee, S., S. H. Yalkowsky, and S. C. Valvani. "Water Solubility and Octanol/Water Partition Coefficients of Organics. Limitations of the Solubility-Partition Coefficient Correlation," *Environ. Sci. Technol.* 14 (10):1227-1229.

Call, D. J., L. T. Brooke, and N. Ahman. 1981. "Toxicity, Bioconcentration, and Metabolism of Selected Chemicals in Aquatic Organisms," U.S. Environmental Protection Agency. Quarterly Progress Reports to EPA, 1979 and 1980.

Hill, E. F., R. G. Heath, J. W. Spann, and J. D. Williams. 1975. "Lethal Dietary Toxicities of Environmental Pollutants to Birds," Special Scientific Report–Wildlife No. 191, U.S. Department of the Interior, Fish and Wildlife Service, Washington, DC.

Johnson, W. W., and M. T. Finley. 1980. "Handbook of Acute Toxicity of Chemicals to Fish and Aquatic Invertebrates," Resource Publication 137, U.S. Department Interior, Fish and Wildlife Service, Washington, DC.

Kenaga, E. E. 1978. "Test Organisms and Methods Useful for Early Assessment of Acute Toxicity of Chemicals," *Environ. Sci. Technol.* 12:1322-1329.

Kenaga, E. E. 1979a. "Aquatic Test Organisms and Methods Useful for Assessment of Chronic Toxicity of Chemicals," in *Analyzing the Hazard Evaluation Process,* K. L. Dickson, A. W. Maki and J. Cairns, Jr., Eds. (Washington, DC: American Fisheries Society).

Kenaga, E. E. 1979b. "Acute and Chronic Toxicity of 75 Pesticides to Various Animal Species," *Down Earth* 35(2):25-32.

Kenaga, E. E. 1979c. "Correlation of Bioconcentration Factors of Chemicals in Aquatic and Terrestrial Organisms with Their Physical and Chemical Properties," *Environ. Sci Technol.* 14:553-556.

Kenaga, E. E. 1980. "Predicted Bioconcentration Factors and Soil Sorption Coefficients of Pesticides and Other Chemicals," *Ecotoxicol. Environ. Saf.* 4:26-38.

Kenaga, E. E. 1981. "Range of Toxicity of 131,596 Chemicals to Plant Seeds," *Ecotoxicol. Environ. Saf.* Vol. 5.

Kenaga, E. E., and C. A. I. Goring. 1980. "Relationship Between Water Solubility, Soil-Sorption, Octanol-Water Partitioning, and Concentration of Chemicals in Biota," in *Aquatic Toxicology,* J. G. Eaton, P. R. Parrish, and A. C. Hendricks, Eds., American Society for Testing and Materials, ASTM STP 707.

Kenaga, E. E., and K. G. Chambers. 1979. "Range of Toxicity of 13,409 Chemicals to Bacteria and Fungi," unpublished report, Health and Environmental Sciences, The Dow Chemical Co., Midland, MI.

Kenaga, E. E., and R. J. Moolenaar. 1979. "Fish and *Daphnia* Toxicity as Surrogates for Aquatic Vascular Plants and Algae," *Environ. Sci. Technol.* 13(12):1479-1480.

LeBlanc, G. A. 1980. "Acute Toxicity of Priority Pollutants to Water Flea (*Daphnia magna*)," *Bull. Environ. Contam. Toxicol.* 24:684-691.

Lichtenberg, J. J., J. W. Eichelberger, R. C. Dressman, and J. E. Longbottom. 1970. "Pesticides in Surface Waters of the United States—A 5-year Summary, 1964-68," *Pestic. Monit. J.* 4(2):71-86.

Manigold, D. B., and J. A. Schulze. 1969. "Pesticides in Selected Western Streams—A Progress Report," *Pestic. Monit. J.* 3(2):124-135.

McCall, P. J., D. A. Laskowski, R. L. Swann, and H. J. Dishburger. 1981. "Estimation of Partitioning of Organic Chemicals in Model Ecosystems," Symposium on Tests for Pesticide Environmental Behavior, 182nd National Meeting, American Chemical Society, New York.

McKim, J. M. 1977. "Evaluation of Tests with Early Life Stages of Fish for Predicting Long-Term Toxicity," *J. Fish. Res. Bd. Can.* 34(8):1148-1154.

Organization for Economic Cooperation and Development (OECD). 1979. "OECD Chemicals Testing Programme. Expert Group Physical Chemistry," Draft Final Report, Berlin, West Germany.

Schulze, J. A., D. B. Manigold, and F. L. Andrews. 1973. "Pesticide in Selected Western Streams—1968-71," *Pestic. Monit. J.* 7(1):73.

U.S. Environmental Protection Agency. 1978. "In-Depth Studies on Health and Environmental Impacts of Selected Water Pollutants," U.S. EPA Contract 68-01-4646.

Yalkowsky, S. H. 1979. "Estimation of Entropies of Fusion of Organic Compounds," *Ind. Eng. Chem. Fundam.* 18(2):108-111.

DISCUSSION

***David Friedman,* EPA:** You had mentioned at the start that we could save a lot of resources if we could use short-term and screening-type tests for determining hazardous materials. I would like to have your opinion on what degree of reproducibility is necessary before these tests are suitable for such purposes. Also, what degree of confidence do we need and what type of validations are necessary before all these things can be used?

E. Kenaga: First of all, I do not believe in using any one factor to evaluate the chemical. You have many factors that you have to consider; for instance, if the chemical is resistant or has the tendency to bioconcentrate and is quite toxic, then it is necessary to study it thoroughly. If the chemical does not have any of those factors, then it can be put aside temporarily. If it fits some place in between, then the predicted factors that I have been talking about come into play.

***Robert Meglen,* University of Colorado:** You indicated that you feel that most compounds at less than one part per million will not show the kinds of toxicity to algae or other organisms that might be of concern. I was wondering if you feel that there is any sort of additive effect. For example, if you have 1000 compounds all at few parts-per billion would it be equal to one compound at one ppm?

E. Kenaga: That is certainly a leading question, isn't it? Well, I used to work with insecticides, and I tried to find things that would synergize insecticides. I work a lot at it, and there are very few compounds that will cause synergism. If I could find one, I would get a patent on it. However, a thousand chemicals is something different. I have the feeling that when you get eight or ten compounds together they tend to neutralize each other more often rather than synergize each other in the end. This is partly brought about by the fact that if you consider what is normal in the environment, you have thousands and hundreds-of-thousands of chemicals that are there as buffering types of chemicals. The environment generally learns how to handle most types of chemicals by reaction in some way or another. Maybe that's not a very good answer, but it appears that most of the predictions in the laboratory are overestimated rather than underestimated. Take toxicity of fish, for example. Nearly always, the field toxicities are an order of magnitude below that predicted from the laboratory toxicity tests. I have a feeling that if there were really bad synergistic effects we would have had big problems a long time ago. That is not to say that we do not have some.

***David Abbott,* Monsanto:** My question comes close to the answer you were just giving. In view of the fact that many of these multispecies tests and field data are off a couple of orders of magnitude, would you comment on the validity or even desire to use single species tests for screening.

E. Kenaga: It is the genetic relations to chemicals in toxicity that are so important ecologically. From the toxicity standpoint, if you take representatives from each of the phyla, you find there is always a very sensitive species, and there is always a very insensitive species among them. The range of toxicity covered by those concentrations is quite often the same. It may vary 3 to 5 orders of magnitude in species of insects, plants, or fish, but the outliers are the cause of that range. The main body are those things within 2 or 3 orders of magnitude.

***Garrie Kingsbury,* Research Triangle Institute:** It's a comment more than a question. I have the feeling that some of the correlations you showed would improve if they were for groups of chemicals that are similar structurally rather than for a large array of chemicals of multiple functional groups. I wondered if you might comment about whether it is a good idea to group chemicals, and do the predictive methods improve when you do?

E. Kenaga: I like your question; I am not sure I know the answer. You noticed Table III was on chlorinated aliphatics. If you are going to be able to make use of any prediction, I think you have to do it with a homologous and analogous series of compounds, which don't have much differential mode of action. If you were to take the DDT series of compounds, you would find a structural difference to toxicity; for example, for compounds that have the same empirical formula, and many of the same physical properties like melting point, there are differences in toxicity. Thus, you are in an entirely different world relative to toxicity, and so I think that straight-chain compounds and things like your homologous series work better than they do in aromatics. Aromatics are terribly unpredictable. I wouldn't want to base very much on those unless the aromatics had substituents that were either just varying the halogen or varying the aliphatic portion or some other specific substitution.

***David Friedman,* EPA:** You mentioned at the beginning that industry has reams of data on chemicals, and as a regulator who is charged with trying to come up with regulations that are environmentally proper and ones that won't be so challenged all the time, how can we get some of this information so that we have some facts to work with?

E. Kenaga: I have given you facts. I realize you don't have the structures available, and I can't do that because of patents and other considerations. I appreciate your problem, too. But I don't think that the conclusions are going to be altered.

SECTION 3
TRANSPORT AND TRANSFORMATION

CHAPTER 12

OVERVIEW

Stephen R. Cordle
Office of Environmental Processes and Effects Research
U.S. Environmental Protection Agency
Washington, DC 20460

To deal effectively with solid wastes, either through regulation or enlightened management, several kinds of information are needed. The scientific community must (1) identify the sources of solid waste and determine their contribution of pollutants to the environment, (2) develop technology for different levels of control and define their costs, (3) estimate the change in concentration and character of pollutants from their entry into the environment to some point of exposure to humans or other biota, (4) develop the technology and costs for cleanup of already polluted environments, and (5) determine the health and environmental effects of exposure to various levels of pollutants derived from solid wastes. This section deals with the third need: to describe the pathways and concentrations of pollutants and their transformation products through the environment to some point where they may have an effect.

Transport and transformation information is needed by regulators and managers to (1) prescribe performance standards for control technology by translating acceptable exposure concentrations into acceptable emission concentrations, (2) prohibit practices which will result in unacceptable environmental damage or health effects, (3) assess the risk to human health and the environment once pollutants have escaped the disposal site, (4) determine the likelihood that a pollutant will persist in the environment (for decisions on the best disposal option for that pollutant), and (5) develop criteria for disposal site selection or rejection.

There are four main pathways that solid-waste-derived pollutants may take through the environment: through the subsurface to groundwater, over land to surface water, through the atmosphere, and through the food chain. Because the major pathway of pollutants from solid waste disposal to humans is generally considered to be through the subsurface, this section has three chapters directed to the better understanding of this route of exposure. One chapter is directed toward delineating the food chain pathway, another the atmosphere, and finally, methods for assessing multiroute exposure.

The plumes of leachate-contaminated groundwater from three municipal landfills in sand aquifiers in Ontario are described by Cherry. Each plume contains a variety of inorganic and organic contaminants. At some distance downflow from the landfill, two of the plumes are overlaid with a zone of uncontaminated groundwater in the upper part of the aquifer. With this and other evidence, Cherry concludes that vertical dispersion in the aquifers has only a weak mixing influence. The stronger effect of longitudinal dispersion is considered to be responsible primarily for the gradual reduction in concentrations of contaminants in the plumes in the direction of groundwater flow.

Water movement and sorption are the primary processes that determine the subsurface movement of inorganic pollutants, according to Hounslow. The sorptive properties of subsurface solids depend not only on their inherent chemical properties but also on their physical properties such as shape, size, and distribution. Some effects of microbiological activity on geochemical reactions are also described.

Hassett, Griffin, and Banwart offer powerful tools for predicting the adsorption of nonpolar organic chemicals by soils. They show that the adsorption of hydrophobic nonpolar organic compounds by soils is a function of the organic matter content of the soil. With the soil organic content and one of two properties of the chemical, the water solubility of the chemical or its *n*-octanol–water partition coefficient, sorption can be predicted. Although considerable data are presented in support of these predictions, the authors do not feel that the need for experimental verification is precluded.

A review of food chain pathways by which hazardous wastes applied to the land could affect humans, agriculture, or the environment is presented by Chaney. A number of heavy metals as well as toxic organic chemicals are considered. His conclusions show that proper development of management practices for land treatment of hazardous wastes could avoid adverse food chain effects for nearly any organic waste and applications of only a few heavy metals would need to be restricted to protect humans.

Recently developed models for predicting the volatilization rates of chemicals from landfills are described by Springer, Thibodeaux, and Chatrathi. Application of the models to polychlorinated biphenyls (PCBs) show that the vapor emission rate of PCBs from landfills can be expected to be quite low. The authors caution, however, that care must be taken in the construction of landfills to limit the amount of biodegradable materials in the same site since the gas generated can either greatly increase the PCB emission rate by sweep action or rupture the cap.

A summary of some multimedia aspects of exposure assessment related to hazardous solid wastes is presented by Falco, Mulkey, and Schaum. The conduct of exposure assessments involves estimation of release rates of toxic chemicals into the environment, estimation of ambient environmental concentrations, and estimation of dose rate.

CHAPTER 13

OCCURRENCE AND MIGRATION OF CONTAMINANTS IN GROUNDWATER AT MUNICIPAL LANDFILLS ON SAND AQUIFERS

John A. Cherry

Department of Earth Sciences
University of Waterloo
Waterloo, Ontario, N2L 3G1

ABSTRACT

Plumes of leachate-contaminated groundwater at three municipal landfills on unconfined sand aquifers have been monitored using networks of multilevel piezometers. The landfills range in age from 15 to 42 years and in area from 4 to 28 ha. The plumes extend over distances ranging from 0.6 to 0.8 km from the landfills, and they occupy much of the vertical thickness of the sand aquifers, except toward the fronts of the plumes where the zones of contamination are much thinner. The concentrations of major contaminants in the plumes decline gradually in the direction of groundwater flow. At two of the sites, distinct vertical zones of contaminated and relatively uncontaminated groundwater persist over large distances in the plumes, which suggests that vertical dispersion does not cause strong vertical mixing.

The concentrations of Cl^-, Na^+, Mg^{2+}, Ca^{2+}, NH_4^+, Fe, Mn, alkalinity, and total dissolved solids in the plumes are considerably above background levels and have spatial distribution patterns with generally similar trends. Because of reducing redox conditions in the landfills and the anoxic status of the plumes, NO_3^- and SO_4^{2-} occur at very low concentrations. An exception occurs at one of the sites where SO_4^{2-}, which is apparently leached from the building demolition debris that contains gypsum and that is abundant in this landfill, is a major constituent in the plume. None of the plumes contain toxic inorganic constituents at levels above the limits specified in the primary drinking water standards. Dissolved organic carbon and organic nitrogen exist at above-background levels throughout the plumes, which indicates that some types of organic compounds are very mobile in the groundwater system.

INTRODUCTION

More than 100,000 municipal landfills, including those that are now inactive, exist in North America. In humid and semihumid regions, leachate is produced in nearly all landfills as a result of the entry of rain or snowmelt into the refuse. Infiltration into landfills is unavoidable during the operation period, and it normally continues even after the landfills receive their final cover for permanent closure. The leachate moves downward through the refuse and enters the groundwater system at the water-table zone near the bottom of the landfill or, if the water table is deeper, it moves farther downward to the groundwater zone in the geological deposits.

At sites where landfills are situated on permeable geological deposits, the movement of leachate in the groundwater zone can cause the development of large zones of contamination. These zones are known as contaminant plumes or contaminant enclaves. This chapter provides brief descriptions of the hydrogeological and hydrochemical nature of contaminant plumes at three landfills situated on unconfined sand aquifers in Ontario. These descriptions are based on the results of ongoing field investigations that the hydrogeology group at the University of Waterloo has been conducting since 1976. Some aspects of these studies are reported on in detail elsewhere (Cherry et al., 1981, 1982; Egboka et al., 1982; Nicholson et al., 1982; Hewetson and Cherry, 1982; Buszka et al., 1982).

The ultimate goal of these studies is to provide a basis for developing an improved methodology for predicting the impact of landfills on water resources in areas of permeable sand terrain. The contaminant plumes at the three Ontario landfills have been monitored in considerable detail by means of networks of multilevel sampling devices. These networks provide a three-dimensional view of the distribution of contaminants in the aquifers and therefore indicate the shapes and the internal hydrochemical characteristics of the plumes.

DESCRIPTION OF THE LANDFILL SITES

The operational history and hydrogeological features of the three Ontario landfills have many differences and a few similarities. The field investigations began in 1976 at the Borden landfill, situated on a military base in southern Ontario, about 75 km northwest of Toronto. In 1979, the program of landfill research was extended to include the landfill that currently serves the city of North Bay, located in northeastern Ontario. In 1980, studies began at the Woolwich landfill, which is situated near Kitchener-Waterloo in southern Ontario.

The Borden landfill, which occupies about 5 ha, was operational during the period of 1940 to 1976. In 1976, the landfill received its final cover layer of sand and was then seeded to grass. The Borden landfill received refuse from the town of Borden, which is a housing development for military personnel, and from the restaurants, offices, garages, and workshops on the Borden military base. The landfill has also received a considerable quantity of demolition debris from old buildings. No evidence exists that the landfill has at any time received hazardous industrial waste or any hazardous chemicals from military operations. The landfill was created by depositing waste beyond the crest of a small sand ridge. The water table in the lowlying land where the landfill was developed was within a meter of the original ground surface. The water table is now just below or just above the bottom of the landfill, depending on the season. The sand aquifer at the Borden site is about 30 m thick beneath the landfill and decreases to 10 m in the direction of groundwater flow.

The landfill at North Bay (28 ha) began operation in 1962 as the main municipal landfill for North Bay, which has a population of about 50,000. The refuse is deposited in excavations created by a sand quarrying operation that existed previously at the site. The refuse is occasionally covered by sand obtained from areas adjacent to the site. When landfilling is complete in a segment of the site, a layer of sand and sometimes sewage sludge or sawdust is used to form the final cover. The water table exists within the bottom 1 or 2 m of the landfill. The thickness of the sand aquifer beneath the landfill ranges from about 2 to 8 m. Beyond the landfill in the direction of groundwater flow, the saturated thickness of the aquifer is between 10 and 15 m. The landfill has received municipal waste, solid waste from local industry and sewage sludge. No hazardous waste landfill exists in the region, and it is thus reasonable to expect that the landfill has received some hazardous liquid wastes.

Landfilling began in a very minor way at the Woolwich site in the mid-1960s. In the early years the landfill mainly received solid wastes from local industry. In 1969, the landfill became a county landfill serving the local rural area and nearby towns. In 1973, the site was brought under the control of the regional municipal government; it has since been operated as a sanitary landfill which receives farm waste, solid wastes from industry, and waste from a small city and several towns. The refuse is deposited in trenches excavated to depths of about 5 m in the sand. The area of active landfilling is covered daily with sand from the trenches. The bottoms of the trenches are situated about 12 m above the water table, which occurs in the sand aquifer that underlies the area. The area that has been landfilled now occupies 4 ha. The saturated thickness of the aquifer varies from about 10 to 20 m.

At all three sites, the aquifers are formed of sand deposited by meltwater streams during the glacial epoch. The sand is layered, with individual layers of sand that is fine-, medium-, or coarse-grained, and that has varying amounts of silt. The sand generally contains less than 1% of clay-sized material. The layers of different textured sand that are visible in cores are generally between a few centimeters and a few tens of centimeters in thickness. The sand at the Borden and Woolwich sites contains abundant calcite and dolomite in addition to quartz and feldspar, whereas the sand at the North Bay site contains almost no calcite or dolomite. At the Woolwich and Borden sites, the aquifers are underlain by relatively impervious clayey or silty strata that effectively isolate the flow in the unconfined sand aquifers from deeper flow system. At the North Bay site, the aquifer is underlain by a thin layer of sand till that rests on Precambrian biotite–gneiss bedrock. Because the unconfined aquifers are laterally extensive and are underlain by relatively impervious zones, groundwater flow is predominantly horizontal at each of the sites. The landfills provide major vertical inputs of contaminated water to the lateral flow regimes in the aquifers.

METHODS OF INVESTIGATION

At each of the landfill sites, numerous boreholes have been drilled using hollow-stem augers to determine the geological conditions and to install groundwater monitoring devices. At the North Bay and Woolwich sites, the groundwater monitoring network is comprised of bundle piezometers and a few conventional standpipe piezometers. A bundle piezometer is an assembly of 0.95- or 1.27-cm-i.d. polyethylene or polypropylene tubes bound to a thickwalled, 1.27-cm.-i.d. PVC center piezometer. The PVC pipe with a short-screened section on the end is the deepest piezometer in the bundle. Each of the tubes has a screened tip that extends to a different depth. The piezometer tips are spaced vertically at intervals ranging from 0.5 to 2 m in the borehole. Each bundle enables a profile of water samples to be collected. The design and installation of bundle piezometers are described by Cherry et al. (1982). At the Borden site, groundwater monitoring has been accomplished by means of a network that includes bundle piezometers and multilevel point samplers of the type described by Pickens et al. (1978).

Groundwater samples were obtained at the Borden and North Bay sites by using narrow-diameter sampling tubes connected to hand-operated or battery-operated vacuum pumps. At the Woolwich site, where the water table is deep, samples were obtained by using the triple-tube gas drive sampler described by Robin et al. (1981). Measurements for Cl^- and electrical conductance were made on unfiltered samples in the laboratory.

Analyses for major ions, minor constituents, and trace inorganic elements were done on samples that were filtered (0.45 μm) in the field immediately after sampling. Preservatives were added to sample splits, except for the subsamples used for constituents such as alkalinity and sulfate. Measurements of pH, redox potential (platinum electrode), dissolved oxygen, and, in many cases, alkalinity were made in the field as soon as each sample was taken. Comprehensive sampling and analysis have been completed at the Borden and North Bay landfills. Only preliminary hydrochemical data have been obtained at the Woolwich landfill. Samples of gas from the landfills were drawn from sampling tubes installed in refuse. The samples were collected using the method described by Reardon et al. (1979).

CONTAMINANT PLUMES

Extensive plumes of contaminated groundwater exist at each of the three landfills. Each of the plumes was delineated by means of chloride concentrations (Figures 1, 2, and 3). Chloride is well suited for this purpose because its mobility is not influenced by geochemical processes such as adsorption, precipitation, oxidation, or reduction, and because it occurs at concentrations of hundreds of milligrams per liter in the leachate and only at a few tens of milligrams per liter or less in the ambient groundwater in the aquifers. The plume configurations delineated using the chloride data are very similar to the configurations obtained using electrical conductance, total dissolved solids, and dissolved organic carbon.

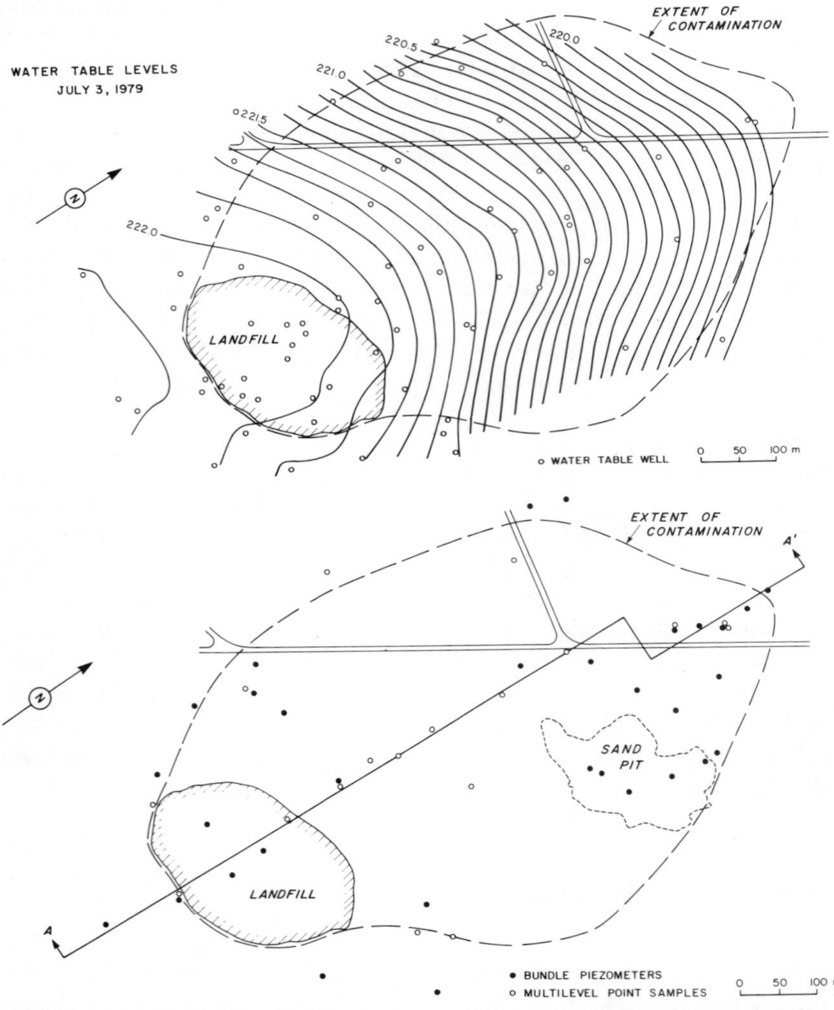

Figure 1. Borden landfill site: (top) outline of the zone of contamination in the sand aquifer and water table contours; (bottom) locations of multilevel sampling devices and of the longitudinal cross section.

At each site, leachate from the landfill has moved along flow paths controlled by the general slope of the water table. At the Borden and Woolwich sites, the plumes are nearly as wide as they are long. The

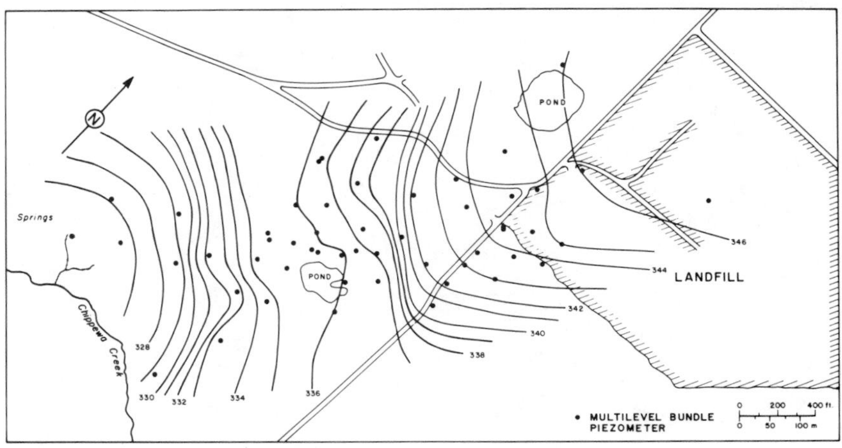

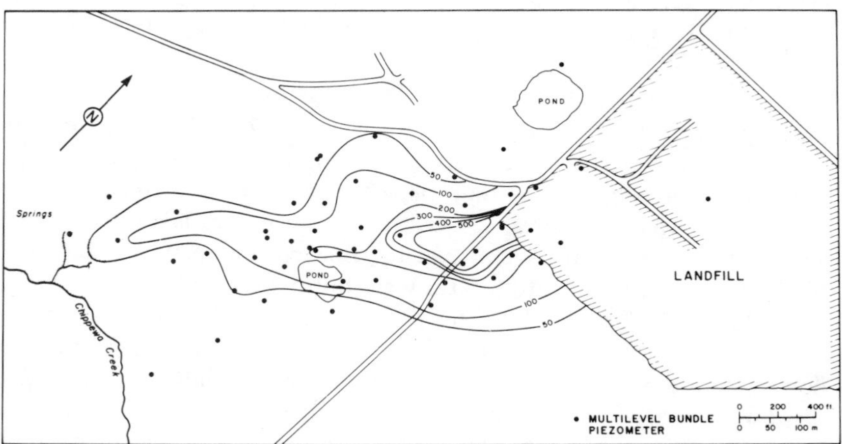

Figure 2. Water table contours (meters above sea level) [top] and chloride distribution (mg/L, maximum values at each multilevel monitoring site) [bottom] in the sand aquifer at the North Bay landfill site.

existence of a slight water table mound beneath the landfills, which results in the divergence of flow lines outward from the landfill areas, appears to be the main cause of the local spreading of the plumes in the directions transverse to main trend of the regional water-table slope.

At the Borden site the front of the plume is approximately 700 m northward of the landfill. At the Woolwich site, the front is at least 625 m from the landfill and may be much farther. At the Borden site, the plume has had 40 years to develop to its present state, whereas the

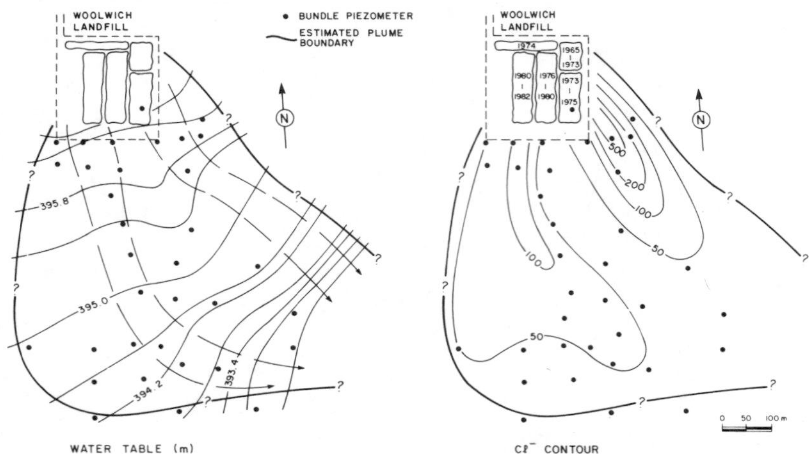

Figure 3. Water table contours (meters above sea level) and chloride distribution (mg/L, maximum values at each multilevel monitoring site) in the sand aquifer at the Woolwich landfill site.

development period has only been 15 years at the Woolwich site. At both of these sites, the plumes are advancing through aquifers that, in the vicinity of the plumes in their present positions, have no connection with surface-water systems such as streams, springs, or lakes. No water-supply wells exist in the vicinity of the Borden plume. The Woolwich plume is advancing toward a farm well, but it may go below the well.

At the North Bay landfill, the plume has spread to the maximum extent possible. It occupies the full extent of the local groundwater flow system from the recharge area where the landfill occurs, to the discharge area where flow paths originating from the landfill emerge in an area of springs situated 650 m southwestward of the landfill (Figure 2). Some of the water in the plume discharges at springs located closer (200 m) to the landfill where the water table intersects the slope of a sand quarry. Southward of the landfill, leachate passes only a short distance through the sand aquifer and discharges into a spruce bog.

The vertical distributions of Cl^- in the plumes at the Borden and North Bay sites are shown on cross sections in Figure 4. Both of these cross sections are positioned along the longitudinal axes of the plumes in the general direction of groundwater flow. At the North Bay site the aquifer is thin beneath the landfill, and it is therefore not surprising that the entire thickness of aquifer is contaminated with leachate. At the Borden site the aquifer is thick beneath the landfill, yet leachate

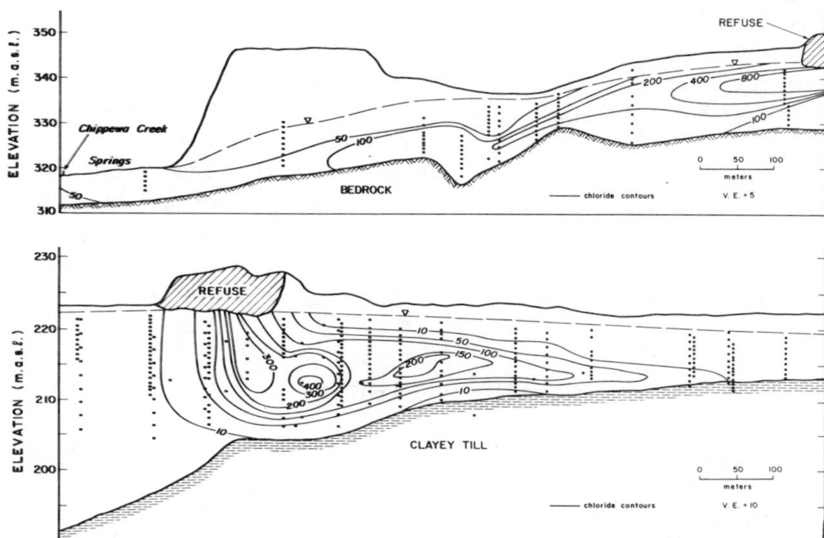

Figure 4. Distributions of chloride (mg/L) along longitudinal cross sections at the North Bay landfill site (top), and at the Borden landfill site (bottom).

has penetrated nearly to the bottom of the aquifer. A similar situation exists in the aquifer beneath the Woolwich landfill. The deep penetration of leachate at these sites appears to be due primarily to downward flow caused by mounding of the water table beneath the landfills. The mounding is a result of a greater recharge rate through the landfill relative to the recharge rate in the surrounding terrain. It is also possible that the greater density of the leachate relative to the natural groundwater contributes to the downward migration of contaminants beneath the landfills. Kimmel and Braids (1980) concluded that the density contrast has caused plumes to move to the bottom of the sand and gravel aquifers beneath two landfills on Long Island, New York.

With distance from the Borden landfill in the direction of groundwater flow, the plume occurs, to an increasing degree, in the bottom part of the aquifer. A zone of uncontaminated recharge water exists above the plume. This zone becomes thicker in the direction of advance of the plume as the cumulative quantity of water that has infiltrated to the aquifer in the area downflow of the aquifer increases. At the North Bay site, the plume generally occupies a greater percentage of the vertical thickness of the groundwater zone in the aquifer than at the Borden site; however, there is some similarity with the Borden case in that there is also a zone of recharge water that has formed about the plume in the area distant from the landfill.

In the direction of flow at the Woolwich site, there is no apparent development of a distinct zone of recharge water above the plume in the sand aquifer. However, the degree of detail of many of the vertical profiles at this site is less than at the other two sites. The vertical zonation of contaminated and uncontaminated water in aquifers at the Borden and North Bay sites illustrates the importance of utilizing multilevel monitoring installations for detecting and delineating zones of contamination at landfills on permeable sandy deposits.

PLUME CHEMISTRY AND PATTERNS

The plumes at the three landfill sites contain a variety of inorganic and organic contaminants. The chemical composition of the most severely contaminated groundwater at monitoring sites beneath or immediately adjacent to each of the landfills is indicated in Table I. Also shown in this table are the maximum concentrations in the plume at representative monitoring sites situated downflow from the landfills. At all three sites, Cl^- and HCO_3^- occur at major concentration levels. At the Borden site, SO_4^{2-} is also a major anion, whereas at the other sites it occurs at levels of a few tens of milligrams per liter or less.

At each of the sites, none of the hazardous inorganic elements such as toxic heavy metals or metalloids are present at levels above the maximum permissible limits specified in the primary drinking water regulations. Some of the constituents that have recommended limits in secondary regulations occur in above-background concentrations in the plumes. These include Fe, Mn, NH_4^+, and organic nitrogen. Concentrations of these constituents at levels that are much above background have also been reported in studies of contaminant plumes at other landfills on sand aquifers (Baedecker and Back, 1979; Golwer et al., 1975; Kimmel and Braids, 1980). Zinc exists in above-background levels in parts of the plumes but below the recommended limit for drinking water.

The concentration patterns of several inorganic species and of dissolved organic carbon at the Borden and North Bay sites are displayed in Figures 5 and 6. The patterns of Ca^{2+} and dissolved organic carbon are similar to the Cl^- pattern, as are the patterns of Na^+, Mg^{2+}, and Mn (and SO_4^{2-} at the Borden site), which are not displayed here. The high concentrations of Ca^{2+} and Mg^{2+} represent the hardness halo that is commonly reported in the literature describing contaminated

Table I. Examples of Chemical Data from the Plumes at the Three Landfills and Comparisons with Limits Specified in Drinking-Water Standards (mg/L, except where specified otherwise)

	Borden[a]		North Bay		Woolwich		Drinking Water Limits[b]
Elect. Cond., μS	3600	2750	5180	2160	6000	750	
TDS, mg/L	3800	2500	2370	1090		600	500
pH	6.91	6.80	6.9	6.7	6.8	7.5	
pE	1.2	5.7	0.9	1.5			
Alk,[c] mg/L as HCO_3	1060	604	2000	1900	3100	250	
Organic N	10	7	14	8			0.15
DOC	16	11	210	90	1300	15	5
Na^+	145	130	600	100	53	32	
K^+	140	54	230	95	10	1	
Ca^{2+}	552	360	330	200	290	85	
Mg^{2+}	184	126	120	35	115	30	
NH_4^{2+}	9	5	184	80			
Cl^-	75	206	550	200	415	45	250
SO_4^{2-}	1640	1000	23	13	100	18	500/250
B^-			1.5	0.5			5*
NO_3^- as N		<0.1	0.5	0.1			10*
Fe, total	15	15	160	300	132	0.01	0.3
Mn	1.5	0.5	21	12		0.7	0.05
Zn	0.75	0.08	0.02	0.07	0.3	0.003	5
Cu		<0.02	<0.01	<0.01	0.04	0.006	1
Ag			0.012			0.02	0.05*
As	0.0005	0.0003	0.003			<0.01	0.05/0.01*
Cd			0.003			0.001	0.005/0.01*
Cr			0.05	0.03		<0.002	0.05*
Pb	<0.03		<0.01			<0.002	0.05*
Ni	<0.03	<0.04	0.06	<0.02		0.003	
Se	<0.001	<0.001	<0.001			<0.01	0.01*
U			<0.05			<0.01	0.02*
H_2S	0.009	0.014					0.05

[a]Maximum ≈ 150 m.
[b]Unless specified otherwise, limits are the Canadian (1978) national values; where two values appear, the first is the Canadian limit, and the second is the Ontario (1978) limit. Values followed by an asterisk (*) are maximum permissible limits; all other values are recommended limits.
[c]Includes alkalinity due to dissolved organic carbon.

groundwater at landfills. Iron and manganese also follow the same trends as Cl^-. The concentrations of these elements are much lower at the Borden and Woolwich sites than at the North Bay site. In all three plumes, the concentrations of dissolved organic carbon are considerably above background levels. At the Borden and North Bay sites the patterns of dissolved organic carbon are generally similar to the chloride patterns.

At all three landfills, the highest concentrations of contaminants in the aquifers occur directly beneath or close to the periphery of the landfills. At the Woolwich site the concentrations are distinctly higher nearest the oldest part of the landfill, which is also the case at the

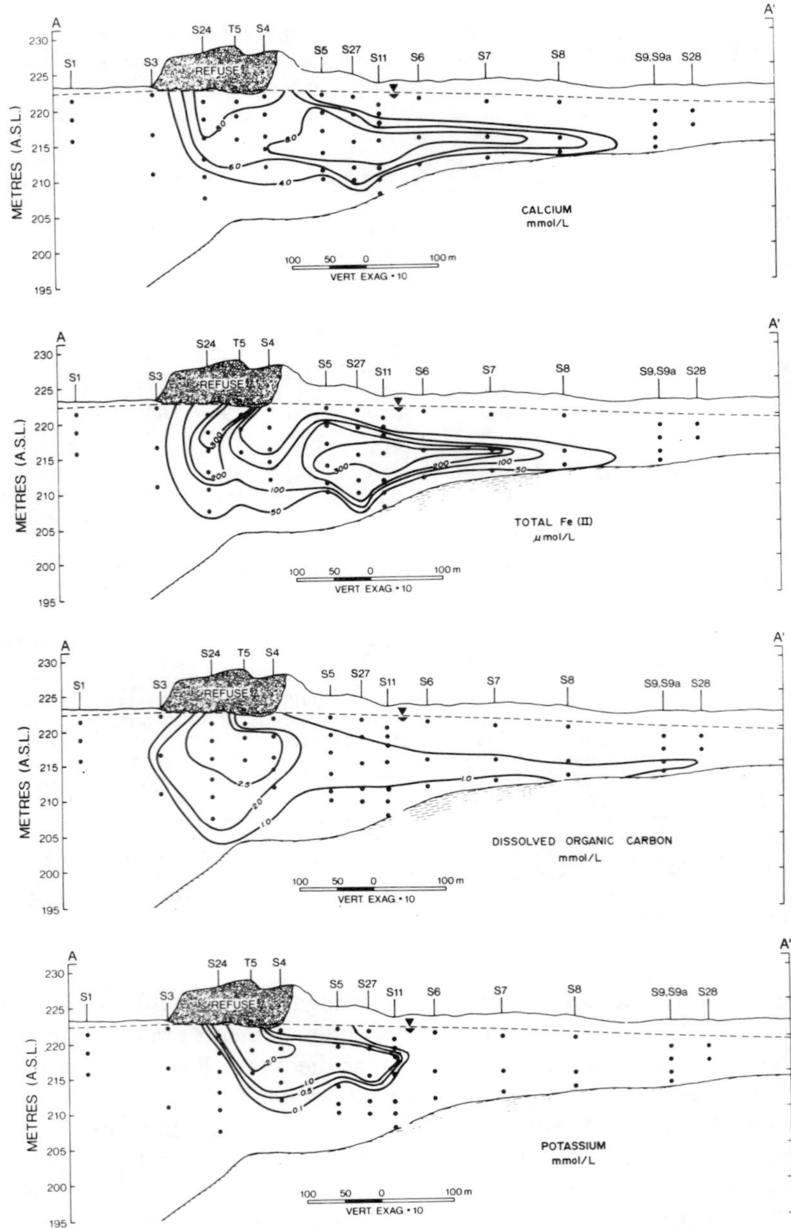

Figure 5. Distributions of Ca^{2+}, total Fe II, dissolved organic carbon, and K^+ on the longitudinal cross section at the Borden landfill (concentrations in millimoles per liter).

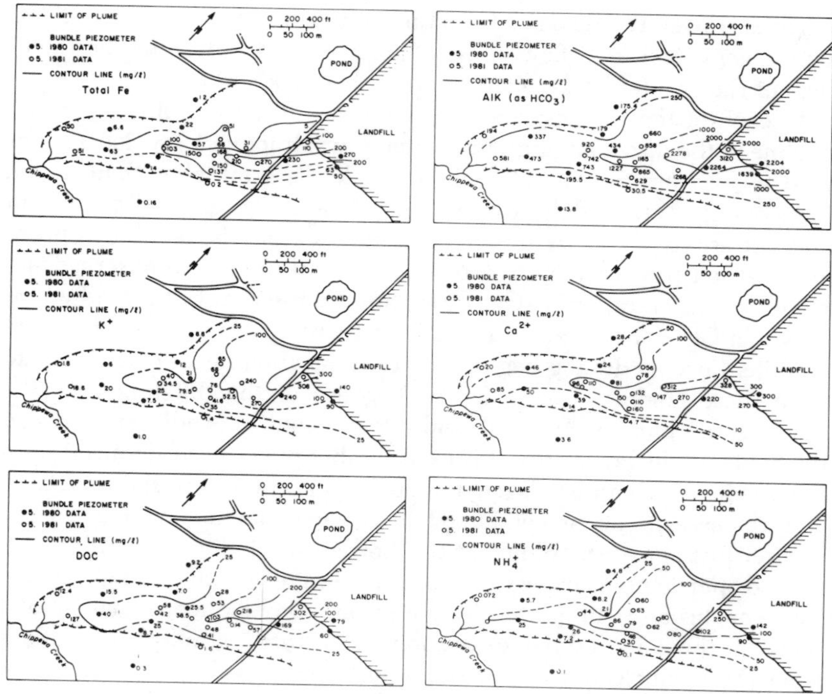

Figure 6. Distributions of total Fe, alkalinity as HCO_3^-, K^+, Ca^{2+}, NH_4^+, and dissolved organic carbon at the North Bay landfill site (concentrations in milligrams per liters).

Borden landfill. There is no evidence suggesting that the generation rate or strength of the leachate at any of the sites is declining appreciably with the aging of the refuse.

DISCUSSION

Physical Processes

The general form of the areal distribution of each plume is predictable from the slope of the water table in the vicinity of and downgradient of the landfills. The Borden and Woolwich plumes have spread out from the landfills in fan-shaped patterns caused by mounding of the water tables beneath the landfills. Farther from the landfills, the plumes advance in response to the slope of regional water table that is uninfluenced by the landfills. At the North Bay site, the plume becomes much narrower at a distance from the landfill. The narrowing is caused by the convergence of flow lines, which in turn is caused by the thickness of the aquifer and the shape of the bedrock surface.

In each of the plumes, the concentrations of Cl^- exhibit a gradual decline in the direction of groundwater flow, although a few local zones exist away from the landfills where the levels are higher than in the immediately surrounding area in the plumes. The gradual decline in concentrations of Cl^- in the direction of flow from the landfills is probably caused primarily by hydrodynamic dispersion. The observed declines in concentration may also be partially the result of a gradual increase in Cl^- concentration in the leachate that has emanated from the landfills since landfilling began. The concentration histories of leachate from these landfills are unknown. Because of the differences in the hydrologic, climatic, chemical, and microbiological conditions between the field and the laboratory, it is unlikely that they can be deduced from the results of laboratory leachate generation reported in the literature.

A tracer experiment conducted in the uncontaminated part of the aquifer just above the plume at the Borden site indicates that longitudinal dispersion is a strong process in the aquifer (Sudicky et al., 1982). This conclusion is supported by the results of one-dimensional simulations of the longitudinal dispersion of tritium (atmospheric fallout from thermonuclear tests) in the aquifer (Egboka et al., 1982). The development of the plume was simulated by means of a two-dimensional numerical model by Sykes and Farquhar (1979), who used large values of longitudinal dispersion to obtain some similarity between the longitudinal field distributions of Cl^- and the simulated distributions. Because of limitations caused by numerical dispersion in the model, they were unable to obtain a realistic representation of vertical distribution of Cl^- in the aquifer.

Dispersion in the vertical direction appears to be relatively weak at the Borden and North Bay sites. This conclusion is based on the observation that (1) the plumes at these sites exhibit vertical profiles that have distinct zones of high concentration which persist over long distances in the lower or middle parts of the aquifer, and (2) the coefficient for vertical dispersion measured by Sudicky et al. (1982) in the tracer test is very small. If vertical dispersion were a strong process, there would be much more vertical mixing in the aquifer as the plumes move away from the landfills. The two landfill plumes in sand and gravel aquifers on Long Island described by Kimmel and Braids (1980) also show little influence of vertical dispersion. There appears to be less vertical zonation of contamination at the Woolwich site, and it is possible that vertical dispersion is a more effective mixing process, particularly in the zone beneath and immediately downflow from the aquifer. The manner in which longitudinal and vertical dispersion influence the behavior of the plumes at the three sites is currently being evaluated using several types of models.

Chemical Processes

The hydrogeochemical processes in the plumes at the Borden and North Bay sites are described in detail by Nicholson et al. (1982) and Buszka et al. (1982), respectively. Chemical attenuation of inorganic contaminants in the plumes occurs from adsorption or ion exchanges, precipitation, oxidation, reduction, and coprecipitation. Major cations such as K^+, Na^+, Mg^{2+}, Ca^{2+}, and NH_4^+ undergo concentration changes primarily due to ion exchange. The cation exchange capacities of the sand at the three landfill sites are small, and these changes are not large, except for K^+, which is significantly retarded due to ion exchange. Because the species indicated above are not included in the primary or secondary drinking water standards, and because their behavior does not appear to have a strong influence on other chemical species in the plumes, their geochemical behavior is not of major practical importance.

The alkalinity and concentrations of dissolved inorganic carbon in the plumes are much higher than that of the background water. This is accounted for by the high partial pressures of CO_2 that exist in the landfills. The high CO_2 conditions provide for high values of H_2CO_3 and HCO_3^-. At the Borden and Woolwich sites, the sand in the aquifer and in the cover layers contains abundant calcite ($CaCO_3$) and dolomite [$CaMg(CO_3)_2$]. These minerals are relatively soluble in the presence of high CO_2 partial pressure.

The pH of the water in the plumes at the three sites is generally between 6.5 and 7.5. This near-neutral pH condition exists even though the gas phase in the landfills has very high partial pressures of CO_2. The concentrations of dissolved inorganic carbon in the plumes are much above background levels. Carbon isotope investigations indicate that much of the dissolved inorganic carbon in the plume at the Borden site is derived from the dissolution of calcite or dolomite in the sand (J. F. Barker, personal communication), whereas at the North Bay site most is derived from microbiological transformations of organic carbon (King et al., 1981). These results are consistent with the observations that the aquifers at the Borden and Woolwich sites contain more than

5 wt % calcite and dolomite and that the North Bay aquifer contains less than 0.01 wt % of these minerals. Also, the concentrations of dissolved organic carbon at the North Bay site are much higher than at the Borden site.

The low SO_4^{2-} concentrations in the plumes at the North Bay and Woolwich landfills suggest that the redox conditions within these two landfills are sufficiently reducing to prevent the formation of SO_4^{2-} from reduced forms of sulfur in the refuse and to cause the reduction of whatever amounts of soluble SO_4^{2-} that may initially exist in refuse. Nearly all zones of contaminated groundwater at landfills have very low SO_4^{2-} concentrations; therefore, the extremely high values of SO_4^{2-} in the Borden plume are exceptional. Based on analyses of the $^{18}O/^{16}O$ ratio in the sulfate, S. Feenstra (personal communication, 1979) concluded that the high SO_4^{2-} at this site is probably caused by dissolution in the landfill of gypsum or anhydrite contained in demolition debris from old buildings. Although the occurrence of methane and hydrogen sulfide indicates that the redox conditions in the landfill are strongly reducing, the SO_4^{2-} manages to persist in the leachate.

Nitrate is not present at concentrations above 1 mg/L in any of the plumes. The lack of NO_3^- and the abundance of NH_4^+ and organic nitrogen are results of the reducing condition in the landfills and of the generally anoxic condition throughout the plumes. In the vicinity of the Woolwich site, the groundwater in the sand aquifer not uncommonly has a few milligrams per liter or more of NO_3^- that probably originates from agricultural sources. The plume is therefore a zone of NO_3^- deficiency relative to the ambient conditions in the aquifer.

The levels of Fe and Mn are very high in the North Bay plume and moderately high in the Borden plume. The extensive occurrence of contaminant levels of these elements throughout the two plumes indicates they are not much influenced by adsorption. Insufficient data are available from the Woolwich site for comparisons to this site to be made. Some of the Fe and Mn in the plumes may be derived from the sand immediately beneath the landfills. The reducing redox conditions beneath the landfills can cause reductive dissolution of oxide coatings of Fe and Mn on the sand grains. The Fe and Mn remains mobile in the plumes because the redox potential remains sufficiently low even at considerable distance from the landfills.

The solubility of siderite ($FeCO_3$) appears to limit the maximum concentrations of Fe in the plumes but with less constraint at the North Bay site than at the Borden site. The values of pH and dissolved inorganic carbon for the two plumes are not much different, but the North Bay plume has an apparent high level of supersaturation with respect to siderite, and the Borden plume does not. The much-higher levels of dissolved organic compounds at the North Bay site may have a significant influence on the concentration levels of Fe and Mn.

Other than Fe and Mn, no other transition metals and no heavy metals or metalloids exist in the plume at levels above the limits specified in the primary or secondary drinking-water standards. These elements do not occur at levels above the drinking-water limits even at sampling points situated in the sand immediately beneath the landfills. It can be concluded, therefore, that these elements are effectively removed from the leachate at the contact between the refuse and the sand below the landfills or that they do not occur at hazardous dissolved concentrations in leachate even within the landfills. From consideration of solubility constraints using equilibrium geochemical models, it is expected that solid-phase solubility controls involving sulfide, hydroxide, and carbonate solid phases severely limit the occurrence in solution of most metals within and beneath the landfills. It is expected that adsorption is also an important mechanism limiting mobility of some of these elements.

Although the plumes do not contain hazardous concentrations of toxic inorganic contaminants, the fact that dissolved organic carbon occurs at elevated levels throughout the plumes is cause for concern. The distribution of dissolved organic carbon indicates that at least some organic compounds are very mobile in the aquifers. Preliminary investigations of organic compounds in the North Bay site indicate that the following compounds occur in the plume: chlorobenzene, dichlorobenzene, numerous substituted benzenes, chloroethane, acetone, dichloroethylene, xylene, and naphthalene (Cherry et al., 1981). The occurrence, movement, and transformations of organic compounds in the plumes at the three landfill sites are currently under investigation.

CONCLUSIONS

Plumes of contaminated groundwater extend many hundreds of meters from each of the three landfills. The lateral and longitudinal extents of the plumes are consistent with the shape of the regional water table in the unconfined sand aquifers and with the occurrence of local transient water table mounds beneath the landfills. The concentrations of mobile contaminants such as Cl^- decline gradually in the direction of lateral groundwater flow. The primary cause of this decline is probably longitudinal dispersion. It is possible that the concentrations of chloride and other major constituents in the leachate at the landfill increased from some period after landfilling began. If increases occurred, they would have contributed to the concentration vs distance trends in the plume. At some distance downflow from the Borden and North Bay landfills, there are distinct vertical zonations in the aquifers, with relatively uncontaminated groundwater occurring in the uppermost part of the aquifers and distinctly contaminated water at greater depths. Because of the persistence of these vertical zones over considerable longitudinal distance, vertical dispersion in the aquifers is regarded as a weak mixing process. This conclusion is supported by the results of a dispersion test using injected tracers, and by interpretations of tritium distributions.

Throughout each of the plumes, Na^+, Mg^{2+}, Ca^{2+}, NH_4^+, Fe, Mn, Cl^-, HCO_3^-, and organic nitrogen occur at levels that are generally much above background. Sulfate is very high in the Borden plume and very low in the other two plumes. Toxic transition metals, heavy metals, metalloids, and all other toxic inorganic constituents do not exist at hazardous levels in any of the plumes. Above-background levels of dissolved organic carbon exist throughout each of the plumes. Preliminary studies of the dissolved organic carbon in the North Bay plume indicate that many specific organic compounds occur in the plume, a number of which are toxic or potentially toxic pollutants.

ACKNOWLEDGMENTS

Many individuals participated in the investigations of the three landfills. Much of the information was from studies by graduate students at the University of Waterloo, including theses or project reports by Paul Buszka, Dave MacFarlane, Ron Nicholson, and Ed Sudicky. Information was also obtained from reports by Abdul Abdul, Janet Hewetson, Scott King, and Gartner Lee Associates. Paul Johnson provided assistance during most phases of the field investigations. Much of the drilling was conducted by Sam Vales using the University of Waterloo drill rig. The chemical analyses of groundwater samples were done by the Laboratory Branch of the Ontario Ministry of the Environment. Financial support for the investigations was provided by the Lottery Fund of the Province of Ontario, Environment Canada, the National Science and Engineering Research Council, and the Regional Municipality of Waterloo.

REFERENCES

Abdul, A. S., and J. A. Cherry, 1982. "Investigations of the Woolwich Landfill: Hydrogeological Conditions and Contaminant Distribution," unpublished report.

Baedecker, M. J., and W. Baek. 1979. "Hydrogeological Processes and Chemical Reactions at a Landfill," *Ground Water* 17(5):429-437.

Buszka, P. M., E. J. Reardon, and J. A. Cherry. 1982. "Investigations of a Municipal Landfill on an Unconfined Sand Aquifer: 2. Hydrochemical Patterns and Processes," in preparation.

Cherry, J. A., J. F. Barker, P. M. Buszka, J. P. Hewetson, and C. I. Mayfield. 1981. "Contaminant Occurrence in an Unconfined Sand Aquifer at a Municipal Landfill," in *Proceedings of the Fourth Annual Madison Conference of Applied Research and Practice on Municipal and Industrial Waste* (in press).

Cherry, J. A., R. W. Gillham, E. G. Anderson, and P. E. Johnson. 1982. "Migration of Contaminants in Groundwater at a Landfill: A Case Study. 2. Groundwater Monitoring Devices," *J. Hydrol.* 63(1-3).

Egboka, B. C. E., J. A. Cherry, R. N. Farvolden, and E. O. Frind. 1982. "Hydrogeological Studies of a Sand Aquifer at an Abandoned Landfill: 3. Bomb Tritium as an Indicator of Dispersion and Recharge," *J. Hydrol.* (in press).

Golwer, A., G. Matthes, and W. Schneider. 1975. "Effects of Waste Deposits on Groundwater Quality," Symposium Proceedings on Groundwater Quality, Moscow, 1971. Int. Assoc. Sci Hydrol. Publ. 102, pp. 159-166.

Hewetson, J. P. and J. A. Cherry. 1982. "Investigations of a Municipal Landfill on an Unconfined Sand Aquifer: 1. Hydrogeological Conditions and Contaminant Migration" (in preparation).

Kimmel, G. E. and O. C. Braids. 1980. "Leachate Plumes in Groundwater from Babylon and Islip Landfills," U.S. Geological Survey, Prof. Paper 1085.

King, K. S., J. F. Barker, J. A. Cherry, P. Fritz, and C. I. Mayfield. 1982. "Carbon Isotope Geochemistry of the North Bay, Ontario Landfill," Joint Annual Meeting of the Geological Association of Canada and Mineralogical Association of Canada. Program with Abstracts, Vol. 7.

MacFarlane, D. S., J. A. Cherry, R. W. Gillham, and E. A. Sudicky. 1982. "Migration of Contaminants in Groundwater at a Landfill: A Case Study. 1. Groundwater Flow and Plume Delineation," *J. Hydrol.* 63 (1-3).

Nicholson, R. V., J. A. Cherry, and E. J. Reardon. 1982. "Migration of Contaminants in Groundwater at a Landfill: A Case Study. 6. Hydrochemical Processes and Patterns," *J. Hydrol.* 63(1-3).

Pickens, J. F., J. A. Cherry, G. E. Grisak, W. F. Merritt, and B. A. Risto. 1978. "A Multilevel Device for Groundwater Sampling and Piezometric Monitoring," *Ground Water* 16(5):322-327.

Reardon, E. J., G. B. Allison, and P. Fritz. 1979. "Seasonal Chemical and Isotopic Variations in Soil CO_2 at Trout Creek," *J. Hydrol.* 43:355-371.

Robin, M. J. L., D. J. Dytynyshyn, and S. J. Sweeney. 1981. "Two Gas-Drive Sampling Devices for Deep Narrow Piezometers," *Groundwater Monit. Rev.* 2(1)63-66.

Sudicky, E. A., J. A. Cherry, and E. O. Frind. 1982. "Migration of Contaminants in Groundwater at a Landfill: A Case Study. 4. A Natural Gradient Tracer Test," *J. Hydrol.* 63 (1-3).

Sykes, J. F., and G. J. Farquhar. 1980. "Modelling of Landfill Leachate Migration," in *Proceedings of the Third International Conference on Finite Elements in Water Resources, Vol. 2,* (University of Mississippi Press), pp. 249-259.

DISCUSSION

***Terry Logan,* Ohio State University:** You indicated that your groundwater plume was biologically active, fairly high in soluble carbon, and absent of nitrate, which would imply to me that possibly you had some denitrification. Is that the case, and if so, is this typical of these types of leachates?

J. Cherry: Well, we don't see nitrates, the oxidized form of nitrogen, in any of our plumes. In fact, there is a series of contour lines of nitrate deficiency in one of the plumes; for example, all the aquifer around it is contaminated with nitrate from agricultural areas, but the plume itself is very low. We think that the generation of the nitrogen is occurring in the reducing environment right in the refuse. However, if nitrate should happen to actually get to the water table, then it would most certainly be reduced. It would be denitrified.

T. Logan: Is that typical?

J. Cherry: All the plumes that we've looked at in our very detailed studies in addition to all the other ones I've looked at in passing have been low in nitrate.

***Harry Freeman,* California Office of Appropriate Technology:** I thought it was a significant finding that you did not find toxic metals such as arsenic and selenium. Do you have any idea what was in the landfill? Did you ever go back and look at the source of the plumes?

J. Cherry: Just let me reiterate one point there. We do not find any of the metals such as selenium in the aquifer. And we are not sure whether these constituents are being attenuated by geologic strata right beneath the refuse or whether they are just not coming out of the refuse. We are not sampling the refuse itself. The Borden landfill, is a military base landfill, and it only receives mess hall waste and construction debris, so it is a very clean landfill, relatively speaking. The other two landfills, however, have received miscellaneous industrial wastes over a period of years. They are not big regional industrial waste sites, but they certainly received industrial waste from local industry. Unfortunately, we do not know what our source term is, so when I say that we are not seeing the metals such as selenium and arsenic, we don't know whether they're geochemically attenuated in the meter or two of sand that leachate passes through before entering the water table, or whether these constituents may not actually come out the bottom of the landfill itself.

John J. Suloway, **Charles T. Main, Inc.:** At what levels were you finding the iron in the second landfill?

J. Cherry: The iron is in the 100- to 200-mg/L range and ends up in the spring areas at the 10- to 100-mg/L range.

John Suloway: One comment. That could not only be an esthetic nuisance but also a nuisance to aquatic organisms. Those precipitates tend to include not only iron but other metals and to adhere to the bodies of fish and aquatic invertebrates, and they are toxic.

David Wilson, **Harwell:** As I said yesterday, we have been involved for a number of years in similar studies of this kind. We have very detailed information on some 20 sites representing a wide variety of geological situations, and I think a couple of the observations we have made might throw some light on discussions this morning. One comment is that we have found generally that underneath landfill sites the leachate is biologically active and it is anaerobic, which would suggest that the nitrate is being used as a food source by the bacteria. We have also used controlled lysimeter experiments to look at migration of heavy metals where we know what the source term is. The extent of the plume after about 3 years was 50 cm maximum in the unsaturated sandstone. In conjunction with the Institute of Geological Sciences, we have done extremely detailed work on just what is happening to the heavy metals. The conclusion is that they are sorbed on the clay particles within the sandstone. Also, we have looked at their precipitation with ferric oxides using very sophisticated analytical techniques. All of this information has been published. There are some 18 reports available, and I can supply reference lists on request.

CHAPTER 14

GEOCHEMISTRY AND SUBSURFACE CHARACTERIZATION RELATED TO THE TRANSPORT AND FATE OF INORGANIC CONTAMINANTS

Arthur W. Hounslow

Department of Geology
Oklahoma State University
Stillwater, Oklahoma 74078

ABSTRACT

The movement and attenuation of heavy metal pollutants in the subsurface depends primarily on groundwater movement and the sorptive properties of the subsurface solids. The principal sorbents are hydrous oxides of iron and manganese, insoluble organic matter, and clay minerals, frequently in that order of importance. They form primarily in soils as a result of weathering but are sufficiently transitory to occur in the underlying sediments. They may influence the porosity of the subsurface medium, and their mode of occurrence depends on the geochemical environment in which they formed, which may affect their sorptive properties.

Groundwater quality results from the interaction of water with the rocks through which it moves. Groundwater composition changes as the geochemistry of the environment changes.

The sorption of heavy metals depends both on the geochemical environment and the form of the sorbent (i.e., whether it is a single phase or a multiphase complex of several materials).

Heavy metals may be desorbed and mobilized by high salt concentrations, changes in pH and redox potential, an influx of natural or synthetic complexing agents, or microbial formation of soluble and toxic alkyl derivatives of heavy metals.

INTRODUCTION

The ultimate aim of subsurface study in this context is to be able to predict the behavior of contaminants in subsurface environments so that these contaminants may be disposed of with minimal effects on both groundwater quality and soils.

The quality of groundwater reflects the mineralogic composition of the rocks with which the water has been in contact. As water moves slowly through the subsurface, its composition gradually changes, reflecting an increasing saturation of some ions or the results of various rock–water interactions. Many of these reactions define the geochemical environment, including such parameters as pH, Eh, and ionic strength, which determine the sorptive properties of the subsurface and the types of microbial processes that may occur, which, in turn, may affect the mobility of many elements.

The most critical mass transport processes that determine the subsurface movement of inorganic pollutants are water movement and sorption. Although many of the stronger adsorbents such as clays, hydrous oxides, and insoluble organic matter form in the soil layer, they often occur in the aquifer but in smaller amounts.

DISCUSSION

A study of the movement and attenuation of pollutants in the subsurface depends on both groundwater movement (in both saturated and unsaturated zones) and the sorptive properties of the subsurface solids. The sorption that may occur depends not only on the inherent chemical properties of these solids but also on the nature, size, distribution of the adsorbents. Because most adsorbing materials in the subsurface are secondary in origin, and to some extent transitory, they also may strongly influence the porosity and permeability of the subsurface medium. Because of this, their physical and chemical properties must be investigated.

Physical Nature of Adsorbent

The physical nature of these subsurface adsorbents as well as their presence and distribution patterns, may be highly significant. For example, humic acids may occur as spheres, sheet-like materials, or bundles of fibers, depending on pH (Schnitzer and Khan, 1978). Grim (1968) describes electron microscope studies showing that kaolinite may often occur as hexagonal flakes, whereas halloysite is elongate and tubular, and illite, chlorite, vermiculite, and some smectites occur as irregular flakes. On the other hand, elongate lathes and fibers are the characteristic form of the attapulgite-sepiolite-palygorskite series and some smectites. The area-to-thickness ratio of smectites is also orders of magnitude greater than that for kaolinite. Further, the movement of water and the sorption of pollutants may be strongly dependent on the shape and size of the voids in the subsurface materials.

Similarly, amorphous sesquioxides often occur as coatings on other minerals, a habit that allows them to exert chemical activity far out of proportion to their total concentration (Jenne, 1968).

Rock-Water Interactions

When water contacts the solid phases in a rock, sediment, or soil several processes take place:

1. The water dissolves certain of the solid constituents that move into the groundwater system. Under normal conditions, this process determines groundwater quality.
2. Insoluble materials such as clay and colloidal sesquioxides may be formed by the chemical alteration of existing minerals. This process characteristically occurs in but is not necessarily restricted to the soil. These secondary materials often have high sorptive capacities and greatly influence the movement of pollutants in the subsurface.
3. Materials may be removed from the aqueous phase by several processes. If the water is saturated with a particular component, or if the geochemical environment changes, new minerals may be precipitated in the pore spaces of the existing rock, soil, or sediment. Another process involves the sorption of inorganic ions or dissolved organic matter onto an existing solid phase. If this is an ion exchange reaction, other ions will be released to the liquid phase.
4. Materials may be added to the liquid phase by desorption reactions. These reactions will often occur when a geochemical barrier is breached. In addition to determining the degree of sorption that occurs, the geochemical parameters determine, to a large extent, the stability of the sorbent itself.

Millot (1979) suggests that the primary factor governing the amount and type of clay at a particular location is the amount of leaching that occurs. This factor is basically climatic.

The hydrous oxides of iron and manganese frequently precipitate at geochemical boundaries such as the presence of dissolved oxygen, the limestone neutralization of acid waters, or the mixing of two groundwaters of differing compositions.

Adsorption of Metals

Some materials with a large surface area, particularly clay minerals, freshly precipitated iron and manganese hydroxides, and organic substances, are capable of sorbing cations from solution.

Clay Minerals

Clay minerals, which are the insoluble products of the chemical weathering of silicate minerals, are usually concentrated in the B soil horizon. These clay minerals have the property of adsorbing certain anions and cations and retaining them in an exchangeable state. These exchangeable irons are held around the outside of the silica-alumina clay-mineral structural unit, and the exchange reaction generally does not affect the structure of the silica-alumina packet. The more common exchangeable cations are Ca^{2+}, Mg^{2+}, H^{+}, K^{+}, NH_4^{+}, and Na^{+}, frequently in about that order of abundance (Grim, 1968). Different clay minerals are formed under different conditions, and although most are capable of sorbing metals to some extent, this ability varies greatly. The principal factors affecting the ability of clay minerals to adsorb metals are particle size, surface area, moisture content, and degree of crystallinity. Further, a specific clay will adsorb different elements to different extents, even under the same conditions. Grim (1968) emphasizes there is no single universal replaceability series.

Generally, the dissolved alkaline earth metals such as Mg, Ca, and Ba are more strongly adsorbed than the univalent alkali metals such as Rb and Cs; however, both groups are adsorbed by clays more frequently than base metals (Levinson, 1974).

Amorphous Hydroxides

In 1968, Jenne suggested that the hydrous oxides of Mn and Fe furnish the principal control on the fixation of Co, Ni, Cu, and Zn in soils and freshwater sediments rather than the clay minerals. He further postulated that the sorption or desorption of these heavy metals occurs in response to the following factors:

1. aqueous concentration of the metal in question,
2. aqueous concentrations of other heavy metals,
3. pH, and
4. amount and strength of organic chelates and other complex ions.

In a study of heavy metal relationships beneath a municipal landfill in central Pennsylvania, Suarez and Langmuir (1976) found that the major source of heavy metals in these soils was the hydrous manganese oxide. The manganese oxide exceeded iron oxide sorption by at least a factor of 10 for some heavy metals, possibly because of the greater degree of crystallinity of the iron oxides as well as a lower pH. They also noted that the metal oxides and trace metals existed predominantly in coatings on quartz grains and were not significantly concentrated in the <15-μm fraction.

These conclusions differ little from well-established geochemical data such as that reported by Rankama and Sahama (1950), namely, that oxidate sediments rich in Mn commonly contain notable amounts of Li, K, Ca, Ba, B, Ti, Co, Ni, Cu, Zu, Tl, Pb, and W.

Organic Matter

Opinions differ greatly as to the role of organic matter in the adsorption of small amounts of heavy metals. That the organic layer of soil accumulates certain metals is not surprising because the exchange capacity of humus can be as high as 500 meq/100 g, whereas clays rarely exceed 150 meq/100 g (Levinson, 1979). Organic matter in swamps and bogs is also likely to show high metal values, although all metals are not necessarily adsorbed equally by organic matter. Horsnail and Elliott (1971) found that, in some swamps in British Columbia, Co and Mo were markedly enriched in organic swamp deposits, whereas Fe, Mn, Co, Ni, and Zn were not.

Takamatsu and Yoshida (1978), studying complexes of the divalent metals Co, Pb, and Cd with a variety of humic acids, found that the bonding is mainly through two –COOH groups such as in the polycarboxylic acids and through –COOH and –OH groups. Saxby (1973) suggested that incorporation of a metal into a sediment may involve several processes, such as a reaction between a metal and organic in solution, followed by adsorption into clay minerals.

Parfitt et al. (1977) showed that the principal mode of interaction between the hydrous oxides of iron and aluminum on one hand, and fulvic and humic acid on the other, is ligand exchange between the humate carboxylate groups and the surface hydroxides of the hydrous oxides.

According to Theng (1979), the isolation of naturally occurring clay organic complexes has met with limited success, principally because interlayer complexes only form at pH greater than 4 and at fairly high concentrations. Further, Kodama and Schnitzer (1968) have shown that the amount of fulvic acid adsorbed by montmorillonite depends on the cation with which the montmorillonite was saturated.

Relative Importance of Adsorbates

Consideration of the data available led Guy and Chakrobarti (1975) to suggest a relative adsorption sequence (highest to lowest) for heavy metals: manganese oxide, humic acid, iron oxide, and clay minerals.

Remobilization of Heavy Metals

Sorption of heavy metals onto organic or inorganic substrates is often only a temporary condition; changes in geochemical parameters or the addition of organic or inorganic complexing agents may release these heavy metals in a slug that may by environmentally catastrophic. Forstner and Wittmann (1979) list five such types of reactions:

1. Elevated salt concentrations could lead to replacement of adsorbed heavy metals by alkali and alkaline earth metals, particularly from clays.
2. Changes in redox conditions, such as a decrease in oxygen potential, could lead to the solubilization of the iron and manganese hydroxides as well as their adsorbed heavy metals.
3. Changes in pH. For example, a decrease in pH could lead to the dissolution of carbonate and hydroxides and decreased sorption because of competition with hydrogen ions.
4. Increased amounts of natural or synthetic complexing agents entering the system could lead to the formation of highly stable heavy metal complexes that would otherwise be adsorbed to solid particles.
5. Microbial activity can lead to the formation of highly soluble and highly toxic alkyl derivatives of a variety of heavy metals, such as methyl mercury.

Geochemical Changes

The three major types of geochemical changes that may take place involve changes in ionic strength, pH, and redox potential. Changes in the salt concentration of water that contacts the solids in an aquifer can be caused by seawater encroachment, contamination by leakage from an upper or lower brine aquifer, and by increased salt contamination from the surface. Because of the preferential adsorption of alkali and alkaline earth metals by clays relative to heavy metals (Levinson, 1974), this group of sorbents is more likely to be affected by this process. However, destabilization of colloids is another potential effect of high salt concentration, which leads to precipitation of other materials such as metal–organic complexes.

Khalid et al. (1977) concluded that low pH and low Eh in sediment-water systems tend to favor the formation of soluble species of many metals, whereas in an oxidized, nonacid system slightly soluble or insoluble forms tend to predominate.

The classic example of increasing the acidity of groundwater and surface water is acid-mine drainage, but this is described later because it is primarily a microbiologically catalyzed reaction. Acid precipitation may also have a similar effect. However, the surface introduction of acids such as pickling liquors to a disposal site may also be a primary cause of changes in the pH of groundwater (Pettyjohn, 1975).

To a first approximation, the major natural changes in redox potential occur with the introduction or depletion of O_2 or H_2S (Hounslow, 1980). The depletion of oxygen would almost certainly lead to the redistribution of Fe and Mn in the subsurface, which in turn would release to the aqueous phase those heavy metals once adsorbed on these hydroxides. However, if H_2S is present, perhaps as a result of microbially catalyzed sulfate reduction, most trace elements are fixed because of their precipitation as sulfides or coprecipitation with iron sulfides.

Dissolved Organic Matter

The dissolved organic matter in aquatic systems consists of the remains of biologically produced compounds of low molecular weight, principally fulvic acid, in addition to a variety of synthetic organic contaminants.

Sholkavity and Copland (1981) have shown that in surface waters Fe, Mn, Cu, Ni, Co, and Cd may be held in the dissolved phase by natural organic substances such as humic acids, between a pH of 3 and 9. However, when such a solution contacts a high concentration of salt, such as in seawater, precipitation of some of these elements may occur due to the destabilization of the colloids. Dissolved organic matter in groundwater may react with and remove dissolved oxygen, causing reduction to occur and possibly dissolving Fe and Mn hydroxides and releasing adsorbed heavy metals. If this reduced water contains sulfate and the appropriate bacteria are present, sulfate reduction can occur, and the generated hydrogen sulfide may react with and cause the precipitation of Fe and the heavy metals.

The introduction of synthetic organic compounds that have strong chelation capabilities, such as NTA, EDTA, and many others, will drastically change the adsorption properties of many heavy metals.

Microbial Activity

Our knowledge of the full extent of the activity of microorganisms in the subsurface, especially the deeper subsurface, is limited. Forstner and Wittmann (1980) postulate three major processes that lead to the mobilization of metals by microbial activity: (1) breakdown of organic matter to more soluble forms, (2) changes in the geochemical parameters of the environment by metabolic activity, and (3) conversion of inorganic compounds into soluble metal-organic complexes.

Theis and Singer (1973,1974) have shown that soluble humic acids produced by the microbiological decay of humic material forms complexes with ferrous iron, which inhibits the formation of insoluble ferric compounds under the appropriate oxidizing conditions.

The production of acid waters by bacteria-catalyzed oxidation of the sulfide components of mine dumps, known as acid mine drainage, is well known. In addition to acid formation, the release and solution of many toxic heavy metals is a major problem. The release of acid and trace metals may continue for decades (Forstner and Wittmann, 1979). In waters with little buffer capacity (i.e., in carbonate-poor areas), dissolved metals may be transported for great distances.

This process is used directly in the bacterial leaching of some metals, particularly for the recovery of Cu and U. In these processes, iron sulfides not only provide a source of acid but also stimulate the bacterial oxidation of other sulfides. It may be used on either low-grade ores or waste dump material.

A more subtle effect of microorganisms, at least in surface waters, is in the alkylation of heavy metals. When uptake of a toxic metal occurs, microorganisms are often able to detoxify their environment through organometallic transformations. This product, however, may be extremely toxic to higher organisms. This is particularly true of methyl mercury. Microorganisms do not seem to require mercury in their diet but deal with it in this manner when it occurs in their food supply.

The mechanism of biologic methylation is also effective in the formation of volatile compounds of Ag, Pb, and Se such as alkyl arsines, tetramethyl lead, and dimethyl selenite. Methylation of tin compounds can be catalyzed by some bacteria. Other metal alkyls, which are stable in water and can be synthesized by methyl-cobalamin reaction, include Te, Pd, Pt, Au, and Tl.

REFERENCES

Forstner, U., and G. T. W. Wittman. 1979. *Metal Pollution in the Aquatic Environment* (New York: Springer-Verlag, Inc.).

Grim, R. E. 1968. *Clay Mineralogy,* 2nd ed. (New York: McGraw-Hill Book Co.).

Guy, R. D., and C. L. Chakrobarti. 1975. "Distribution of Metal Ions Between Soluble and Particulate Forms," Abstract, International Conference on Heavy Metals in the Environment, Toronto, Ontario, Canada.

Horsnail, R. F., and I. L. Elliott. 1971. "Some Environmental Influences on the Secondary Dispersion of Molybdenum and Copper in Western Canada," in *Geochemical Exploration,* CIM Special Vol. 11, pp. 165-175.

Hounslow, A. W. 1980. "Ground-Water Geochemistry: Arsenic in Landfills," *Ground Water* 18:331-333.

Jenne, E. A. 1968. "Controls on Mg, Fe, Co, Ni, Cu and Zn Concentrations in Soils and Water: the Significant Role of Hydrous Mn and Fe Oxides," in *Trace Organics in Water,* American Chemical Society Adv. Chem. Ser. 73, pp. 337-387.

Khalid, R. A., R. P. Gambrell, M. G. Verloo, and W. H. Patrick, Jr. 1977. "Transformations of Heavy Metals and Plant Nutrients in Dredged Sediments as Affected by Oxidation Reduction Potential and pH, Vol. 1, Literature Review," Contract Report D-77-4 for Office, Chief of Engineers," United States.

Kodama, H., and M. Schnitzer. 1968. "Effects of Interlayer Cations on the Adsorption of a Soil Humic Compound by Montmorillonite," *Soil Sci.* 106:73-74.

Levinson, A. A. 1974. *Introduction to Exploration Geochemistry* (Applied Publishing, Ltd.).

Millot, G. 1979. "Clay," *Sci. Am.* 240:109-118.

Parfitt, R. L., A. R. Fraser, and V. C. Farmer. 1977. "Adsorption on Hydrous Oxides. Vol. III. Fulvic Acid and Humic Acid on Goethite, Gibbsite and Imogolite," *J. Soil Sci.* 28:289-296.

Pettyjohn, W. A. 1975. "Pickling Liquors, Strip Mines and Ground-Water Pollution," *Ground Water* 13:4-10.

Rankama, K., and T. G. Sahama. 1950. *Geochemistry* (Chicago: The University of Chicago Press).

Saxby, J. D. 1973. "Digenesis of Metal Organic Complexes in Sediments: Formation of Metal Sulfides from Cysteine Complexes," *Chem. Geol.* 12: 241-288.

Schnitzer, M., and S. U. Khan. 1978. *Soil Organic Matter* (New York: Elsevier).

Sholkavitz, E. R., and D. Copland. 1981. "The Coagulation, Solubility, and Adsorption Properties of Fe, Mn, Cu, Ni, Cd, Co and Humic Acids in a River Water," *Geochim. Cosmochim. Acta* 45:181-189.

Suarez, D. L., and D. Langmuir. 1976. "Heavy Metal Relationships in a Pennsylvania Soil," *Geochim. Cosmochim. Acta* 40:589-598.

Takamatsu, T., and T. Yoshida. 1978. "Determination of Stability and Constants of Metal-Humic and Complexes by Potentiometric Saturation and Nonselective Electrodes," *Soil Sci.* 125:377-386.

Theis, T. L., and P. C. Singer. 1973. "The Stabilization of Ferrous Ions by Organic Compounds in Natural Waters," in *Trace Metals and Metal-Organic Interactions in Natural Waters*, P. C. Singer, Ed. (Ann Arbor, MI: Ann Arbor Science Publishers, Inc.).

Theis, T. L., and P. C. Singer. 1974. "Complication of Iron (III) by Organic Matter and Its Effect on Iron (II) Oxygenation," *Environ. Sci. Technol.* 8:569.

Theng, B. K. G. 1979. "Formation and Properties of Clay-Polymer Complexes," *Devel. Soil Sci.* 9:362.

DISCUSSION

***John Brown*, CEGB:** I can appreciate your message that if we disturb the system nothing is safe or irreversible, but for most of the situations if the metals are "fixed" under a reducing environment you are really back to the situation that existed in the earth. I would anticipate that in a real situation you may know what the waste is and know in a relative sense the state of the aquifer. If you have immobilized the heavy metals, then you've not just adsorbed them, you may have precipitated them and fixed them chemically for time immemorial, if you like. So I would appreciate your comments on that.

A. Hounslow: I agree that in some cases if you leave the system alone it may remain safe for a generation or so; however, thinking geologically, I would not agree with your "time immemorial." A common situation that must be considered is that a dump exists, was operated efficiently and safely, covered with clay, and the land sold. If the new owner builds roads and buildings on the site and in so doing disturbs the cap or lining, infiltration will probably occur and leakage will most likely result. I agree that if you can say "Let's leave it alone forever, put a sign on it, a fence around it, and some monitoring wells in it," you can assume that once immobilized the metals will probably stay that way for the foreseeable future. My only concern is that an abandoned dump is seldom left undisturbed.

John Brown: I appreciate that too. The trouble is that if you want to make an application for license to dispose, your argument can be used to prevent almost any disposal scheme on the grounds that the licensing authority is a little unsure about what is going to happen in the future. And all I am trying to do is redress the balance by saying that I can see some schemes where it is impossible to predict the future.

A. Hounslow: That is something that the regulating body has to decide—what percentage safety or permanence is adequate. I am not really saying that we shouldn't do any of this: I am pointing out that there are some major hazards if you don't look far enough ahead.

***David Wilson,* Harwell:** You said that if you disturb the site by building a road across it or whatever, you may immobilize your heavy metals, then mobilize them again. If you take the alternative management strategy, and you keep your materials in the site by lining it, and you build a road across it, you have even greater problems because your materials are there. You are going to be interfering with them directly; you are going to have direct toxicity problems, etc.

A. Hounslow: I am not quite sure that I understand the question. I am somewhat against building on hazardous waste sites. I think that if you have a site you should leave it as a site and forget about it. Put some grass on it maybe but don't try to build a city on it. But that's just a personal opinion.

***Ralph Franklin,* Department of Energy:** I am concerned that these modeling generalizations might be used by a regulatory agency to make decisions related to specific disposal situations. Although you acknowledged that the problem was complex and that generalizations did not always hold true, the models are so oversimplified as to be meaningless, for example, Fe oxide $<$ Mn oxide, which only holds for certain solutes. You really need to consider the specific reaction mechanisms which apply to the surface-solute interaction; for example, the valence of inorganic ions, the functional groups on organic solutes, and whether special affinites or reactions are involved (such as for phosphates with iron oxide) are examples of factors that control the relative affinity of the various surfaces for the various solutes.

A. Hounslow: I agree. I am not arguing that this could be used by anybody except maybe to lead you in a direction of the correct type of monitoring and research needed to study a site. I think one frequently goes to a site without really knowing what to look for. The object of this talk was to give an indication of the interactions that you may get and what you should be looking for at a particular site.

CHAPTER 15

CORRELATION OF COMPOUND PROPERTIES WITH SORPTION CHARACTERISTICS OF NONPOLAR COMPOUNDS BY SOILS AND SEDIMENTS: CONCEPTS AND LIMITATIONS

John J. Hassett and W. L. Banwart
Department of Agronomy
University of Illinois
Urbana, Illinois 61801

Robert A. Griffin
Illinois State Geological Survey
368 Natural Resources Building
Champaign, Illinois 61820

ABSTRACT

Sorption constants for the sorption of nonpolar (hydrophobic) organic compounds are highly correlated with the organic carbon content of the sorbing phase (soil or sediment) and with water solubility and the octanol-water partition coefficient of the compound.

The purposes of this chapter are (1) to review and expand the concept of hydrophobic sorption; (2) to show the relationships between molar Freundlich adsorption constants and soil, and sediment organic carbon contents; (3) to develop regression equations relating sorption to the organic compounds water solubilities and octanol-water partition coefficients; and (4) to discuss the use and limitations of these concepts.

INTRODUCTION

The current emphasis in the United States on the environmental impact of the production and use of organic chemicals has resulted in a need to make predictions of their environmental behavior from basic physicochemical properties of the compounds. Significant quantities of organic chemicals are released to soils each year as a result of their manufacture, transport, use, and disposal. Hence, organic chemicals are subject to possible transport through soil media. There are, however, more than 40,000 compounds currently in commercial production in the United States, and it is impractical to conduct extensive soil-adsorption and attenuation studies on each compound on every soil type. It would be highly desirable to predict soil adsorption and mobility of organic compounds from some basic physicochemical properties. Although it may be difficult to generalize about the sorption of organics by soils if all classes of organic compounds are considered, recent research by a number of investigators (i.e., Chiou et al., 1979; Hassett et al., 1980; Karickhoff et al., 1979; Means et al., 1979; McCall et al., 1980) has demonstrated that generalizations about the sorption of nonpolar (hydrophobic) organic compounds by soils may be possible.

SORPTION OF NONPOLAR ORGANICS

Sorption of many nonpolar (hydrophobic) organic compounds by soils has produced isotherms that are well described by the Freundlich equation:

$$C_s = K_f \cdot C_w{}^{1/n} \tag{1}$$

where C_s is the amount sorbed per unit weight of adsorbent, C_w is the equilibrium solution concentration, and K_f and $1/n$ are empirical constants. When $1/n$ values are >1, the amount of sorption tends to increase without limit at a nearly constant C_w value; this is suggestive of precipitation. When values of $1/n$ are <1, sorption appears to approach a limit. When values of $1/n$ are $= 1$, linear isotherms are produced, and K_f reduces to a simple partition coefficient. Most nonpolar organic compounds produce linear isotherms over the concentration ranges found in the environment (Chiou et al., 1979; Griffin and Chian, 1979; Hassett et al., 1980; Means et al., 1980).

The units in which C_s and C_w and hence K_f are expressed are critical, if $1/n$ is $\neq 1$, and must be considered when comparing sorption constants from different sources and when averaging K_f values. Osgerby

(1970) developed a simple relationship between K_f values expressed on a mass basis and a molar basis:

$$K_{molar} = [K_{mass} \times (\text{mol. wt.})^{1/n}]/(\text{mol. wt.}) \quad (2)$$

This relationship can be expanded to allow conversion between K_f values determined in different concentration ranges:

$$K_{\mu g} = [K_{ng} \times (1000)^{1/n}]/1000. \quad (3)$$

Table I shows the effect of 1/n of K_f values determined in different concentration ranges. A similar effect of 1/n values on the value of K_f has been demonstrated by Bowman (1981) for the sorption of several different organic compounds. When $1/n = 1$, the units of C_s and C_w have no effect on the value of K_f. If 1/n is $\neq 1$, molar K_f values should be used.

Table I. Effect of 1/n Values on the Value of K_f Determined for Different Concentration Ranges

1/n	$K_{\mu g}$	K_{ng}	K_{pg}
0.62	500	6,906	95,386
0.90	500	998	1,992
1.00	500	500	500
1.21	500	117	28

CORRELATION OF K_f VALUES WITH SOIL AND COMPOUND PROPERTIES

Adsorption experiments with nonpolar organic compounds and different soils produce a different K_f value for each soil. Correlation of K_f against soil properties has shown that most of the variation in K_f values between soils can be explained by differences in the organic carbon (humus) contents of the soils. Figure 1 illustrates the relation of K_f values and the organic carbon contents of several soils for four representative compounds. Sorption appears to increase linearly with an increase in the organic carbon content of the soil. The slopes of these lines represent the sorption constant (K_{oc}) for the respective compounds normalized for the organic carbon content of the respective soil.

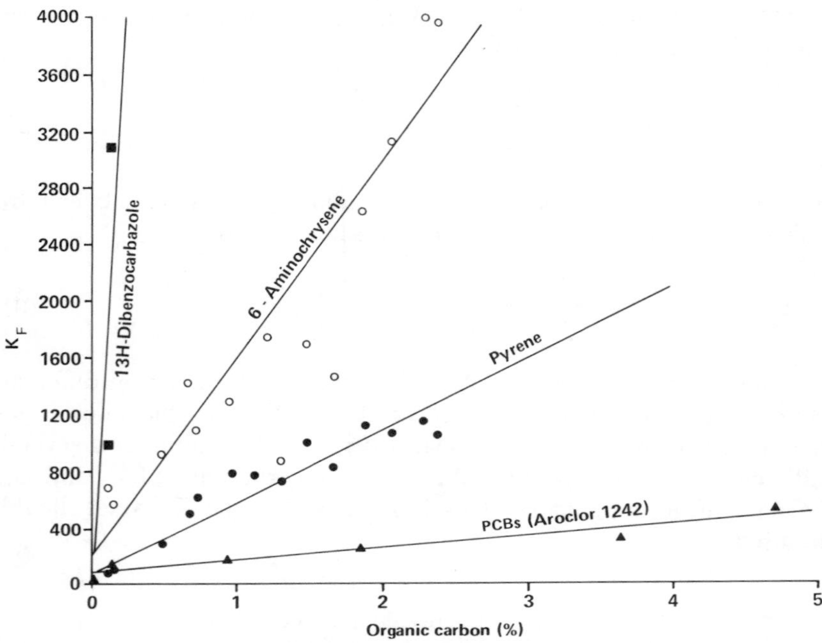

Figure 1. Relation between adsorption constant (K_f) and the organic carbon content of soils .

Sorption constant values (K_f) can be adjusted for the organic carbon content of the soil in two basic ways. The first method is to use individual values of K_f and the respective percent OC contents of the soils.

$$K_{oc} = (K_f \times 100)/\text{percent organic carbon} \qquad (4)$$

The second method is to obtain K_{oc} from a regression of K_f vs percent OC for a group of soils. The latter method is illustrated in Figure 1 and is the method preferred by us. This method minimizes the effects of experimental errors or spurious data points that could affect a K_{oc} value obtained from the average of a few soils. A small positive intercept is usually obtained from a regression relation of this type. This has been interpreted to be caused by the adsorption that can be expected from the mineral fraction of the soil material (Griffin and Chian, 1979) from hydrophobic sites present on soil minerals such as clays (Chen, 1976).

Values of K_{oc} are compound properties, not soil properties, and have been related to other compound properties such as water solubilities

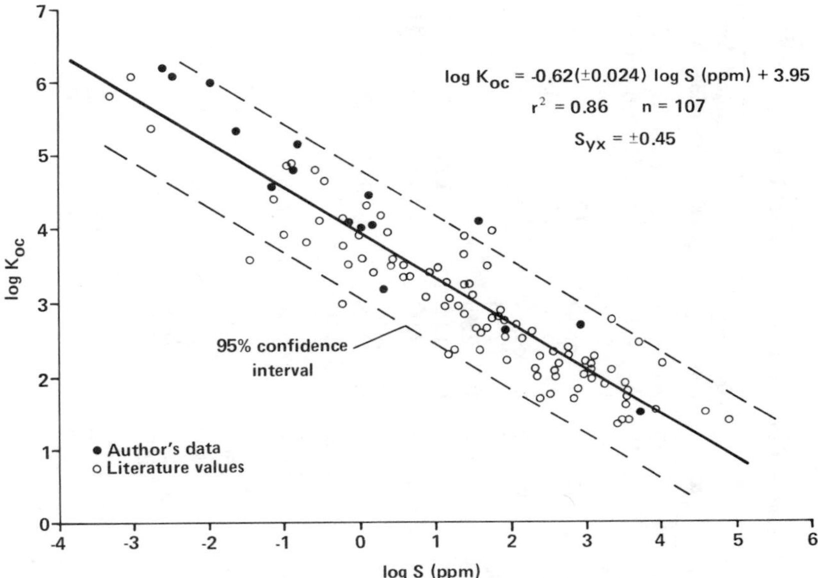

Figure 2. Relation of normalized sorption constant (K_{oc}) and the aqueous solubility (S) of the compounds.

(S) and *n*-octanol/water partition coefficients (K_{ow}). Figure 2 shows the K_{oc}-S relationship for 91 compounds for which data from the literature could be obtained, and 16 compounds studied by the authors. Data for these compounds are listed in Table II. Literature values for K_{oc} were often based on a single K_f value (Kenaga, 1980; Griffin, unpublished; Briggs, 1973; Banerjee et al., 1980; Mills and Biggar, 1969). The linear regression obtained was as follows:

$$\log K_{oc} = 3.95 - 0.62 \log S \text{ (ppm)} \quad (5)$$
$$r^2 = 0.86; \quad n = 107$$

The 95% confidence intervals as well as the standard error of the slope and the standard error of the estimate of K_{oc} were computed according to Steele and Torrie (1960).

The data indicate that a good first approximation of sorption by soils can be made from the water solubility of a compound. The K_{oc} can be predicted from the above relationship with 95% confidence that the true or experimentally measured K_{oc} will be within ±0.90 log units, or

Table II. Water Solubility, Soil Adsorption Constant, and Octanol–Water Partition Coefficient Data

	Compound	Solubility (ppm)	Reference[a]	log S	K_{oc}	Reference	log K_{oc}	K_{ow}	Reference	log K_{ow}
1	Acetophenone	5,440	A	3.74	35	A	1.54	38.6	A	1.59
2	Acridine	38.4	A	1.58	12,910	A	4.11	4,200	A	3.62
3	Alachlor (Lasso)	242	B	2.38	190	B	2.28			
4	2-Aminoanthracene	1.30	A	0.11	28,129	A	4.45	13,400	A	4.13
5	6-Aminochrysene	0.155	A	−0.81	143,355	A	5.16	96,600	A	4.98
6	Anthracene	0.073	C	−1.14	26,000	C	4.41	35,000	C	4.54
7	Anthracene-9-carboxylic acid	85.0	A	1.93	422	A	2.63	1,300	A	3.11
8	Asulam	5,000	B	3.70	300	B	2.48			
9	Benzene	1,780	C	3.25	83	C	1.92	130	C	2.11
10	2,2′-Biquinoline	1.02	A	0.01	10,404	A	4.02	20,200	A	4.31
11	Bromacil	815	B	2.91	72	B	1.86			
12	Bromobenzene	446	B	2.65	150	B	2.18	977	E	2.99
13	Butralin (Amex)	1	B	0.00	8,200	B	3.91			
14	Carbaryl	40	B	1.60	230	B	2.36			
15	Carbofuran (Furadan)	415	B	2.62	105	F	2.02			
16	Carbophenothion (Trithion)	0.34	B	−0.47	45,400	B	4.66			
17	Chloramben, methyl ester	120	B	2.08	507	B	2.71			
18	Chlorobromuron (Maloran)	50	B	1.70	460	B	2.66			
19	Chloroneb	8	B	0.90	1,200	B	3.08			
20	Chloroxuron (Teneran)	2.7	B	0.43	3,200	B	3.51			
21	Chlorpropham (Chloro-IPC)	88	B	1.94	590	B	2.77			
22	Chlorpyrifos (Lorsban)	0.3	B	−0.52	13,600	B	4.13	128,825	E	5.11
23	Chlorpyrifos-methyl	4.0	B	0.60	3,300	B	3.52	20,417	E	4.31
24	Chlorthiamid (Prefix)	950	B	2.98	107	B	2.03			

25	Crotoxyphos (Ciodrin)	1,000	B	3.00	170	B	2.23			
26	Cycloate (Ro-Neet)	85	B	1.93	345	B	2.54			
27	DBCP	1,230	B	3.09	129	B	2.11			
28	DDT	0.0017	B	−2.77	238,000	B	5.38	1,548,817	E	6.19
29	Di-allate (Avadex)	14	B	1.15	1,900	B	3.28			
30	Diamidphos (Nellite)	50,000	B	4.70	32	B	1.51			
31	Dibenz (a,h) anthracene	0.00249	A	−2.60	1,668,800	A	6.22	3,170,000	A	6.50
32	13H-Dibenzo (a,i) carbazole	0.0104	A	−1.98	1,055,926	A	6.02	2,514,000	A	6.40
33	Dibenzothiophene	1.47	A	0.17	11,230	A	4.05	24,000	A	4.38
34	1,2-Dibromo-3-chloropropane	1,230	G	3.09	143	A	2.16			
35	1,2-Dibromoethane	3,520	G	3.55	69	A	1.84			
36	Dichlobenil (Casoron)	18	B	1.26	235	B	2.37			
37	1,2-Dichlorobenzene	148	G	2.17	343	A	2.54	2,510	H	3.40
38	1,2-Dichloroethane	8,450	G	3.93	36	A	1.56	28.2	H	1.45
39	1,2-Dichloropropane	3,570	G	3.55	51	A	1.71			
40	*cis*-1,3-Dichloropropene	2,700	B	3.43	23	B	1.36			
41	*trans*-1,3-Dichloropropene	2,800	B	3.45	26	B	1.41			
42	Diflubenzuron	0.2	B	−0.70	6,790	B	3.83			
43	7,12-Dimethylbenz (a) anthracene	0.0244	A	−1.61	225,308	A	5.35	953,000	A	5.98
44	Dinitramine	1.1	B	0.04	4,000	B	3.60			
45	Dipropetryn (Sancap)	16	B	1.20	1,170	B	3.07			
46	Disulfoton	25	B	1.40	1,780	B	3.25			
47	Diuron	42	B	1.62	400	B	2.60	172	D	2.24
48	EPTC	365	B	2.56	240	B	2.38			
49	Ethion	2	B	0.30	15,400	B	4.19			
50	Ethylene dibromide	3,370	B	3.53	44	B	1.64			
51	Fenuron	3,850	B	3.59	27	B	1.43			
52	Fluometuron	90	B	1.95	175	B	2.24			

Table II, continued

Compound	Solubility (ppm)	Reference[a]	log S	K_{oc}	Reference	log K_{oc}	K_{ow}	Reference	log K_{ow}
53 Hexachlorobenzene	0.035	B	−1.45	3,914	B	3.59			
54 2,4,5,2′,4′,5′-Hexachlorobiphenyl	0.00095	C	−3.02	1,200,000	C	6.08	2,200,000	C	6.34
55 Hexachlorocyclopentadiene	2.0	D	0.30	1,493	D	3.17			
56 Isocil	2,150	B	3.33	130	B	2.11			
57 Isopropalin (Paarlan)	0.11	B	−0.96	75,250	B	4.88			
58 Leptophos (Phosvel)	2.4	B	0.38	9,300	B	3.97			
59 γ-Lindane	8.5	I	0.93	2,627	I	3.42			
60 β-Lindane	2.7	I	0.43	3,619	I	3.56			
61 Linuron	75	B	1.88	820	B	2.91			
62 Methazole (Probe)	1.5	B	0.18	2,620	B	3.42			
63 Methomyl	10,000	B	4.00	160	B	2.20			
64 Methoxchlor	0.120	C	−0.92	80,000	C	4.90	120,000	C	5.08
65 2-Methoxy-3,5,6-trichloropyridine	20.9	B	1.32	920	B	2.96			
66 Methyl 1 parathion	57	B	1.76	9,800	B	3.99			
67 9-Methylanthracene	0.261	C	−0.58	65,000	C	4.81	117,000	C	5.07
68 3-Methylchloanthrene	0.00323	A	−2.49	1,244,046	A	6.09	2,632,000	A	6.42
69 2-Methylnaphthalene	25.4	C	1.40	8,500	C	3.93	13,000	C	4.11
70 Metobromuron (Patoran)	330	B	2.52	60	B	1.78			
71 Metribuzin	1,220	B	3.09	95	B	1.98			
72 Mirex	0.6	B	−0.22	5,800	B	3.76			
73 Monolinuron (Aresin)	580	B	2.76	200	B	2.30			
74 Monuron	230	B	2.36	100	B	2.00			
75 Naphthalene	31.7	C	1.50	1,300	C	3.11	2,300	C	3.36
76 1-Naphthol	866	A	2.94	522	A	2.72	700	A	2.85

77	Napropamide (Devranol)	73	B	1.86	680	B	2.83			
78	Neburon	4.8	B	0.68	2,300	B	3.36			
79	Nitralin (Planavin)	0.6	B	−0.22	960	B	2.98			
80	Nitrapyrin (N-Serve)	40	B	1.60	420	B	2.62			
81	Norflurazon	28	B	1.45	1,914	B	3.28			
82	Oxadiazon (Ronsfar)	0.7	B	−0.15	3,241	B	3.51			
83	Parathion	24	B	1.38	4,800	B	3.68	6,457	E	3.81
84	Pebulate (Tillam)	60	B	1.78	630	B	2.80			
85	2,2′,4,5,5′-Pentachlorobiphenyl (Aroclor 1254)	0.07	J	−1.15	42,500	B	4.63	1,288,250	E	6.11
86	Pentachlorophenol (PCP)	14.0	B	1.15	900	B	2.95			
87	Phenanthrene	1.29	C	0.11	23,000	C	4.36	37,000	C	4.57
88	Phenol	82,000	B	4.91	27	B	1.43			
89	Phorate	50	B	1.70	3,200	B	3.51	5,623	D	3.75
90	Profluralin (Tolban)	0.1	B	−1.00	8,600	B	3.93			
91	Pronamide (Kerb)	15	B	1.18	200	B	2.30			
92	Propachlor	580	B	2.76	265	B	2.42			
93	Propham (IPC)	250	B	2.40	51	B	1.71			
94	Pyrazon (Alicep)	400	B	2.60	120	B	2.08			
95	Pyrene	0.135	A	−0.87	63,400	A	4.80	124,000	A	5.09
96	Pyroxychlor	11.3	B	1.05	3,000	B	3.48			
97	Tebuthiuron (Spike)	2,300	B	3.36	620	B	2.79			
98	Terbacil	710	B	2.85	51	B	1.71			
99	Terbutryne (Igran)	25	B	1.40	700	B	2.85			
100	Tetracene	0.0005	C	−3.30	650,000	C	5.81	800,000	C	5.90
101	1,1,2,2-Tetrachloroethane	3,230	G	3.51	88	A	1.94	245	H	2.39
102	Tetrachloroethene (tetrachloroethylene)	200	G	2.30	400	A	2.60			

Table II, continued

Compound	Solubility (ppm)	Reference[a]	log S	K_{oc}	Reference	log K_{oc}	K_{ow}	Reference	log K_{ow}
103 Tri-allate(Far-Go)	4	B	0.60	2,220	B	3.35			
104 Trichlorobiphenyl(Aroclor 1242)	0.703	J	−0.15	12,400	J	4.09			
105 Trichloroethane	1,360	G	3.13	198	A	2.30	295	H	2.47
106 3,5,6-Trichloro-2-pyridinol	220	B	2.34	130	B	2.11			
107 Trifluralin	0.6	B	−0.22	13,700	B	4.14			

[a]References: see list at end of this chapter for the complete reference.

A Hassett et al. (1980)
B Kenaga (1980)
C Karickhoff et al. (1979)
D Unpublished Data
E Chiou et al. (1977)
F McCall et al. (1980)
G Chiou et al. (1979)
H Banerjee et al. (1980)
I Mills and Biggar (1969)
J Griffin and Chian (1979)

less than one order of magnitude, from the predicted value. The high linear correlation ($r^2 = 0.86$) is notable for such a large group of organic compounds covering nine decades of water solubility. This is especially true when one considers the variability in methodology and presentation of the data used by the various researchers, and that for many of the compounds the solubility was obtained from one reference and the adsorption data from a second reference.

The sorption of nonionic organics has also been related to the partitioning of an organic compound between an organic solvent (*n*-octanol) and water (Karickhoff et al., 1979; Kenaga and Goring, 1980; Briggs, 1973). The octanol–water partition coefficient (K_{ow}) measures the tendency of a compound to partition between an organic solvent and water. It follows therefore that K_{ow} should be related to the tendency of the same compound to partition between water and soil organic matter. Figure 3 illustrates the relationship between K_{oc} and K_{ow}. There is a high linear correlation yielding the regression equation:

$$\log K_{oc} = 0.088 + 0.909\ K_{ow} \qquad (6)$$
$$r^2 = 0.93;\ n = 34$$

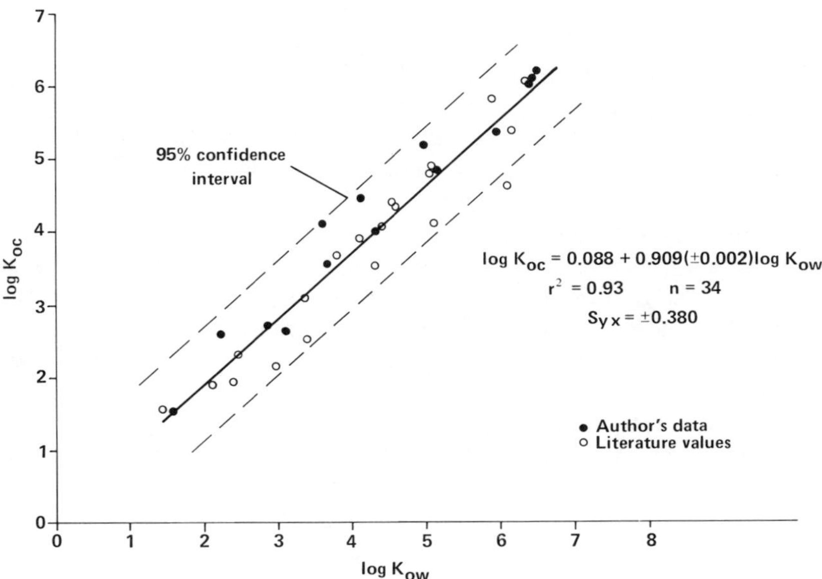

Figure 3. Relation of normalized sorption constant (K_{oc}) and the *n*-octanol–water partition coefficient (K_{ow}) of the compounds.

Again, the 95% confidence intervals and the standard errors are given in the figure, and the solid data points represent our data. The open points are data obtained from the literature. Note the nearly 1:1 correspondence between K_{oc} and K_{ow} for rule-of-thumb predictions. The errors for the estimate of K_{oc} from K_{ow} data are slightly better than from solubility data but appear to offer no important advantages over water solubility data for predictions of K_{oc} values. Both relationships, S and K_{ow}, offer powerful tools for prediction of K_{oc} values for organic compounds.

The octanol–water partition coefficient is usually obtained by measuring the concentration of the compound in equilibrium with *n*-octanol and water (Hassett et al., 1980). However, a second method is to predict K_{ow} from the retention time of the compound on a reverse-phase, high-performance liquid chromatographic (RP-HPLC) column (Veith et al., 1979). An extension of this is to use the retention time to predict K_{oc} directly. This technique has been successfully applied by McCall et al. (1980) and appears to offer a convenient method for future predictions of the soil adsorption characteristics of organic compounds.

THEORETICAL CONSIDERATIONS

Hydrophobic sorption is the partitioning of the nonpolar solute out of the polar aqueous phase onto the hydrophobic surfaces in the soil. The hydrophobic surfaces are primarily associated with soil organic matter but may also include ≡Si-O-Si≡ bonds at mineral surfaces (Chen, 1976). The primary feature of hydrophobic sorption is the very weak interaction between the solute and the solvent. As the term hydrophobic suggests, the solvent water plays a major role in the interaction. Although not strictly the case, the interaction may be so weak as to acquire the nature of a repulsion of the organic molecule from the water. The more hydrophobic the organic molecule, the greater the repulsion from the water and the greater the sorption on a hydrophobic surface.

In a stricter treatment, hydrophobic sorption is thought to be an entropy-driven process governed by the basic thermodynamic relation:

$$\Delta G = \Delta H - T\Delta S \tag{7}$$

where ΔG is the free-energy change of the system, ΔH is the heat of adsorption (enthalpy), ΔS is the entropy of the system, and T is temperature. Sorption results when there is a negative free-energy change. This may be due to either a large negative enthalpy of sorption and a relatively small entropy contribution, either favoring or opposing the sorption, or it may be due to a large positive entropy change with a small enthalpy change opposing or favoring the sorption. The sorption of polar compounds by soils appears to be an enthalpy-driven process with relatively large heats of adsorption (Green, 1974; Yamane and Green, 1972). Hydrophobic sorption appears to be an entropy-driven process (Hassett et al., 1981) with small heats of adsorption (Chiou et al., 1979; Haque and Coshow, 1971). This is illustrated by the sorption of α-naphthol and phenanthrene (Table III) by soils where hydrophobic sorption may or may not be the major sorption mechanism. For phenanthrene, which appears to be strictly hydrophobically sorbed regardless of the soil studied, there is little temperature dependence of the measured K_{oc} values and hence a small enthalpy change. Sorption of α-naphthol, a more polar compound, also showed little effect of temperature on sorption for those soils that contained moderate to high amounts of organic matter. For these soils, measured K_{oc} values agreed well with the predicted K_{oc} value. But for soil 15, where there was a high montmorillonite-to-carbon ratio, which caused greater amounts of sorption than predicted by K_{oc}-S correlation, there was a marked temperature effect and hence a large heat of sorption.

The primary driving force in hydrophobic sorption appears to be the large entropy change resulting from the removal of the organic molecule from solution and sorption onto a hydrophobic surface (Nemethy and Scherara, 1962; Horvath and Melander, 1978). The entropy change is primarily due to the destruction of the cavity occupied by the solute in the solvent and the destruction of the highly structured water shell surrounding the solvated organic compound (Frank and Evans, 1945; Horvath and Melander, 1978).

Table III. Effect of Equilibrium Temperature and Soil Properties on Measured K_{oc} Values

Compound	Soil[a]	Measured K_{oc} 15°C	25°C	35°C	Predicted K_{oc}
Phenanthrene	5	14,380	13,325	14,899	15,744
	15	12,310	15,907	13,268	
α-Naphthol	5	237	239	338	181
	15	1,973	2,634	3,302	

[a]Soil 5 contains 2.28% OC and 31% clay; Soil 15 contains 0.95% OC and 36% clay.

The role of the solute–solvent interaction in determining the degree of sorption of hydrophobic compounds was demonstrated by a soil thin-layer chromatography study using solutes and solvents (mobile phases) of different polarities (Hassett et al., 1981). When the mobile phase was 100% water (i.e., polar), the polar compound was weakly sorbed and moved with the solvent front, whereas the less-polar compounds were strongly sorbed and showed little movement. As the mobile phase was made less polar by additions of ethanol, it became a better solvent for the nonpolar compounds and their sorption decreased. With ethanol additions, the mobile phase became a poorer solvent for the polar compound and its sorption increased and movement decreased.

LIMITATIONS

Several major limitations must be discussed as they relate to the use of predicted K_{oc} values from water-solubilities and *n*-octanol–water partition coefficients.

1. The K_{oc}-K_{ow} and K_{oc}-S correlations are log-log relationships; hence, small deviations from the regression line can represent significant deviations in actual and predicted K_{oc} values.
2. Several sorption mechanisms are possible for the sorption of organic compounds. In those cases where more than one mechanism is involved, the K_{oc}-S correlation may accurately predict the hydrophobic contribution to sorption (Zierath et al., 1980) but may not correctly predict a given soils total K_f value for the compound. In general, the more polar the compound the greater the potential for more than one sorption mechanism.
3. Soils may contain hydrophobic surfaces other than those associated with soil organic matter. This appears to be the case in soils whose contents contain low organic carbon in combination with medium to high montmorillonite (Griffin and Chian, 1979; Hassett et al., 1981). In soils with medium to high organic matter, the hydrophobic surfaces of montmorillonite appear to be coated by humus.
4. Changes in solvent properties that affect solubility may decrease or increase sorption.

CONCLUSIONS

Several conclusions are apparent. The adsorption of hydrophobic nonpolar organic compounds by soils is a function of the organic matter content of the soil. Individual adsorption constants converge on a common value, K_{oc}, when adjusted for the soil organic carbon content.

Values of K_{oc} were negatively correlated with the water solubility (S) and were positively correlated with the *n*-octanol/water partition coefficient (K_{ow}). The relations correlating K_{ow} and S data with K_{oc} offer powerful tools for making predictions of the K_{oc} values of nonpolar organics by soils; K_{oc} values can also be used to predict sorption on individual soils if organic carbon contents of the soil materials are known.

Although these relationships offer the potential for prediction of sorption from simple physicochemical properties, they do not preclude the need for experimental verification of predictions.

ACKNOWLEDGMENTS

Research support from U.S. EPA Grants No. R-804684 and R-808417 is gratefully acknowledged.

REFERENCES

Banerjee, S., S. H. Yalkowsky, and S. C. Valvani. 1980. "Water Solubility and Octanol/Water Partition Coefficients or Organics. Limitations of the Solubility-Partition Coefficient Correlation," *Environ. Sci. Technol.* 14: 1227-1229.

Briggs, G. G. 1973. "A Simple Relationship Between Soil Adsorption of Organic Chemicals and Their Octanol/Water Partition Coefficients," Proceedings Seventh British Insecticide and Fungicide Conference.

Bowman, B. T., 1981. "Anomalies in the Log Freundlich Equation Resulting in Deviations in Adsorption K Values of Pesticides and Other Organic Compounds When the System of Units is Changed," *J. Environ. Sci. Health* B16:113-123.

Chen, N. Y. 1976. "Hydrophobic Properties of Zeolites," *J. Phys. Chem.* 80:60-64.

Chiou, C. T., V. H. Freed, D. W. Schmedding, and R. L. Kohnert. 1977. "Partition Coefficient and Bioaccumulation of Selected Organic Chemicals," *Environ. Sci. Technol.* 11:475-478.

Chiou, C. T., L. J. Peters, and V. H. Freed. 1979. "A Physical Concept of Soil-Water Equilibrium for Nonionic Organic Compounds," *Science* 206: 831-832.

Frank, H. S., and M. W. Evans. 1945. "Free Volume and Entropy in Condensed Systems III. Entropy in Binary Liquid Mixtures; Partial Molal Entropy in Dilute Solutions; Structure and Thermodynamics vs Aqueous Electrolytes," *J. Chem. Phys.* 13:507-532.

Green, R. E. 1974. "Pesticide-Clay-Water Interactions," *Pesticides in Soil and Water*, W. D. Guenzi, Ed. (Madison, WI: Soil Science Society of America) pp. 3-37.

Griffin, R. A., and E. S. K. Chian. 1979. "Attenuation of Water-Soluble Polychlorinated Biphenyls by Earth Materials," Environ. Geol. Notes 89, Ill. State Geol. Surv.

Griffin, R. A. Unpublished data. Ill. State Geol. Surv.

Hassett, J. J., W. L. Banwart, S. G. Wood, and J. C. Means. 1981. "Sorption of α-Naphthol: Implications Concerning the Limits of Hydrophobic Sorption," *Soil Sci. Soc. Am. J.* 45:38-42.

Hassett, J. J., J. C. Means, W. L. Banwart, and S. G. Wood. "Sorption Properties of Sediments and Energy-Related Pollutants," EPA-600/3-80-041, U.S. Environmental Protection Agency.

Haque, R., and W. R. Coshow. 1971. "Adsorption of Isocil and Bromocil from Aqueous Solution onto Some Mineral Surfaces," *Environ. Sci. Technol.* 5:139-141.

Horvath, C., and W. Melander. 1978. "Reversed-Phase Chromatography and the Hydrophobic Effect," *Am. Lab.* 10:17-36.

Karickhoff, S. W., D. S. Brown, and T. A. Scott. 1979. "Sorption of Hydrophobic Pollutants on Natural Sediments," *Water Res.* 13:241-248.

Kenaga, E. E. 1980. "Correlation of Bioconcentration Factors of Chemicals in Aquatic and Terrestrial Organisms with Their Physical and Chemical Properties," *Environ. Sci. Technol.* 14:553-556.

Kenaga, E. E., and C. A. I Goring. 1980. "Relationship Between Water Solubility, Soil Sorption, Octanol-Water Partitioning, and Concentration of Chemicals in Biota," *Aquatic Toxicology,* ASTM STP 707, J. G. Eaton, P. R. Parrish, and A. C. Hendricks, Eds., American Society for Testing and Materials, p. 78-115.

McCall, P. J., R. L. Swan, D. A. Laskowski, S. M. Unger, S. A. Vrona, and H. J. Dishburger. 1980. "Estimation of Chemical Mobility in Soil from Liquid Chromatographic Retention Times," *Bull. Environ. Contam. Toxicol.* 24:190-195.

Means, J. C., J. J. Hassett, S. G. Wood, and W. L. Banwart. 1979. "Sorption Properties of Energy-Related Pollutants and Sediments," *Polynuclear Aromatic Hydrocarbons,* P. W. Jones and P. Leber, Eds. (Ann Arbor, MI: Ann Arbor Science Publishers, Inc.) pp. 327-340.

Means, J. C., S. G. Wood, J. J. Hassett, and W. L. Banwart. 1980. "Sorption of Polynuclear Aromatic Hydrocarbons by Sediments and Soils," *Environ. Sci. Technol.* 14:1524-1528.

Mills, A. C., and J. W. Biggar. 1969. "Solubility-Temperature Effect on the Adsorption of Gamma- and Beta-BHC from Aqueous and Hexane Solutions by Soil Materials," *Soil Sci. Soc. Am. Proc.* 33:210-216.

Nemethy, G., and H. A. Scherara. 1962. "The Structure of Water and Hydrophobic Bonding in Proteins, Vol. III. The Theromodynamic Properties of Hydrophobic Bonds in Proteins," *J. Phys. Chem.* 66:1773-1789.

Osgerby, J. M. 1970. "Sorption of Un-ionized Pesticides by Soils," in *Sorption and Transport Processes in Soils,* Soc. Chem. Ind. London, Monogr. 37 pp. 63–78.

Steel, R. G. D., and J. H. Torrie. 1960. *Principles and Procedure of Statistics* (New York: McGraw-Hill, Book Co.) p. 175.

Veith, G. D., N. M. Austin, and R. T. Morris. 1979. "A Rapid Method for Estimating log P for Organic Chemicals," *Water Res.* 13:43-47.

Yamane, V. K., and R. E. Green. 1972. "Adsorption of Ametryne and Atrazine on an Oxisol, Montmorillonite, and Charcoal in Relation to pH and Solubility Effects," *Soil Sci. Soc. Am. Proc.* 36:58-64.

Zierath, D. L., J. J. Hassett, and W. L. Banwart. 1980. "Sorption of Benzidine by Sediments and Soils," *Soil Sci.* 129:277-281.

DISCUSSION

***John Cherry*, University of Waterloo:** Some of the early isotherms you showed for several organic compounds were determined over a rather narrow concentration range. Do you have any comments on how those linear isotherms behave when a larger concentration range is used? What form do the isotherms take, and on what basis did you choose the test range?

J. Hassett: They are over a fairly narrow concentration range. We ran all of our isotherms from 50% of water solubility down. That was our initial solution concentration. If you go above 50% water solubility for almost all of the compounds, the isotherms turn up. That would suggest again that the compound rather than being absorbed is precipitating out of solution. Chemists like to call it incipient phase formation, but it is probably a precipitation process. It says that as you approach the water solubility of the compound, you reach a point where the water solubility of the compound is beginning to dominate due to precipitation, not adsorption. Occasionally, when we used the more polar compounds, the isotherms would level out and approach a limit. That was only for the more polar compounds, not for nonpolar compounds.

***John Brown*, CEGB:** Do you know what the entropy change is in the process, because what you get looks to be pretty well ordered to me?

J. Hassett: We are currently trying to measure changes in entropy. The problem with entropy is that it is not directly measured in sorption experiments. We do have measured entropy values, and they can be quite large. In fact, Osgerby in his work calculated some entropy values, and they went up to 10,000 kcal. There is one point I want to make. What we are looking at is the partitioning of organic compounds out of solution onto a hydrophobic surface. It doesn't matter what that hydrophobic surface is. Competitive hydrophobic surfaces are the hydrophobic sites on a bacteria, or an algae, or a fish. All these membranes have hydrophobic sites. When you study a metal and say that it is strongly adsorbed, it almost always implies a low biological availability. In contrast, strong adsorption with nonpolar organic compounds may not mean low biological availability. Availability will be determined by the relation of the competition of the two phases—partition to the membranes of the organism vs partition to the soil. Whichever K is the largest is going to decide which way the compound partitions. Kenaga has shown tremendous correlations between octanol-water partitions and biological availability.

***T. Tamura*, ORNL:** Could you tell me in a soil profile how the organic matter might change with depth and, therefore, at what point we might be dealing with inorganic substrate in a particular landfill-type operation below the surface?

J. Hassett: The surface horizons, the A horizons, and the B horizons as well, will have fairly high organic matter contents, depending on their development. In B horizons you still have a fair amount of organic matter, but its placement tends to be on the surface of the peds, for example, on the surface of the particles. And so, even in the B horizons you might have low amounts of organic matter that quite often dominate sorption of organic compounds. Most landfills are not in soil but in the regolith; thus, the organic carbon contents are not high and the mineral material begins to dominate sorption. The work that Dr. Griffin did with clays showed that he got strong adsorption of organics by clays. But he still had a high degree of correlation with the trace amounts of organic carbon and clay. Therefore, hydrophobic sorption appears to have a tremendous effect even in very low organic clays.

CHAPTER 16

FOOD CHAIN PATHWAYS FOR TOXIC METALS AND TOXIC ORGANICS IN WASTES

Rufus L. Chaney

Biological Waste Management and Organic Resources Laboratory
Agricultural Research Service
U.S. Department of Agriculture
Beltsville, Maryland 20705

ABSTRACT

Pathways to and potential effects on the food chain of microelements and toxic organic compounds in wastes applied to land treatment sites are considered in detail. Wasteborne toxic chemicals can enter the food chain in (1) ingested wastes, wastes adhering to forages, or waste-amended soil or (2) plants that have increased levels of the toxic chemical when grown on waste-amended soil. Some wasteborne elements fail to injure animals even when the waste is ingested (e.g., Ti, Cr^{3+}, Zr, Si, Sn, Au, Ag), either because animals tolerate high exposures or because the element is bound very strongly by the ingested soil or waste. Other chemicals can injure animals that ingest the waste or soil but not when animals ingest plants grown on the amended soil (Fe, Pb, Ni, Zn, Cu, Be, As, and bioaccumulated toxic organic compounds) because phytotoxicity kills the plant before the metal concentration in the plant reaches toxic levels for the animal. On the other hand, risk may result from ingestion of forage crops that contain phytotoxic levels of Co or excessive nonphytotoxic levels of Se, Mo, or Cd. Cumulative applications of only a few elements (Cd and Pb) must be restricted to protect humans from worst-case situations. If land treatment sites were not allowed to be converted to normal, unregulated land-use patterns until toxic organic compounds were degraded, management practices that avoided adverse food-chain effects could be developed for nearly any organic waste.

INTRODUCTION

Ultimate disposal of hazardous wastes is an important problem of modern industrial societies. Methods are needed which are of low risk to society and the environment, yet are less expensive than hazardous waste landfills. Methods that treat (degrade to nontoxic materials) hazardous chemicals are preferred over those that simply store the wastes.

One method that appears to satisfy the need is land treatment (previously called landfarming or land cultivation of wastes). The U.S. Environmental Protection Agency (U.S. EPA, 1980) has published regulations for land treatment of hazardous wastes.

Persistent negative impacts on humans, agriculture, or environment would be the basis for prohibiting treatment of a waste or a management or closure technique for a land treatment site. This chapter considers food-chain pathways for effects of hazardous waste land treatment, based on long experience with natural toxicants and recent experience with land application of sewage sludge and other wastes. Concerns fall into three groups: during site operation, during the closure period, and persistent beyond closure. Many wastes can be managed at land treatment sites if the wastes are sufficiently characterized and if site personnel are appropriately trained.

FOOD CHAIN PATHWAYS FOR WASTE-BORNE TOXIC CHEMICALS

Liquid sludges can be spray-applied to cropland and tilled into the soil. Alternatively, liquid sludge can be sprayed onto forage or pastureland where it can contact plants and remain on the soil surface. Dewatered or dried sludges or composted wastes can be applied and mixed with the soil or can be allowed to remain on the soil surface. These management options allow substantially different quantities of toxic chemicals to enter the food chain by quite different routes. Some options allow animals to directly ingest sludges, whereas other options use reactions in soils and properties of plants to largely prevent exposure.

Sludge Adherence to Existing Crops

When liquid sludges (0 to 10% solids) are sprayed on pastures or forage crops, a thin film of the sludge coats the plant foliage. Research has determined that some wastes dry and adhere strongly, whereas others dry and flake off upon weathering. The first records of organic waste adherence came from a study of land application of high-copper, pig-manure slurry (Batey et al., 1972); forage grasses were enriched in copper from adhering manure.

After these findings were reported, research was begun on sewage sludge adherence to forage crops and the effects on grazing cattle. Chaney and Lloyd (1979) found that after liquid digested sludge had dried on tall fescue forage, it was not readily washed off by rainfall. Growth of the crop biomass diluted the sludge percentage in harvested forage. Sludge adherence was greater at higher application rates. Jones et al. (1979) found that sludge could be washed off forages before, but not after, it dried. They also found that the amount of adhering sludge was approximately a linear function of the percent solids of the applied liquid sludge.

Sludge has adhered to all crops studied. Sludge adherence is easily characterized because the levels of some microelements in sludge-contaminated forage are much greater than levels ordinarily possible from uptake and translocation by forage plants. Plant uptake and translocation to shoots of Cu, Pb, Cr, Fe, and other elements are so limited that high levels of these elements indicate direct sludge contamination (Chaney and Lloyd, 1979). Many early reports on uptake of microelements from surface-applied sludges presumed uptake when, in fact, sludge adherence fully explains the observations (Boswell, 1975; Fitzgerald, 1978). We have recently observed that industrial aerobic sludges adhere to forages in a manner similar to that of sewage sludge.

Another route by which microelements enter the food chain is through farm equipment. Studies with pig manure indicate that organic wastes on the soil surface can be lifted and mixed into baled hay (Dalgarno and Mills, 1975).

When increased levels of microelements in forage indicate sludge adherence, all constituents present in the sludge contaminate the forage. Not only microelements, but also macroelements, pathogens

(Brown et al., 1980), and toxic organics (Fitzgerald, 1978) are increased.

Ingestion of Sludge-Amended Soil or Sludge on the Soil Surface

Several research programs have established that grazing animals consume soil as a part of the normal grazing process. Teeth of sheep and cattle wear out more rapidly when the forage is contaminated with soil (Healy and Ludgwig, 1965). Study of the teeth-wear problem led Healy (1968) to more fully develop Field and Purves' (1964) method of soil ingestion measurement, in which the Ti level in forage and feces is compared with that of soil. Titanium present in soil is not appreciably absorbed and translocated by plants. Forage Ti level thus becomes a label for soil in or on forages. Healy et al. (1974) found that wet weather and excessive stocking rates caused forages to be trampled into the soil, thereby increasing soil adherence to forages. Although soil was normally 1 to 2% of sheep's diet, it reached 24% in the worst cases. In other research, Mayland et al. (1975, 1977) found that cattle grazing on crested wheatgrass grown on dryland consumed considerable quantities of soil. Because the cattle consumed plants complete with soil-laden roots, the ingested diet contained 20% soil. Fries et al. (1982) recently reviewed soil ingestion by dairy cattle.

Ingested soil can cause lead poisoning of livestock when cattle graze soil naturally high in lead (Egan and O'Cuill, 1970; Harbourne et al., 1968; Thornton and Kinniburgh, 1978). Even after closure of a smelter lead-enriched crop residues remain on the soil surface, exposing cattle to possible lead poisoning. Reclamation of rangeland polluted by a lead smelter required incorporation of the organic sward thatch into the soil to prevent ingestion by cattle (Edwards and Clay, 1977).

Similarly, sewage or composted sludge is ingested from the soil surface. Decker et al. (1980) found 6.5% (1977) and 2.0% (1978) compost in feces of cattle grazing on pastures fertilized by sludge compost. Compost did not adhere to the plant surfaces but lay on the soil surface.

Soil ingestion can also expose humans to waste-applied microelements in land treatment sites subsequently developed for housing. Some children and adults deliberately consume soil in a practice called pica. If the soil is high in Pb (>500 to 1000 ppm), individuals may absorb excessive amounts of Pb (Wedeen et al., 1978; Shellshear et al., 1975). Children also ingest soil and dust by hand-to-mouth play activities and by mouthing of toys, and other articles (Lepow et al., 1975; Sayre et al., 1974; Roels et al., 1980). High-Pb soils carried into homes on dusty clothing and shoes causes house dust to become rich in Pb (Jordan and Hogan, 1975). Several research programs have shown that soil Pb and house dust Pb contribute to Pb exposure of children (Angle and McIntire, 1979; Charney et al., 1979; Galke et al., 1977; Roels et al., 1980, Barltrop et al., 1974; Hammond et al., 1980). Paint, air, soil, and house dust Pb constitute a multiple source model for Pb exposure of children.

Perhaps the clearest example of the possibility of human health effects from inadvertent ingestion of high-Pb house dust is found in recent cases in which children of Pb-industry workers experienced excessive blood Pb. The only identifiable source of Pb was the industrial dusts carried into the house on the workers' clothing and shoes (Baker et al., 1977; Giguere et al., 1977; Rice et al., 1978).

Soil Pb can only cause excessive blood Pb if it is at least somewhat bioavailable to humans. Studies have shown that soil Pb is largely bioavailable to rats (Dacre and Ter Haar, 1977; Stara et al., 1973), but that soil properties can reduce Pb bioavailability (Chaney et al., 1980). A recent National Research Council (NRC) review considered the potential importance of soil and dust ingestion in Pb poisoning of children (NRC, 1980a).

Soil or sludge ingestion can be an important process that allows a sludge-borne microelement into the food chain, especially when the element is normally not absorbed by plants (plant level $\leq$ soil level). For some elements (e.g., Zn, Cd, Mn, and Se), plant levels often exceed soil levels, and plant uptake is a more important process than soil ingestion. However, soil ingestion is a potential route for allowing excessive Pb, Fe, Cu, F, As, Hg, Cu, Co, Mo, Se, and other elements into the food chain. Further, soil ingestion can interfere with availability of microelements in plants to animals. Individual elements will be discussed below.

Soil ingestion is an especially important pathway for persistent lipophilic toxic organic compounds. Harrison et al. (1970) found increased levels of DDT in sheep-grazing pastures where DDT was on the soil surface. They also studied lindane (Harrison et al., 1969; Collett and Harrison, 1968). Bergh and Peoples (1977) noted PCB movement from surface-applied dewatered sludge to milk of a grazing cow, but they did not estimate sludge ingestion. Hansen et al. (1981) noted PCB retention by swine grazing in a field where the surface soil was largely sewage sludge.

SOIL-PLANT BARRIER TO MICROELEMENTS IN THE FOOD CHAIN

Some elements (e.g., Zn, Cd, Mn, Mo, Se, B) are easily absorbed and translocated to food-chain plant tissues, whereas others are not. These other elements (e.g., Fe, Pb, Hg, Al, Ti, Cr^{3+}, Ag, Au, Sn, Si, and Zr) which are strongly bound to soil or retained in plant roots, are not translocated to plant foliage in injurious amounts, even when soils are greatly enriched. Even though an element may be easily or relatively easily absorbed and translocated to plant foliage, phytotoxicity may limit plant levels of an element (e.g., Zn, Cu, Ni, Mn, As, and B) to levels that are safe for animals.

These concepts were developed during the past 40 years by many researchers. Important reviews of the research supporting these concepts have been prepared but have not named the general theory (Underwood, 1977; Allaway, 1968,1977a,b; Bowen, 1966,1979; Baker and Chesnin, 1976; Chaney, 1980; Lisk, 1972; Kienholz, 1980; Loneragen, 1975; Reid and Horvath, 1980; Cataldo and Wildung, 1978; Leeper, 1978; Ammerman et al., 1977; Shacklette et al., 1978; Page, 1974; Walsh, et al., 1976). Chaney (1980) introduced the term soil-plant barrier to describe these concepts when considering waste-soil-plant-animal relationships of toxic microelements. A soil-plant barrier protects the food chain from toxicity of a microelement when one or more processes limit maximum levels of that element in edible plant tissues to levels safe for animals: (1) insolubility of the element in soil prevents uptake; (2) immobility of an element in fibrous roots prevents translocation to edible plant tissues; or (3) phytotoxicity of the element occurs at concentrations of the element in edible plant tissues below that injurious to animals.

Unfortunately, the soil-plant barrier does not protect animals from toxicities of all elements. The exceptions important in assessing risk from land application of municipal sludge are Cd, Se, and Mo; a few more elements (e.g., Be, Co) may have to be considered for land application of industrial wastes. Ingestion of amended soil or sludge circumvents the soil-plant barrier. These concepts will be applied in considering risk to the food chain from individual microelements applied in wastes.

Influence of Dietary and Sludge Constituents on Microelement Availability

Evaluation of the potential impact of microelements on animals through consumption of sludge, sludge-amended soil, or crops grown on sludge-amended soil is complex. Animal species differ in their tolerance to microelements. Tolerance to microelements is also influenced by age; younger animals are generally more sensitive than older ones. Crop species absorb unequal amounts of microelements. Total and relative microelement uptake is affected by crop species and cultivar, soil pH, organic matter, soil temperature, and other factors. Wastes differ in levels of elements and ratios among elements. Individual, potentially toxic elements interact with other elements in the diet, often reciprocally. Because these interactions are often the basis for physiological toxicity, they are highly important in assessing risk.

Interactions affecting Cu deficiency in ruminant animals, which were among the first studied, have been intensely examined because of their practical significance. Animals can experience simple Cu deficiency or Mo-, sulfate-, Zn-, Cd-, or Fe-induced Cu deficiency. Among the most complex is the three-way Cu-Mo-S interaction. Dietary sulfate is reduced to sulfide in the rumen; sulfide reacts with Mo to form a thiomolybdate. Thiomolybdate reacts with Cu to form an insoluble compound, which is unavailable and is excreted; this loss leads to depletion of liver Cu reserves and, subsequently, to clinical Cu deficiency (Mills et al., 1978; Bremner, 1979; Spence et al., 1980). Copper is of lower bioavailability in young forage plants than in mature plants and in fresh forages than in dried hay (Hartmans and Bosman, 1970). Soil consumed with forages reduces Cu absorption by sheep, perhaps because of soil Mo, Zn, or Fe, but probably because of Cu sorption by soil constituents, which reduces Cu absorption in the intestine (Suttle et al., 1975).

After the Cu-Mo-S interaction in ruminants was identified, it became clear that Zn, Cd, and Fe also interact with Cu bioavailability to both ruminants and monogastric animals (Bunn and Matrone, 1966; Hill et al., 1963; Matrone, 1974; McGhee et al., 1965; Mills, 1974,1978; Standish et al., 1971; Standish and Ammerman, 1971; Suttle and Mills, 1966). Reciprocally, high dietary Cu interacts to reduce absorption and toxicity of Zn, Fe, and Cd (Bunn and Matrone, 1966; Grant-Frost and Underwood, 1958; Cox and Harris, 1960; Lee and Matrone, 1969; L'Estrange, 1979; McGhee et al., 1965). Other elemental interactions have been studied and were found to be important in assessing risks (Underwood, 1977; Matrone, 1974; Levander, 1979; NRC, 1980b; Mills and Dalgarno, 1972; Mahaffey and Vanderveen, 1979; Fox 1974,1979; Fox et al., 1979; Bremner, 1979.)

In many cases, food chain toxicity results from a microelement imbalance as much as from an increased supply of one potentially toxic element. When one element is increased to a level such that the ratio of it to other elements or dietary constituents is great enough to induce a deficiency of another element, then animal weight gain declines and a health effect is observed. Chaney (1980) noted that domestic sewage sludge contains a mixture of potentially toxic elements. Consumption of sludge or sludge-amended soil is a different case for risk assessment than standard toxicological studies, where a soluble salt of one element is added at rates to cause health effects (and often to purified rather than practical diets). With sludge ingestion, increased levels of dietary Zn are balanced by increased levels of Cu and Fe. Industrial wastes and sludge differ somewhat from domestic sewage sludge because they are more likely to be enriched in only one or a few elements. The potential for microelement imbalances is greater with sludges rich in one element. Also, a wider range of elemental interactions must be considered to evaluate food chain impact of industrial wastes. A few elements are considered below as an example of sludge-soil-plant-animal interactions influencing the food chain. Many other elements are discussed in a recent report to EPA (Chaney et al., 1981).

EVALUATION OF POTENTIAL FOR FOOD CHAIN IMPACTS OF INDIVIDUAL MICROELEMENTS APPLIED IN INDUSTRIAL WASTES

Phytotoxicity and other aspects of the soil-plant barrier must be considered in relation to maximum plant foliage levels of the microelement in question. Interactions should be considered through the complete sludge–soil–plant–animal system.

An important reference for tolerance of microelements by animals was recently published by the NRC (1980b). The NRC committee considered increased levels of only the element being evaluated, although interactions are discussed. Tolerance levels are shown in Table I. Unfortunately, these levels may not be valid for sludge-fertilized crops or for ingestion of sludge or soil.

Zinc

Zinc is an essential element for plants and animals. Phytotoxicity (about 25% yield reduction) occurs in most plants at about 500 mg Zn/kg dry foliage (Chapman, 1966; Boawn, 1971; Boawn and Rasmussen, 1971; Walsh et al., 1972b). Leafy vegetables such as chard and spinach (but not lettuce) may be more tolerant of foliar Zn; chard does

Table I. Maximum Tolerable Levels of Dietary Minerals for Domestic Livestock in Comparison with Levels in Forages

Element	Soil-Plant Barrier	Level in Plant Foliage[a] (mg/kg dry foliage)		Maximum Levels Chronically Tolerated[b] (mg/kg dry diet)			
		Normal	Phytotoxic	Cattle	Sheep	Swine	Chicken
As, inorganic	Yes	0.01-1	3-10	50	50	50	50
B	Yes	7-75	75	150	(150)	(150)	(150)
Cd[c]	Fails	0.1-1	5-700	0.5	0.5	0.5	0.5
Cr^{3+}, oxides	Yes	0.1-1	20	(3000)	(3000)	(3000)	3000
Co	Fail?	0.01-0.3	25-100	10	10	10	10
Cu	Yes	3-20	25-40	100	25	250	300
F	Yes?	1-5	–	40	60	150	200
Fe	Yes	30-300	–	1000	500	3000	1000
Mn	?	15-150	400-2000	1000	1000	400	2000
Mo	Fails	0.1-3.0	100	10	10	20	100
Ni	Yes	0.1-5	50-100	50	(50)	(100)	(300)
Pb[c]	Yes	2-5	–	30	30	30	30
Se	Fails	0.1-2	100	(2)	(2)	2	2
V	Yes?	0.1-1	10	50	50	(10)	10
Zn	Yes	15-150	500-1500	500	300	1000	1000

[a]Based on literature summarized in Chaney et al. (1981).

[b]Based on NRC (1980b). Continuous long-term feeding of minerals at the maximum tolerable levels may cause adverse effects. Levels in parentheses were derived by interspecific extrapolation by NRC.

[c]Maximum levels tolerated based on human food residue consideration, rather than toxicity to livestock.

not show phototoxicity in acid soils until foliar Zn is about 1500 mg/kg (dry) (Baxter et al., 1974). Zinc toxicity to domestic livestock occurs to 300 to 1000 mg/kg dry diet (NRC, 1980b). This Zn toxicity appears to result from Zn-induced Cu deficiency (Grant-Frost and Underwood, 1958; Lee and Matrone, 1969; Campbell and Mills, 1979; Bremner, 1979). Sheep are more sensitive to excessive dietary Zn if Cu in the feed is marginal to deficient (Campbell and Mills, 1979).

Several factors influence risk of soil Zn. Crop uptake of Zn is strongly reduced as soil pH increases. Forage crops are poor accumulators of Zn compared with vegetables and crop plants. Forage crops grown on sewage–sludge-amended soil would have a normal to somewhat enriched Cu concentration to counteract the Zn; however, industrial wastes high in Zn but low in Cu would comprise a greater Zn risk to the food chain than sewage sludge.

Ruminant animals could consume Zn-rich sludges from the soil surface. For alkaline soils, sludge and soil ingestion could supply more Zn than the forage crops grown on the amended soil; for acidic soils, crop uptake would be the predominant Zn source. For animals consuming domestic sewage sludge, dietary Cu and Fe would be substantially increased along with Zn and would counteract the risk of Zn toxicity. For high-Zn sludges or industrial wastes, these protective factors might not be increased.

Forage crops present the worst case for evaluation. Under conditions of high-Zn supply, grain and fruit contain lower Zn concentrations than do leaves. Research by Ott et al. (1966), Campbell and Mills (1979), L'Estrange (1979), and Bremner et al. (1976) indicates that crop uptake of Zn is unlikely to cause Zn poisoning of cattle or sheep. Although humans could consume appreciable amounts of high-Zn leafy vegetables (acid soils, phytoxicity), Zn toxicity would be unlikely to occur in humans, even in this worst case. Leafy vegetables supply appreciable Cu and Fe, especially if grown on sludge-amended soils, and these counteract the Zn risk. Individuals are unable to consume high-Zn leafy vegetables as the bulk of their diet the whole year. Many humans now consume low or deficient amounts of Zn (Hambridge et al., 1972); increased food Zn will be beneficial in many cases.

Wildlife constitute a greater risk case than do domestic animals because wildlife could chronically consume Zn-phytotoxic foliage as most of their diet. High-Zn sludges and industrial wastes could possibly cause adverse health effects (from Zn-induced Cu deficiency) in wildlife.

In summary, land application and treatment of sewage sludges and industrial wastes are unlikely to cause Zn toxicity in domestic livestock or humans, even under worst-case conditions. Wastes unusually high in Zn but low in Cu may cause Zn toxicity to animals if they are mismanaged.

Iron

Iron is an essential element for plants and animals. Phytotoxicity of Fe is not an agronomic problem, except when rice is grown on some reduced soils (Foy et al., 1978). Leaves of most crops contain 30 to 300 mg Fe/kg dry weight. Fertilizing with sludges does not cause foliar Fe to rise above these normal levels; sludge can increase foliar Fe concentration from deficient levels (30 to 40 ppm in chlorotic leaves) to normal and may be a valuable Fe fertilizer (McClaslin and Rodriguez, 1978).

Plant Fe can reach unusually high levels from interactions with other plant micronutrients (Olsen, 1972). Deficiencies of Cu (DeKock et al., 1974), Mn, or Zn (Ambler and Brown, 1969; Warnock, 1970) can cause foliar Fe to exceed 100 ppm. Excessive Fe in leaves is stored as phytoferritin (Seckbach, 1968,1969).

The levels noted above are for leaves free of soil. Soil dust ($>1\%$ dry weight) on leaves can supply more total Fe than internal plant Fe. Trampled forages become "soiled" (Healy et al., 1974). Spray-applied sludges may greatly increase Fe in and on forages (Chaney and Lloyd, 1979), allowing forage Fe to exceed even 1% dry weight (Decker et al., 1980). This external Fe can injure domestic animals, but internal plant Fe has shown no evidence of toxicity to animals.

Animals tolerate higher levels of Fe than normally occur in feedstuffs (NRC, 1980b). However, chronic Fe toxicity is complex, and toxicity of sludge-borne Fe depends on many interactions. Dietary Fe interacts strongly with Cu and Zn; a forage diet with 1000 ppm Fe added reduced Cu in sheep and cattle liver (Standish et al., 1969,1971; Standish and Ammerman, 1971; Grun et al., 1978). On low-Cu diets, Fe toxicity is expressed as an Fe-induced Cu deficiency. However, when sewage sludge supplies the excessive Fe, it also supplies Cu, which would tend to correct Cu deficiency, and Zn, Cd, S, Mo, and fiber, which could further interfere with Cu utilization (Anke, 1973; Bremner, 1979).

The chemical form of Fe influences Fe bioavailability. Iron in $FeSO_4$ is three to five times more available to cattle than Fe in ferric oxide (Ammerman et al., 1967). Much of the Fe in anaerobically digested sludge is in the ferrous form; this quickly oxidizes as the sludge becomes aerobic. Iron in soils (and sludges on the soil surface) should have different bioavailability, depending on the degree of crystallinity of the ferric oxides present.

Iron toxicity was suspected in cattle grazing on pastures sprayed with a digested sewage sludge containing 11% Fe (Decker et al., 1980). That sludge was low in Cu. Liver Cu was at marginal or deficient levels in the cattle on sludge treatments. Liver Cu has been low in most studies of sludge feeding (Kienholz, 1980). In one study, with sludge Cu at 1000 ppm, the cattle showed Cu balance or a slight increase in liver Cu (Kienholz, 1980; Baxter et al., 1980).

Thus, high-Fe wastes that are low in Cu can promote Fe-induced Cu deficiency and high Fe accumulation in liver, spleen, and intestine. Even the erosion of cartilage in joints of cattle (Decker et al., 1980) may be caused by a severe Cu deficiency. Zinc-induced Cu deficiency in horses (Willoughby et al., 1972; Willoughby and Oyaert, 1973) caused erosion of cartilage in joints similar to that observed in cattle grazing on pastures fertilized with high-Fe sludge.

Industrial sludges and other wastes high in Fe may cause Fe-induced Cu deficiency if they are surface-applied to pasture land. Zinc and Mo in the waste can also contribute to the toxicity, whereas Cu in the waste can counteract it. Supplemental Cu may be needed for ruminants grazing on pastures where high-Fe wastes remain on the soil surface.

Molybdenum

Molybdenum, which is essential for both plants and animals, is present in soils and plants as anionic molybdate. This form causes Mo to differ markedly in its reactions with soils and plants from those of the potentially toxic cations for which liming to alkaline pH reduces plant uptake. Soils adsorb Mo more strongly under acidic conditions; inorganic compounds of Mo have not been identified in soils (Vlek and Lindsay, 1977a). The strong Mo sorption at low soil pH holds Mo against leaching and reduces plant Mo uptake. Alkaline soil pH weakens soil sorption of Mo, promoting plant uptake of Mo and allowing Mo to leach through the root zone with leaching water. Sorption of Mo is further decreased by reducing conditions (low E_h). Molybdenum uptake is greatest in alkaline, poorly drained soils with Mo enrichment (Kubota, 1977). The chemical or biological basis for greater Mo concentration in plants growing on poorly drained soils remains unclear. [A higher percentage of plant roots in the surface soil, greater Mo solubility (lower sorption) in reduced soils, or reduced plant biomass diluting absorbed Mo may explain these higher concentrations

(Allaway, 1977c; Kubota et al., 1961,1963; Vlek and Lindsay, 1977b).]

Plants tolerate very high levels of Mo and translocate Mo to edible plant tissues. Legume crops accumulate substantially more Mo than do grasses (Chapman, 1966; Allaway, 1977c; Jensen and Lesperance, 1971; Kubota, 1977). Vegetable crops are also Mo accumulators (Hornick et al., 1977; Gupta et al., 1978).

Ruminant animals are especially susceptible to Mo toxicity (molybdenosis) because Mo interacts to induce a Cu deficiency (Mills et al., 1978; Underwood, 1977). Dietary Mo interacts also with sulfate (increased sulfate competitively reduces Mo absorption), Mn, Zn, Fe, tungstate, sulfur-containing amino acids, etc. Dietary sulfate and Cu are the primary factors interacting with excessive dietary Mo. Normal forages contain 0.1 to 3.0 ppm Mo (dry weight). Levels >10 ppm Mo are generally high enough to cause molybdenosis; lower levels can cause molybdenosis depending on dietary Cu availability. Plants grown on poorly drained, alkaline, Mo-enriched soils can exceed 100 ppm Mo, far surpassing the levels necessary for molybdenosis. Industrial air pollution from Mo smelters, oil refineries, and steel mills has caused molybdenosis (Hornick et al., 1977; Alary et al., 1981; Gardner and Hall-Patch, 1962,1968; Buxton and Allcroft, 1955; Parker and Rose, 1955). The soil-plant barrier clearly does not protect the ruminant-animal food chain from excessive soil Mo.

The few soils in the United States that have naturally high Mo levels cause molybdenosis in cattle and sheep (Kubota, 1977). Soil and crop management can reduce the problem; for example, draining the area and growing grasses rather than legumes reduces potential molybdenosis considerably. Land treatment sites for Mo-rich wastes can be managed to remove the Mo from the surface soil before ruminants are allowed to graze the area; draining and crop selection are important. Adequate or high Cu levels in crops and surface soil reduces. Mo risk. Animals can be supplemented with Cu by injection or salt licks containing Cu.

Sewage sludges are seldom high in Mo. Historically, sludge use has not been reported to cause molybdenosis in ruminants through either plant uptake or direct ingestion of sludge. Sludge Cu would counteract any effects of considerable amounts of Mo in ingested sludges. Fly ash is richer in Mo than sludges and has been evaluated as a Mo fertilizer; fly ash Mo is about as available to plants as Na molybdate (Doran and Martens, 1972). Molybdenosis could result if high rates of fly ash are applied on alkaline pastures (Gutenmann et al., 1979).

Monogastric animals are protected from excess soil Mo by the soil-plant barrier; Mo concentration in grain is lower than in leaves; monogastric animals are more tolerant of Mo than ruminants; and monogastric animals are seldom fed grain crops grown only on a restricted area of land. Humans could conceivably (but very unlikely) be at some risk of molybdenosis by consuming large amounts of garden crops grown on high-Mo alkaline soils that had previously been used for land treatment of Mo-rich wastes. Sites managed at lower soil pH to avoid excessive crop Mo can subsequently cause molybdenosis after limestone is applied (Hornick et al., 1977).

EVALUATION OF POTENTIAL FOR FOOD CHAIN IMPACTS OF TOXIC ORGANIC COMPOUNDS APPLIED IN INDUSTRIAL WASTES

Animals can be exposed to toxic organic compounds (TOs) present in wastes by (1) direct ingestion of wastes, wastes adhering to forages, or wastes lying on the soil surface; (2) ingestion of plant tissues that are increased in TO content after plant uptake or volatilization from the soil to the plant; or (3) consumption of animal products enriched in TO by other routes. The chemical and physical properties of a TO control adsorption by soil, volatilization, plant uptake and translocation, biodegradation (in soil, plant, or animal), and accumulation in animal tissues. Because each TO is chemically and pharmacologically unique, each compound will have its unique behavior in waste-soil-plant-animal systems (Fries, 1982; Majeta and Clark, 1981).

Although much research has been conducted on insecticides, fungicides, and herbicides, insufficient information is available to assess food chain risk of waste-borne TOs. Environmentally relevant research on TOs is quite limited, even among pesticides. Little is known about fate and potential for food chain effects of industrial TO wastes and by-products that may be considered for application to land treatment sites.

Thus, the processes that influence movement of TOs in waste-soil-plant-animal food chains are discussed in this section. It should illustrate the research needs for assessing potential impacts of a TO or a TO-enriched waste. PCBs in sewage sludge provide a particularly relevant example because regulations were developed from the available research (U.S. EPA, 1979).

Bioavailability of Waste-Borne Toxic Organics

Lipophilic toxic organics in ingested sludges and soil are bioavailable. DDT and lindane in soil were absorbed by sheep and stored in their fat (Harrison et al., 1969, 1970; Collett and Harrison, 1968). PCBs and other compounds in ingested sludge were absorbed and stored in the fat of cattle (Kienholz, 1980; Baxter et al., 1980; Fitzgerald, 1978, 1980), cow's milk, (Bergh and Peoples, 1977), and the fat of swine (Hansen et al., 1981). In general, PCB residues in fat reached fivefold levels in feed. Based on these studies and basic research on bioaccumulation of PCBs, Fries (1982) concluded that PCBs should not exceed 2.0 mg/kg dry sludge if milk cows are to be allowed to graze pastures under worst-case conditions, which allow 14% sludge in their diet. This conclusion was based on a biomagnification from diet to milk fat of four- to fivefold, and Food and Drug Administration (FDA) tolerances of 1.5 mg PCB/kg milk fat (FDA, 1979). Forages have PCB residues that are about one-tenth that of the soil, or lower. Good management practices (delay grazing for 30 d after surface application of sludge and supply feed concentrates during periods of low-forage availability) reduce sludge ingestion so that more than 10 ppm PCBs could be allowed in sludge surface applied at 10 t/ha·yr. Injection of sludge below the soil surface would further reduce exposure.

A seldom considered concentration step involves soil fauna. Earthworms accumulate Cd and lipophilic toxic organics. Beyer and Gish (1980) noted substantial residues of DDT, dieldrin, and heptachlor in earthworms many years after application. Birds and shrews, which consume appreciable earthworm biomass, are exposed to Cd and pesticides. More study is needed to assess the importance of this unusual food-chain pathway in relation to land treatment of industrial wastes.

Plant Uptake of Toxic Organics in the Soil

Toxic organics can enter edible parts of plants by two processes: (1) uptake from the soil solution, with translocation from roots to shoots, or (2) adsorption by roots or shoots to TOs volatilized from the soil. "Systemic" acting pesticides are those that are applied to the soil, absorbed and translocated by the plant, and act to protect the plant leaves. These compounds, which are quite water soluble, would probably not appear in industrial wastewater treatment sludges at appreciable levels. Most systemic TOs are prohibited from use on food crops (other than seed protectants) because residues of the compound or its metabolites on or in food may be unacceptable. The EPA-approved label for each compound lists acceptable uses.

The lipophilic halogenated pesticides represent the case for water-insoluble compounds, which are largely sorbed by plants from the soil air or the pesticide-enriched air near the soil surface. Beall and Nash

(1971) developed a method to discriminate between movement of a TO through the plant vascular system (uptake and translocation) vs vapor-phase movement. They found that soybean shoots were contaminated by dieldrin, endrin, and heptachlor largely by uptake and translocation. Vapor transport predominated for DDT and was equal to uptake and translocation for endrin. Using the method of Beall and Nash, Fries and Marrow (1981) found that PCBs reached shoots via vapor transport, whereas the less-volatile PBBs did not contaminate plant shoots by either process (Chou et al., 1978; Jacobs, Chou, and Tiedje, 1976). Suzuki et al. (1977) found that PCBs with a low number of chlorines could be absorbed and translocated at low rates by soybean seedlings from sand treated with high levels of PCBs.

Root crops are especially susceptible to contamination by the vapor transport route. Carrots have a lipid-rich peel layer that serves as a sink for volatile lipophilic TOs. Depending on the water solubility and vapor pressure of the individual compound, the TO may reside nearly exclusively in the peel layer of carrots or penetrate the storage root several millimeters (Lichtenstein et al., 1964,1965; Jacobs, et al., 1976; Lichtenstein and Schulz, 1965; Iwata and Gunther, 1976; Iwata et al., 1974; Fox et al., 1964; Landrigan et al., 1978).

Carrot cultivars differ in uptake and in peel vs pulp distribution of the chlorinated hydrocarbon pesticides, endrin and heptachlor (Lichtenstein, Myrdal, et al., 1965; Hermanson et al., 1970). Other root crops (sugarbeet, onion, turnip, and rutabaga) are much less effective in accumulating lipophilic TOs in their edible roots, possibly because the surface of the peel is lower in lipids (Moza et al., 1976,1979; Moza et al., 1976; Fox et al., 1964; Chou et al., 1978; Lichtenstein and Schulz, 1965).

The level of chlorinated hydrocarbon in carrots is sharply reduced by increased organic matter in soil. The increased organic matter adsorbs the TOs, preventing them from being released to the soil solution or soil air (Filonow et al., 1976; Weber and Mrozek, 1979; Chou et al., 1978; Strek et al., 1981). Added sewage sludge increased the ability of soils to adsorb PCBs (Fairbanks and O'Connor, 1980). At some low level of PCBs in sludge, the increased sorption capacity may fully counteract the increased PCB (Lee et al., 1980).

Assessing risk environmental exposure to PCBs or other TOs is difficult. The residue of PCBs in waste products is depleted of the lower-chlorinated compounds, which are relatively more volatile, but most research is conducted with the commercial mixture. Plant contamination by the higher chlorinated compounds is clearly much less than that by lower chlorinated ones (Iwata and Gunther, 1976; Suzuki et al.,

1977; Moza, et al., 1976; Moza et al., 1979; Fries and Marrow, 1981). Recently, research has begun with the individual ^{14}C-labeled PCBs; risk evaluation should focus on the penta-, hexa- and more highly chlorinated compounds that remain in wastes and soils. As long as carrots are peeled, the only significant exposure to these higher-chlorinated PCBs is to grazing ruminants through soil ingestion.

Another research effort centered on assessing risk from polycyclic aromatic hydrocarbons (PAHs). Some PAHs are carcinogenic [e.g., benzo(a)pyrene]. Researchers found PAHs in composted municipal refuse and noted that carrot roots (but not mushrooms) accumulated many PAHs from compost-amended soils (Muller, 1976; Linne and Martens, 1978; Wagner and Siddiqi, 1971; Siegfried, 1975; Neudecker, 1978; Ellwardt, 1977; Borneff et al., 1973). The level of 3,4-benzypyrene in carrot roots declined with successive cropping of compost-amended soil. Multigeneration feeding studies of control and compost-grown carrots found no risk to rats (Neudecker, 1978).

Many other carcinogenic or toxic compounds may be present in wastes and may contaminate the food chain through plant uptake, volatile contamination of crop root or shoots, or soil ingestion. Little information is available on these substances. Nitrosamines have been found in sewage wastes (Yoneyama, 1981; Green et al., 1981) and are accumulated from soil by plants (Brewer et al., 1980; Dean-Raymond and Alexander, 1976). However, nitrosamines appear to be rapidly degraded in soils and plants. Research on N-nitrosodimethylamine and N-nitrosodiethylamine found rapid degradation in soil; plant uptake did occur, but these compounds were rapidly degraded there (Dressel, 1976a,b). Traces of nitrosamines are found in nitroanaline-based herbicides. These compounds are rapidly degraded, and no detectable nitrosamine was found in soybean shoots (Kearney et al., 1980a). An IUPAC committee, which assessed the environmental consequences of these trace nitrosamines, found no risk to the food chain (Kearney et al., 1980b).

Aflatoxin presents another useful example of the fate of toxic organic compounds. Aflatoxin-contaminated agricultural wastes are usually tilled into cropland. Aflatoxin is readily decomposed or transformed to nonextractable forms in soil, although detectable aflatoxin remained for about 50 d when 2 ppm was applied (Angle and Wagner, 1980). If present in nutrient solution or freshly amended soil, aflatoxin can be absorbed by corn or lettuce (Mertz et al., 1980, 1981). Thus, although plants can absorb aflatoxin from aflatoxin-amended land treatment sites, no aflatoxin would remain after closure, and little would remain at the time of crop growth after preparing the soil for seeding.

Bioassay techniques for plant residues of soil-borne TOs include (1) field studies, (2) controlled pot studies, (3) vapor barrier pot studies, and (4) small-pot-intensive uptake (many plants to observe maximum possible uptake) techniques. Results from growing many plants in a small volume of pesticide-treated soil (Fuhremann and Lichtenstein, 1980) provide valuable information on possible pathways of TO movement. However, little information obtained is relevant to actual environmental exposures.

Land treatment sites can be managed to avoid all unacceptable effects on the food chain from waste-borne TOs. Wastes can be injected below the soil surface. Mechanically harvesting fresh forages or grain crops avoids soil contamination of the food chain. Crops can be returned to the treatment site soil until plant-absorbable TOs no longer reach unacceptable concentrations. Knowledge of the biodegradation and transport route for compounds in industrial wastes is required to prepare hazardous waste land treatment site management and closure plans (EPA, 1980b).

ACKNOWLEDGMENTS

This study was performed under Interagency Agreement No. AD-12-F-0-055-0 between the U.S. Department of Agriculture and the U.S. Environmental Protection Agency. Ms. M. M. Leach, Technical Information Specialist, assisted in searching and retrieval of relevant literature. The guidance and support of Dr. Richard J. Mahler, EPA Project Officer; Mr. Carlton Wiles, EPA-MERL; and my colleagues in the Biological Waste Management and Organic Resources Laboratory were helpful in completing this review.

REFERENCES

Alary, J., P. Bourbon, J. Esclassan, J. C. Lepert, J. Vandaele, J. M. Lecuire, and F. Klein. 1981. "Environmental Molybdenum Levels in Industrial Molybdenosis of Grazing Cattle," *Sci. Total Environ.* 19:111-119.

Allaway, W. H. 1968. "Agronomic Controls over the Environmental Cycling of Trace Elements," *Adv. Agron.* 20:235-274.

Allaway, W. H. 1977a. "Soil and Plant Aspects of the Cycling of Chromium, Molybdenum, and Selenium," *Proc. Intern. Conf. Heavy Metals in the Environment* 1:35-47.

Allaway, W. H. 1977b. "Food Chain Aspects of the Use of Organic Residues," in *Soils for Management of Organic Wastes and Wastewaters,*

L. F. Elliott and F. J. Stevenson, Eds. (Madison, WI: American Society of Agronomy), pp. 282-298.

Allaway, W. H. 1977c. "Perspectives on Molybdenum in Soils and Plants," in *Molybdenum in the Environment, Vol. 2,* W. R. Chappell and K. K. Petersen, Eds. (New York: Marcel Dekker, Inc.), pp. 317-339.

Ambler, J. E., and J. C. Brown. 1969. "Cause of Differential Susceptibility to Zinc Deficiency in Two Varieties of Navy Beans (*Phaseolus vulgaris L.*)," *Agron. J.* 61:41-43.

Ammerman, C. B., J. M. Wing, B. G. Dunavant, W. K. Robertson, J. P. Feaster, and L. R. Arrington. 1967. "Utilization of Inorganic Iron by Ruminants as Influenced by Form of Iron and Iron Status of the Animal," *J. Anim. Sci.* 26:404-410.

Ammerman, C. B., S. M. Miller, K. R. Fick, and S. L. Hansard, III. 1977. Contaminating Elements in Mineral Supplements and Their Potential Toxicity: A Review," *J. Animal Sci.* 44:485-508.

Angle, C. R., and M. S. McIntire. 1979. "Environmental Lead and Children: the Omaha, Nebraska Study," *J. Toxicol. Environ. Health* 5:855-870.

Angle, J. S., and G. H. Wagner. 1980. "Decomposition of Aflatoxin in Soil," *Soil Sci. Soc. Am. J.* 44:1237-1240.

Anke, M. 1973. "Disorders Due to Copper Deficiency in Sheep and Cattle," *Monat. Vet. Med.* 28:294-298.

Baker, D. E., and L. Chesnin. 1976. "Chemical Monitoring of Soils for Environmental Quality, and Animal and Human Health," *J. Agric. Sci.* 84: 249-254.

Baker, E. L., Jr., D. S. Folland, T. A. Taylor, M. Frank, W. Peterson, G. Lovejoy, D. Cox, J. Housworth, and P. J. Landrigan. 1977. "Lead Poisoning in Children of Lead Workers. Home Contamination with Industrial Dust," *New England J. Med.* 296:260-261.

Barltrop, D., C. D. Strehlow, I. Thornton, and J. S. Webb. 1974. "Significance of High Soil Lead Concentrations for Childhood Lead Burdens," *Environ. Health Perspect.* 7:75-82.

Batey, T., C. Berryman, and C. Line. 1972. "The Disposal of Copper-Enriched Pig-Manure Slurry on Grassland," *J. Br. Grassland Soc.* 27:139-143.

Baxter, J. C., R. L. Chaney, and C. S. Kinlaw. 1974. "Reversion of Zn and Cd in Sassafras Sandy Loam as Measured by Several Extractants and by Swiss Chard," *Agron. Abstr.* 1974:23.

Baxter, J. C., D. E. Johnson, and E. W. Kienholz. 1980. "Uptake of Trace Metals and Persistent Organics into Bovine Tissues from Sewage Sludge—Denver Project," in *Sludge—Health Risks of Land Application,* G. Bitton et al., Eds. (Ann Arbor, MI: Ann Arbor Science Publishers, Inc.).

Beall, M. L., Jr., and R. G. Nash. 1971. "Organochlorine Insecticide Residues in Soybean Plant Tops: Root vs. Vapor Sorption," *Agron J.* 63:460-464.

Bergh, A. K., and R. S. Peoples. 1977. "Distribution of Polychlorinated Biphenyls in a Municipal Wastewater Treatment Plant and Environs," *Sci. Total Environ.* 8:197-204.

Beyer, W. N., and C. D. Gish. 1980. "Persistence in Earthworms and Potential Hazards to Birds of Soil Applied DDT, Dieldrin, and Heptachlor," *J. Appl. Ecol.* 17:295-307.

Boawn, L. C. 1971. "Zinc Accumulation Characteristics of Some Leafy Vegetables," *Commun. Soil Sci. Plant Anal.* 2:31-36.

Boawn, L. C., and P. E. Rasmussen. 1971. "Crop Response to Excessive Zinc Fertilization of Alkaline Soil," *Agron. J.* 63:874-876.

Borneff, J., G. Farkasdi, H. Glathe, and H. Kunte. 1973. "The Fate of Polycyclic Aromatic Hydrocarbons in Experiments Using Sewage Sludge-Garbage Composts as Fertilizers," *Zbl. Bakt. Hyg.* 157:151-164.

Boswell, F. C. 1975. "Municipal Sewage Sludge and Selected Element Applications to Soil: Effect on Soil and Fescue," *J. Environ. Qual.* 4:267-273.

Bowen, H. J. M. 1966. *Trace Elements in Biochemistry* (New York: Academic Press).

Bowen, H. J. M. 1979. *Environmental Chemistry of the Elements* (New York: Academic Press).

Bremner, I. 1979. "The Toxicity of Cadmium, Zinc, and Molybdenum and Their Effects on Copper Metabolism," *Proc. Nutr. Soc.* 38:235-242.

Bremner, I., B. W. Young, and C. F. Mills. 1976. "Protective Effect of Zinc Supplementation Against Copper Toxicosis in Sheep," *Brit. J. Nutr.* 36:551-561.

Brewer, W. S., A. C. Draper, III, and S. S. Wey. 1980. "The Detection of Dimethylnitrosamine and Diethylnitrosamine in Municipal Sewage Sludge Applied to Agricultural Soils," *Environ. Poll.* 81:37-43.

Brown, K. W., S. G. Jones, and K. C. Donnelly. 1980. "The Influence of Simulated Rainfall on Residual Bacteria and Virus on Grass Treated with Sewage Sludge," *J. Environ. Qual.* 9:261-265.

Bunn, C. R., and G. Matrone. 1966. "*In Vivo* Interactions of Cadmium, Copper, Zinc and Iron in the Mouse and Rat," *J. Nutr.* 90:395-399.

Buxton, J. C., and R. Allcroft. 1955. "Industrial Molybdenosis of Grazing Cattle," *Vet. Rec.* 67:273-276.

Campbell, J. K., and C. F. Mills. 1979. The Toxicity of Zinc to Pregnant Sheep," *Environ. Res.* 20:1-13.

Cataldo, D. A., and R. E. Wildung. 1978. "Soil and Plant Factors Influencing the Accumulation of Heavy Metals by Plants," *Environ. Health Perspect.* 27:145-149.

Chaney, R. L. 1980. "Health Risks Associated with Toxic Metals in Municipal Sludge," in *Sludge–Health Risks of Land Application*, G. Bitton et al., Eds. (Ann Arbor, MI: Ann Arbor Science Publishers, Inc.), pp. 59-83.

Chaney, R. L., D. D. Kaufman, S. B. Hornick, J. F. Parr, L. J. Sikora, W. D. Burge, P. B. Marsh, G. B. Willson, and R. H. Fisher. 1981. "Review of Information Relevant to Land Treatment of Hazardous Wastes," Draft Report to U.S. Environmental Protection Agency.

Chaney, R. L., O. A. Levander, S. O. Welsh, and H. W. Mielke. 1980. "Effect of Soil on Availability of Pb to Rats, and Bioavailability of Pb in Urban Garden Soils," unpublished results, U.S. Department of Agriculture, Beltsville, MD.

Chaney, R. L., and C. A. Lloyd. 1979. "Adherence of Spray-Applied Liquid Digested Sewage Sludge to Tall Fescue," *J. Environ. Qual.* 8:407-411.

Chapman, H. D. 1966. "Diagnostic Criteria for Plants and Soils," University of California, Division of Agricultural Science, Riverside.

Charney, E., J. W. Sayre, and M. Coulter. 1979. "Increased Lead Adsorption in Inner City Children: Where Does the Lead Come From?," *Proceedings of the Second International Symposium on Environmetal Lead Research*, in press.

Chou, S. F., L. W. Jacobs, D. Penner, and J. M. Tiedje. 1978. "Absence of Plant Uptake and Translocation of Polybrominated Biphenyls (PBB's)," *Environ. Health Perspect.* 23:9-12.

Collett, J. N., and D. L. Harrison. 1968. "Lindane Residues on Pasture and in the Fat of Sheep Grazing Pasture Treated with Lindane Prills," *N.Z.J. Agr. Res.* 11:589-600.

Cox, D. H., and D. L. Harris. 1960. "Effect of Excess Dietary Zn on Fe and Cu in the Rat," *J. Nutr.* 70:514-520.

Dacre, J. C., and G. L. Ter Harr. 1977. "Lead Levels in Tissues from Rats Fed Soils Containing Lead," *Arch. Environ. Contam. Toxicol.* 6:111-119.

Dalgarno, A. C., and C. F. Mills. 1975. "Retention by Sheep of Copper from Aerobic Digests of Pig Faecal Slurry," *J. Agric. Sci.* 85:11-18.

Decker, A. M., R. L. Chaney, J. P. Davidson, T. S. Rumsey, S. B. Mohanty, and R. C. Hammond. 1980. "Animal Performance on Pastures Topdressed with Liquid Sewage Sludge and Sludge Compost," in *Proceedings of the National Conference on Municipal and Industrial Sludge Utilization and Disposal* (Silver Spring, MD: Information Transfer, Inc.), pp. 37-41.

Dean-Raymond, D., and M. Alexander. 1976. "Plant Uptake and Leaching of Dimethylnitrosamines," *Nature* 262:394-396.

DeKock, P. C., M. V. Cheshire, and A. Hall. 1971. "Comparison of the Effect of Phosphorus and Nitrogen on Copper-Deficient and -Sufficient Oats," *J. Sci. Food Agric.* 22:437-440.

Doran, J. W., and D. C. Martens. 1972. "Molybdenum Availability as Influenced by Applications of Fly Ash to Soil," *J. Environ. Qual.* 1:186-189.

Dressel, J. 1976a. "Dependence of Nitrogen Containing Constituents Which Influence Plant Quality by the Intensity of Fertilization," *Landwirtsch. Forsch. Sondern.* 33:326-334.

Dressel, J. 1976b. "Relationship Between Nitrate, Nitrite and Nitrosamines in Plants and Soil at Intensive Nitrogen Fertilization," *Qual. Plant. Plant Foods Hum. Nutr.* 25:381-390.

Edwards, W. C., and B. R. Clay. 1977. "Reclamation of Rangeland Following a Lead Poisoning Incident in Livestock from Industrial Airborne Contamination of Forage," *Vet. Human Toxicol.* 19:247-249.

Egan, D. A., and T. O'Cuill. 1970. "Cumulative Lead Poisoning in Horses in a Mining Area Contaminated with Galena," *Vet. Rec.* 86:736-738.

Ellwardt, P.-C. 1977. "Variation in Content of Polycyclic Aromatic Hydrocarbons in Soil and Plants by Using Municipal Waste Composts in Agriculture," in *Soil Organic Matter Studies, Vol. II*, pp. 291-298.

Fairbanks, B. C., and G. A. O'Connor. 1980. "Adsorption of Polychlorinated

Biphenyls (PCB's) by Sewage Sludge Amended Soil," in *Sludge–Health Risks of Land Application,* G. Bitton et al., Eds. (Ann Arbor, MI: Ann Arbor Science Publishers, Inc.), p. 346.

Field, A. C., and D. Purves. 1964. "The Intake of Soil by the Grazing Sheep," *Proc. Nutr. Soc.* (London) 23:24-25.

Filonow, A. B., L. W. Jacobs, and M. M. Mortland. 1976. "Fate of Polybrominated Biphenyls (PBB's) in Soils. Retention of Hexabromobiphenyl in Four Michigan Soils," *J. Agric. Food Chem.* 24:1201-1204.

Fitzgerald, P. R. 1978. "Toxicology of Heavy Metals in Sludges Applied to the Land," in *Proceedings of the Fifth National Conference on Acceptable Sludge Disposal Techniques* (Rockville, Md: Information Transfer, Inc.), pp. 106-116.

Fitzgerald, P. R. 1980. "Observations on the Health of Some Animals Exposed to Anaerobically Digested Sludge Originating in the Metropolitan Sanitary District of Greater Chicago System," in *Sludge–Health Risks of Land Application,* G. Bitton et al., Eds. (Ann Arbor, MI: Ann Arbor Science Publishers), pp. 267-284.

Food and Drug Administration. 1979. "Polychlorinated Biphenyls (PCB's); Reduction of Tolerances," *Fed. Regist.* 44:38330-38340.

Fox, M. R. S. 1974. "Effect of Essential Minerals on Cadmium Toxicity. A Review," *J. Food Sci.* 39:321-324.

Fox, M. R. S. 1979. "Nutritional Influences on Metal Toxicity: Cadmium as a Model Toxic Element," *Environ. Health Perspect.* 29:95-104.

Fox, M. R. S., R. M. Jacobs, A. O. L. Jones, and B. E. Fry, Jr. 1979. "Effects of Nutritional Factors on Metabolism of Dietary Cadmium at Levels Similar to Those of Man," *Environ. Health Perspect.* 28:107-114.

Fox, C. J. S., D. Chisholm, and D. K. R. Stewart. 1964. "Effect of Consecutive Treatments of Aldrin and Heptachlor on Residues in Rutabagas and Carrots and on Certain Soil Arthropods and Yield," *Can. J. Plant Sci.* 44: 149-156.

Foy, C. D., R. L. Chaney, and M. C. White. 1978. "The Physiology of Metal Toxicity in Plants," *Ann. Rev. Plant Physiol.* 29:511-566.

Fries, G. F. 1982. "Potential Polychlorinated Biphenyl Residues in Animal Products from Application of Contaminated Sewage Sludge to Land," *J. Environ. Qual.* 11:14-20.

Fries, G. F. and G. S. Marrow. 1981. "Chlorobiphenyl Movement from Soil to Soybean Plants," *J. Agric. Food Chem.* 29:757-759.

Fries, G. F., G. S. Marrow, and P. A. Snow. 1982. "Soil Ingestion by Dairy Cattle," *J. Dairy Sci.* 65:611-618.

Fuhremann, T. W., and E. P. Lichtenstein. 1980. "A Comparative Study of the Persistence, Movement and Metabolism of Six Carbon-14 Insecticides in Soils and Plants," *J. Agric. Food Chem.* 28:446-452.

Galke, W. A., D. I. Hammer, J. E. Keil, and S. W. Lawrence. 1977. "Environmental Determinants of Lead Burdens in Children," *Proc. Int. Conf. Heavy Metals in the Environment, Vol. III* pp. 53-74.

Gardner, A. W., and P. K. Hall-Patch. 1968. "Molybdenosis in Cattle Grazing Downwind from an Oil Refinery Unit," *Vet. Rec.* 82:86-87.

Giguere, C. G., A. B. Howes, M. McBean, W. N. Watson, and L. E. Witherell. 1977. "Increased Lead Absorption in Children of Lead Workers–Vermont," *Morb. Mortal. Wkly. Rep.* 26(8):61-62.

Grant-Frost, D. B., and E. J. Underwood. 1958. "Zn Toxicity in the Rat and Its Interrelation with Cu." *Aust. J. Exp. Biol. Med. Sci.* 36:339-346.

Green, S., M. Alexander, and D. Leggett. 1981. "Formation of N-Nitrosodimethylamine During Treatment of Municipal Waste Water by Simulated Land Application," *J. Environ. Qual.* 10:416-421.

Grun, M., M. Anke, A. Hennig, W. Seffner, M. Partshefeld, G. Flachowsky, and B. Groppel. 1978. "Excessive Oral Iron Application to Sheep. 2. The Influence on the Level of Iron, Copper, Zinc, and Manganese in Different Organs" (in German), *Arch. Tierernaehr.* 28:341-347.

Gupta, U. C., E. W. Chipman, and D. C. Mackay. 1978. "Effects of Molybdenum and Lime on the Yield and Molybdenum Concentration of Vegetable Crops Grown on Acid Sphagnum Peat Soil," *Can. J. Plant Sci.* 58:983-992.

Gutenmann, W. H., I. S. Pakkala, D. J. Churey, W. C. Kelly, and D. J. Lisk. 1979. "Arsenic, Boron, Molybdenum, and Selenium in Successive Cuttings of Forage Crops Field Grown on Fly Ash Amended Soil," *J. Agric. Food Chem.* 27:1393-1395.

Hambridge, K. M., C. Hambridge, M. Jacobs, and J. D. Baum. 1972. "Low Levels of Zinc in Hair, Anorexia, Poor Growth, and Hypogeusia in Children," *Pediatr. Res.* 6:868-874.

Hammond, P. B., C. S. Clark, P. S. Gartside, O. Berger, A. Walker, and L. W. Michael. 1980. "Fecal Lead Excretion in Young Children as Related to Sources of Lead in their Environments," *Int. Arch. Occup. Environ. Health* 46:191-202.

Hansen, L. G., P. K. Washko, L. G. M. T. Tuinstra, S. B. Dorn, and T. D. Hinesly. 1981. "Polychlorinated Biphenyl, Pesticide, and Heavy Metal Residues in Swine Foraging on Sewage Sludge Amended Soils," *J. Agric. Food Chem.* 29:1012-1017.

Harbourne, J. F., C. T. McCrea, and J. Watkinson. 1968. "An Unusual Outbreak of Lead Poisoning in Calves," *Vet. Rec.* 83:515-517.

Harrison, D. L., J. C. M. Mol, and W. B. Healy. 1970. "DDT Residues in Sheep from the Ingestion of Soil," *N.Z. J. Agric. Res.* 13:664-672.

Harrison, D. L., J. C. M. Mol, and J. E. Rudman. 1969. "DDT and Lindane: New Aspects of Stock Residues Derived from a Farm Environment," *N.Z. J. Agric. Res.* 12:553-574.

Hartmans, J., and M. S. M. Bosman. 1970. "Differences in the Copper Status of Grazing and Housed Cattle and Their Biochemical Backgrounds," *Trace Element Metabolism in Animals–1*, C. F. Mills, Ed. Livingstone, Edinburgh, pp. 362-366.

Healy, W. B. 1968. "Ingestion of Soil by Dairy Cows," *N.Z. J. Agr. Res.* 11:487-499.

Healy, W. B., and T. G. Ludwig. 1965. "Wear of Sheeps' Teeth. 1. The Role of Ingested Soil," *N.Z. J. Agr. Res.* 8:737-752.

Healy, W. B., P. C. Rankin, and H. M. Watts. 1974. "Effect of Soil Contamination on the Element Composition of Herbage," *N.Z. J. Agr. Res.* 17: 59-61.

Hermanson, H. P., L. D. Anderson, and F. A. Gunther. 1970. "Effects of Variety and Maturity of Carrots upon Uptake of Endrin Residues from Soil," *J. Econ. Entomol.* 63:1651-1654.

Hill, C. H., G. Matrone, W. L. Payne, and C. W. Barber. 1963. "*In Vivo* Interactions of Cadmium with Copper, Zinc, and Iron," *J. Nutr.* 80:227-235.

Hornick, S. B., D. E. Baker, and S. B. Guss. 1977. Crop Production and Animal Health Problems Associated with High Soil Molybdenum," in *Molybdenum in the Environment, Vol. 2,* W. R. Chappell and K. K. Peterson, Eds. (New York: Marcel Dekker), pp. 665-684.

Iwata, Y., F. A. Gunther, and W. E. Westlake. 1974. "Uptake of a PCB (Arochlor 1254) from Soil by Carrots Under Field Conditions," *Bull. Environ. Contam. Toxicol.* 11:523-52.

Iwata, Y., and F. A. Gunther. 1976. "Translocation of the Polychlorinated Biphenyl Arochlor 1254 from Soil into Carrots Under Field Conditions," *Arch. Environ. Contam. Toxicol.* 4:44-59.

Jacobs, L. W., S. F. Chou, and J. M. Tiedje. 1976. "Fate of Polybrominated Biphenyls (PBB's) in Soils. Persistence and Plant Uptake," *J. Agr. Food Chem.* 24:1198-1201.

Jensen, E. H., and A. L. Lesperance. 1971. "Molybdenum Accumulation by Forage Plants," *Agron. J.* 63:201-204.

Jones, S. G., K. W. Brown, L. E. Deuel, and K. C. Donnelly. 1979. "Influence of Rainfall on the Retention of Sludge Heavy Metals by the Leaves of Forage Crops," *J. Environ. Qual.* 8:69-72.

Jordan, L. D., and D. J. Hogan. 1975. "Survey of Lead in Christchurch Soils," *N.Z. J. Sci.* 18:253-260.

Kearney, P. C., J. E. Oliver, A. Kontson, W. Fiddler, and J. W. Pensabene. 1980a. "Plant Uptake of Dinitroaniline Herbicide-Related Nitrosamines," *J. Agric. Food Chem.* 28:633-635.

Kearney, P. C., M. E. Amundson, K. I. Beynon, N. Drescher, G. J. Marco, J. Miyamoto, J. R. Murphy, and J. E. Oliver. 1980b. "Nitrosamines and Pesticides. A Special Report on the Occurrence of Nitrosamines as Terminal Residues Resulting from Agricultural Use of Certain Pesticides," *Pure and Appl. Chem.* 52:499-526.

Kienholz, E. W. 1980. "Effect of Toxic Chemicals Present in Sewage Sludge on Animal Health," *Sludge—Health Risks of Land Application,* G. Bitton et al., Eds. (Ann Arbor, MI: Ann Arbor Science Publishers, Inc.), pp. 153-171.

Kubota, J. 1977. "Molybdenum Status of United States Soils and Plants," *Molybdenum in the Environment, Vol. 2,* W. R. Chappel and K. K. Peterson, Eds. (New York: Marcel Dekker, Inc.), pp. 555-581.

Kubota, J., E. R. Lemon, and W. H. Allaway. 1963. "The Effect of Soil Moisture Content on the Uptake of Molybdenum, Copper, and Cobalt by Alsike Clover," *Soil Sci. Soc. Am. Proc.* 27:679-683.

Kubota, J., V. A. Lazar, L. W. Langan, and K. C. Beeson. 1961. "The Relationship of Soils to Molybdenum Toxicity in Cattle in Nevada," *Soil Sci. Soc. Am. Proc.* 25:227-232.

Landrigan, P. J., C. W. Heath, Jr., J. A. Liddle, and D. D. Bayse. 1978. "Exposure to Polychlorinated Biphenyls in Bloomington, Indiana," Report EPI-77-35-2, Public Health Service, Center for Disease Control, Atlanta.

Lee, C. Y., W. F. Shipe, Jr., L. W. Naylor, C. A. Bache, P. C. Wszolek, W. H. Gutenmann and D. J. Lisk. 1980. "The Effect of a Domestic Sewage Sludge Amendment to Soil on Heavy Metals, Vitamins, and Flavor in Vegetables," *Nutr. Rep. Int.* 21:733-738.

Lee, D., and G. Matrone. 1969. "Fe and Cu Effect on Serum Ceruloplasmin Activity of Rats with Zn-Induced Cu Deficiency," *Proc. Soc. Exp. Biol. Med.* 130:1190-1194.

Lepper, G. W. 1978. *Managing the Heavy Metals on the Land* (New York: Marcel Dekker, Inc.).

Lepow, M. L., L. Bruckman, M. Gillette, S. Markowitz, R. Robino, and J. Kapish. 1975. "Investigations into Sources of Lead in the Environment of Urban Children," *Environ. Res.* 10:415-426.

L'Estrange, J. L. 1979. "The Performance and Carcass Fat Characteristics of Lambs Fattened on Concentrate Diets. 4. Effects of Barley Fed Whole, Ground or Pelleted and of a High Level of Zinc Supplementation," *Ir. J. Agric. Res.* 18:173-182.

Levander, O. A. 1979. "Lead Toxicity and Nutritional Deficiencies," *Environ. Health Perspect.* 29:115-125.

Lichtenstein, E. P., and K. R. Schulz. 1965. "Residues of Aldrin and Heptachlor in Soils and Their Translocation into Various Crops," *J. Agric. Food Chem.* 13:57-63.

Lichtenstein, E. P., G. R. Myrdal, and K. R. Schulz. 1964. "Effect of Formulation and Mode of Application of Aldrin on the Loss of Aldrin and Its Epoxide from Soils and Their Translocation into Carrots," *J. Econ. Entomol.* 57:133-136.

Lichtenstein, E. P., G. R. Myrdal, and K. R. Schulz. 1965. "Absorption of Insecticidal Residues from Contaminated Soils into Five Carrot Varieties," *J. Agric. Food Chem.* 13:126-133.

Linne, C., and R. Martens. 1978. "Examination of the Risk of Contamination by Polycyclic Aromatic Hydrocarbons in the Harvested Crops of Carrots and Mushrooms after the Application of Composted Municipal Refuse," (in German), *Z. Pflanzenernaehr. Bodenk.* 141:265-274.

Lisk, D. J. 1972. "Trace Metals in Soils, Plants, and Animals," *Adv. Agron.* 24:267-325.

Loneragen, J. F. 1975. "The Availability and Absorption of Trace Elements in Soil-Plant Systems and Their Relation to Movement and Concentrations of Trace Elements in Plants," in *Trace Elements in Soil-Plant-Animals Systems*, D. J. D. Nicholas and A. R. Egan, Eds. (New York: Academic Press, Inc.), pp. 109-134.

Mahaffey, K. R., and J. E. Vanderveen. 1979. "Nutrient-Toxicant Interactions: Susceptible Populations," *Environ. Health Perspect.* 29:81-87.

Majeti, V. A., and C. S. Clark. 1981. "Health Risks of Organics in Land Application," *J. Environ. Eng. Div. Proc. ASCE* 107:339-357.

Matrone, G. 1974. "Chemical Parameters in Trace-Element Antagonisms," in *Trace Element Metabolism in Animals-2,* W. G. Hoekstra et al., Eds. (Baltimore, MD: University Park Press), pp. 91-103.

Mayland, H. F., A. R. Florence, R. C. Rosenau, V. A. Lazar, and H. A. Turner. 1975. "Soil Ingestion by Cattle on Semiarid Range as Reflected by Titanium Analysis of Feces," *J. Range Manage.* 28:448-452.

Mayland, H. F., G. E. Shewmaker, and R. C. Bull. 1977. "Soil Ingestion by Cattle Grazing Crested Wheatgrass," *J. Range Manage.* 30:264-265.

McClaslin, B. D., and V. L. Rodriguez. 1978. "Gamma Irradiated Digested Sewage Sludge as Micronutrient Fertilizer on Calcareous Soil," *Agron. Abstr.* 1978:30-31.

McGhee, F., C. R. Creger, and J. R. Couch. 1965. "Copper and Iron Toxicity," *Polut. Sci.* 44:310-312.

Mertz, D., T. Edward, D. Lee, and M. Zuber. 1981. "Absorption of Aflatoxin by Lettuce Seedlings Grown in Soil Adulterated with Aflatoxin B_1," *J. Agric. Food Chem.* 29:1168-1170.

Mertz, D., D. Lee, M. Zuber, and E. Lillehoj. 1980. "Uptake and Metabolism of Aflatoxin by *Zea mays*," *J. Agric. Food Chem.* 28:963-966.

Mills, C. F. 1974 "Trace-Element Interactions: Effects of Dietary Composition on the Development of Imbalance and Toxicity," *Trace Element Metabolism in Animals-2,* W. G. Hoekstra, Ed. (Baltimore, MD: Universite Park Press), pp. 79-90.

Mills, C. F. 1978. "Heavy Metal Toxicity and Trace Element Imbalance in Farm Animals," *World Congr. Anim. Feed* 7:275-281.

Mills, C. F., and A. C. Dalgarno. 1972. "Copper and Zinc Status of Ewes and Lambs Receiving Increased Dietary Concentrations of Cadmium," *Nature* 239:171-173.

Mills, C. F., I. Bremner, T. T. El-Gallad, A. C. Dalgarno, and B. W. Young. 1978. "Mechanisms of the Molybdenum/Sulphur Antagonism of Copper Utilization by Ruminants," *Trace Element Metabolism in Man and Animals*-3, M. Kirchgessner, Ed. (Baltimore, MD: University Park Press), pp. 150-158.

Moza, P., I. Schuenert, W. Klein, and F. Korte. 1979. "Studies with 2,4′,5-Trichlorobiphenyl-^{14}C and 2,2′,4,4′,6-Pentachlorobiphenyl-^{14}C in Carrots, Sugar Beets and Soil," *J. Agric. Food Chem.* 27:1120-1124.

Moza, P., I. Weisgerber, and W. Klein. 1976. "Fate of 2,2′-Dichlorobiphenyl-^{14}C in Carrots, Sugar Beets and Soil Under Outdoor Conditions," *J. Agric. Food Chem.* 24:881.

Muller, H. 1976. "Aufnahme von 3,4-Benzpyren Durch Nahrungspflanzen aus Kunstlich Angereicherten Substraten," *Z. Pflanzenernaehr. Bodenk.* 139:685-695.

National Research Council. 1980a. *Lead in the Human Environment* (Washington, DC: National Academy of Sciences).

National Research Council. 1980b. *Mineral Tolerance of Domestic Animals* (Washington, DC: National Academy of Sciences).

Neudecker, C. 1978. "Toxicological Long-Term Animal Feeding Studies on Carrots Cultivated with Composted Garbage or Sewage Sludge" (in German), *Qual. Plant.* 28:119-134.

Olsen, S. R. 1972. "Micronutrient Interactions," *Micronutrients in Agricul-*

ture, J. J. Mortvedt, P. M. Giordano, and W. L. Lindsay, Eds. (Madison, WI: Soil Science Society of America), pp. 243-264.

Ott, E. A., W. H. Smith, R. B. Harrington, and W. M. Beeson. 1966. "Zinc Toxicity in Ruminants. II. Effect of High Levels of Dietary Zinc on Gains, Feed Consumption, and Feed Efficiency of Beef Cattle," *J. Anim. Sci.* 25:419-423.

Page, A. L. 1974. "Fate and Effects of Trace Elements in Sewage Sludge When Applied to Agricultural Lands. A Literature Review Study," EPA-670/2-74-005. U.S. Environmental Protection Agency.

Parker, W. H., and T. H. Rose. 1955. "Molybdenum Poisoning (Teart) Due to Aerial Contamination of Pastures," *Vet. Rec.* 67:276-279.

Reid, R. L., and D. J. Horvath. 1980. "Soil Chemistry and Mineral Problems in Farm Livestock: A Review," *Anim. Feed Sci. Technol.* 5:95-167.

Rice, C., A. Fischbein, R. Lilis, L. Sarkozi, S. Kon, and I. J. Selikoff. 1978. "Lead Contamination in the Homes of Employees of Secondary Lead Smelters," *Environ. Res.* 15:375-380.

Roels, H. A., J. P. Buchet, R. R. Lauwreys, P. Bruaux, F. Claeys-Thoreau, A. Lafontaine, and G. Verduyn. 1980. "Exposure to Lead by the Oral and the Pulmonary Routes of Children Living in the Vicinity of a Primary Lead Smelter," *Environ. Res.* 22:81-94.

Sayre, J. W., E. Charney, J. Vostal, and I. B. Pless. 1974. "House and Hand Dust as a Potential Source of Childhood Lead Exposure," *Am. J. Dis. Child.* 127:167-170.

Seckbach, J. 1968. "Studies on the Deposition of Plant Ferritin as Influenced by Iron Supply to Iron-Deficient Beans," *J. Ultrastruct. Res.* 22:413-423.

Seckbach, J. 1969. "Iron Content and Ferritin in Leaves of Iron Treated *Xanthium pennsylvanicum* Plants," *Plant Physiol.* 44:816-820.

Shacklette, H. T., J. A. Erdman, T. F. Harms, and C. S. E. Papp. 1978. "Trace Elements in Plant Foodstuffs," in *Toxicity of Heavy Metals in the Environment*, F. W. Oehme, Ed. (New York: Marcel Dekker, Inc.), pp. 25-68.

Shellshear, I. D., L. D. Jordan, D. J. Hogan, and F. T. Shannon. 1975. "Environmental Lead Exposure in Christchurch Children: Soil Lead a Potential Hazard," *N.Z. Med. J.* 81:382-386.

Siegfried, R. 1975. "The Influence of Refuse Compost on the 3,4-Benzpyrene Content of Carrots and Cabbage," (in German), *Naturwissenschaften* 62:300.

Spence, J. A., N. F. Suttle, G. Wendham, T. El-Gallad, and I. Bremner. 1980. "A Sequential Study of the Skeletal Abnormalities Which Develop in Rats Given a Small Dietary Supplement of Ammonium Tetrathiomolybdate," *J. Comp. Path.* 90:139-153.

Standish, J. F., and C. B. Ammerman. 1971. "Effect of Excess Dietary Iron as Ferrous Sulfate and Ferric Citrate on Tissue Mineral Composition of Sheep," *J. Anim. Sci.* 33:481-484.

Standish, J. F., C. B. Ammerman, C. F. Simpson, F. C. Neal, and A. Z. Palmer. 1969. "Influence of Graded Levels of Dietary Iron, as Ferrous Sulfate, on Performance and Tissue Mineral Composition of Steers," *J. Anim. Sci.* 29:496-503.

Standish, J. F., C. B. Ammerman, A. Z. Palmer, and C. F. Simpson. 1971. "Influence of Dietary Iron and Phosphorus on Performance, Tissue Mineral Composition and Mineral Absorption in Steers," *J. Anim. Sci.* 33: 171-178.

Stara, J., W. Moore, M. Richards, N. Barkley, S. Neiheisel, and K. Bridbord. 1973. "Environmentally Bound Lead. III. Effects of Source on Blood and Tissue Levels of Rats," in *EPA Environmental Health Effects Research Series,* A-670/1-73-036, pp. 28-29.

Strek, H. J., J. B. Weber, P. J. Shea, E. Mrozek, Jr., and M. R. Overcash. 1981. "Reduction of Polychlorinated Biphenyl Toxicity and Uptake of Carbon-14 Activity by Plants Through the Use of Activated Carbon," *J. Agric. Food Chem.* 29:288-293.

Suttle, N. F., and C. F. Mills. 1966. "Studies of the Toxicity of Copper to Pigs. I. Effects of Oral Supplement of Zinc and Iron Salts on the Development of Copper Toxicosis," *Br. J. Nutr.* 20:135-148.

Suttle, N. F., B. J. Alloway, and I. Thornton. 1975. "An Effect of Soil Ingestion on the Utilization of Dietary Copper by Sheep," *J. Agric. Sci.* 84: 249-254.

Suzuki, M., N. Aizawa, G. Okano, and T. Takahashi. 1977. "Translocation of Polychlorobiphenyls in Soil into Plants: A Study by a Method of Culture of Soybean Sprouts," *Arch. Environ. Contam. Toxicol.* 5:343-352.

Thornton, I., and D. G. Kinniburgh. 1978. "Intake of Lead, Copper, and Zinc by Cattle from Soil and Pasture," in *Trace Element Metabolism in Man and Animals,* M. Kirchgessner, Ed. 3:499.

Underwood, E. J. 1977. *Trace Elements in Human and Animal Nutrition*, 4th ed. (New York: Academic Press).

U.S. Environmental Protection Agency. 1979. "Criteria for Classification of Solid Waste Disposal Facilities and Practices," *Fed. Regist.* 44(179): 53438-53464.

U. S. Environmental Protection Agency. 1980. "Hazardous Waste Management System: Standards for Owners and Operators of Hazardous Waste Treatment, Storage, and Disposal Facilities," *Fed. Regist.* 45(98):33154-33258.

Vlek, P. L. G., and W. L. Lindsay. 1977a. "Thermodynamic Stability and Solubility of Molybdenum Minerals in Soils," *Soil Sci. Soc. Am. J.* 41: 42-46.

Vlek, P. L. G., and W. L. Lindsay. 1977b. "Molybdenum Contamination in Colorado Pasture Soils," *Molybdenum in the Environment, Vol. 2,* W. R. Chappell and K. K. Petersen, Eds. (New York: Marcel Dekker, Inc.), pp. 619-650.

Wagner, K. H., and I. Siddiqi. 1971. "Die Speicherung von 3,4-Benzfluoranthen in Sommerweizen und Sommerrogen," *Z. Pflanzenernaehr. Bodenk.* 127:211-218.

Walsh, L. M., D. R. Stevens, H. D. Seibel, and G. G. Weis. 1972b. "Effects of High Rates of Zinc on Several Crops Grown on an Irrigated Plainfield Sand," *Commun. Soil Sci. Plant Anal.* 3:187-195.

Walsh, L. M., M. E. Sumner, and R. B. Corey. 1976. "Consideration of Soils for Accepting Plant Nutrients and Potentially Toxic Nonessential Elements," in *Land Application of Waste Materials* (Ankeny, IA: Soil Conservation Society of America), pp. 22-47.

Warnock, R. E. 1970. "Micronutrient Uptake and Mobility Within Corn Plants (*Zea mays* L.) in Relation to Phosphorous-Induced Zinc Deficiency," *Soil Sci. Soc. Am. Proc.* 34:765-769.

Weber, J. B., and E. Mrozek, Jr. 1979. "Polychlorinated Biphenyls: Phytotoxicity, Absorption, and Translocation by Plants, and Inactivation by Activated Charcoal," *Bull. Environ. Contam. Toxicol.* 23:412-417.

Wedeen, R. P., D. K. Mallik, V. Batuman, and J. D. Bogden. 1978. "Geophagic Lead Nephropathy: Case Report," *Environ. Res.* 17:409-415.

Willoughby, R. A., E. MacDonald, B. J. McSherry, and G. Brown. 1972. "Lead and Zinc Poisoning and the Interaction Between Pb and Zn Poisoning in the Foal," *Can. J. Comp. Med.* 36:348-359.

Willoughby, R. A., and W. Oyaert. 1973. "Zinc Poisoning in Foals," (in Dutch) *Vlaams Diergeneeskd. Tijdschr.* 42:134-143.

Yoneyama, T. 1981. "Detection of N-Nitrosodimethylamine in Soils Amended with Sludges," *Soil Sci. Plant Nutr.* 27:249-253.

DISCUSSION

***Terry Logan*, Ohio State:** Rufus, are you implying that perhaps in our analytical scheme for sludges, we ought to be looking more closely at selenium and molybdenum? We really haven't done that routinely.

R. Chaney: It has always been a recommendation from our laboratory that on the once-a-year kind of sludge analysis, the more comprehensive analysis, one had a analyze for things like selenium, molybdenum, and fluoride—elements and organics which are industrial pollutants but are very rarely found at high levels in sludge. There has been one high-molybdenum sludge found in Ontario; of all that have been reported, I think there are only one or two high in molybdenum.

Terry Logan: That would have to be on a composited sample then.

R. Chaney: Yes, on a once-a-year basis. And, then, when the once-a-year sample starts showing that you have a pollution source causing a high sludge residue for something that could potentially be a problem in the way you're managing your sludge, you have to change your plans. But on the other hand, we don't want to tell people to be spending a fortune analyzing selenium and molybdenum, etc. when we have so many domestic sludges.

J. Hansen: Two questions. One, considering the worst cadmium case, could then calcium, zinc, or other trace-element deficiencies occur at a later life stage, which in turn would cause a cadmium toxicity problem? And two, considering nickel uptake in plants, is not skin disease rather than toxicity the real problem, and then at a lower intake level?

R. Chaney: Nickel deficiency does not occur in practical agriculture or food chains. Nickel is always above deficient levels.

J. Hansen: It was not a deficiency of nickel that was a problem but rather too high nickel intake. You mentioned toxicity. Isn't nickel toxicity related to a skin disease arising from high intake of nickel?

R. Chaney: Not from dietary. The skin disease is topical application like chromate eczema. The cadmium question—as far as we can tell, and I think the data are getting more and more conclusive—there is no evidence of any health effect until you get to the point where the proximal tubules are injured and you start seeing low-molecular-weight proteins. If anything you have done across your lifetime caused you to reach that amount of cadmium, then you are in trouble. I know of no evidence showing that later Ca or Zn deficiency has any affect on previously adsorbed (and retained) Cd.

CHAPTER 17

SIMULATION STUDY OF THE VOLATILIZATION OF POLYCHLORINATED BIPHENYLS FROM LANDFILL DISPOSAL SITES

Charles Springer, Louis J. Thibodeaux, and Shrikrishna Chatrathi

Department of Chemical Engineering
University of Arkansas
Fayetteville, Arkansas 72701

ABSTRACT

Computer simulation studies were made of the emission of polychlorinated biphenyls from landfill disposal sites. The model included mechanisms for the migration of vapor through the landfill covering (cap) by (1) diffusion within the pore spaces, (2) barometric pressure pumping because of fluctuating barometric pressure, and (3) sweeping flow of gases produced by biodegradation of other waste materials.

Increased cap depth and decrease cap porosity have the expected result on the vapor emission rate in the absence of biologically produced sweep gas, but the emission rate is approximately proportional to the rate of gas production.

The vapor emission rate of polychlorinated biphenyls from landfills may be expected to be quite low in any event, but the accompanying deposition of biodegradable materials may have undesirable results caused by the formation of gases from biological decomposition processes.

INTRODUCTION

Landfill disposal of polychlorinated biphenyls (PCB) has been used in the past for disposal of condensers, transformers, or other electrical gear containing PCB as a dielectric fluid. Likewise, landfilling has been used to dispose of PCB-contaminated oils and other substances.

Although satisfactory chemical destruction methods now exist for PCB and combustible oils contaminated with PCB, alternatives to landfilling for noncombustible, contaminated materials are few. Specifically, contaminated sludge from rivers and estuaries might conceivably be dredged and landfilled to facilitate cleanup of the affected waterways (Tofflemire and Quinn, 1979).

The purpose of this study was to estimate by computer simulation the volatilization rate from the landfill disposal of PCB as a "pure" material and the landfill disposal of a contaminated riverbed silt.

The results of simulations of this type can be of material assistance in planning future action and in establishing protocols for effective and safe disposal.

MATHEMATICAL MODEL

Thibodeaux et al. (1981) have recently developed models of the processes by which vapors can be evolved from landfills. Three processes are considered in the modeling: (1) vapor phase molecular diffusion, (2) diffusion enhancement by fluctuating barometric pressure (barometric pumping), and (3) sweeping by biologically generated gas venting. If little or no biodegradable material is present with the volatilizing chemical, then the third mechanism will be inoperative.

For a uniform composition of gas within a landfill cell, an equation of continuity (conservation of mass) can be written:

$$\epsilon_c \frac{d\rho}{dt} = -\frac{\rho v}{h_c} + r_g \tag{1}$$

where ϵ_c = cell porosity
h_c = cell depth, cm
ρ = cell gas density, g/cm^3
v = superficial outward velocity, cm/s
t = time, s
r_g = rate of gas generation in the cell, g/cm^{-3}s^{-1}

An equation to describe the unidirectional motion of a gas flowing through porous media can be obtained from Darcy's law:

$$v = \frac{K}{\mu L}(P - \pi) \tag{2}$$

where
K = permeability of the covering material, darcies ($cm^2 cP s^{-1} atm^{-1}$)
μ = gas viscosity, cP
L = thickness of the covering, cm
P = pressure within the cell, atm
π = barometric pressure

The biological gas generation rate is usually expressed as r_b, a gas generation rate per unit mass of material in the cell, $cm^3/g^{-1} \cdot s^{-1}$. This may be converted to volumetric gas rate, r_g, for specific cell conditions by:

$$r_g = r_b \rho_c \rho \tag{3}$$

where ρ is the bulk density of material in the cell.

Combining Equations 1, 2, and 3 while solving for the cell gas density by the ideal gas law results in an expression for the rate of change of cell pressure:

$$\frac{dP}{dt} = \frac{r_b \rho_{wp}}{\epsilon_c} - \frac{KP(P - \pi)}{\epsilon_c h_c L \mu} \tag{4}$$

Since the barometric pressure π varies in time, the cell pressure P will also vary, but it will lag because of the capacity of the cell. The variation of cell pressure with atmospheric pressure will not be linear, even if the biogas generation is zero.

The rate of transfer of a single component, A, of a mixture for unidirectional flow without reaction or adsorption on the pore space walls, can be related to the diffusion coefficient of the component by the following continuity equation:

$$\frac{v d\rho_A}{dy} = D_{A,P} \frac{d^2 \rho_A}{dy^2} \tag{5}$$

where A = component A; thus, ρ_A = concentration of component A in the vapor phase in the cell, g/cm^3
$D_{A,P}$ = diffusivity of component A in the pore spaces of the covering material, cm^2/s
y = vertical distance into the cover

Thibodeaux (1981) has shown the integration of Equation 5 as

$$N_A = \frac{D_{A,P}}{L} \rho_A^* \left(\frac{R \exp R}{\exp R - 1}\right) \tag{6}$$

where $R = Lv/D_{A,P}$
N_A = outward flux of component A, $g/cm^2 \cdot s$
ρ_A^* = equilibrium concentration of the component in the cell, g/cm^3

The model assumes that the vapor within the cell is saturated with respect to component *A* at all times.

The simulation was performed by simultaneously solving Equations 2, 4, and 6 to determine the flux of the component of interest. The simulation was carried out by using the IBM Continuous Systems Modeling Program.

SIMULATION PARAMETERS

Two PCB disposal scenarios were assumed in this study. In the first scenario, a quantity of liquid PCB, perhaps 1 or 2 gal, is disposed of at the bottom of an excavation. The liquid is assumed to be in a container or in a piece of discarded electrical gear, and the excavation is backfilled with a medium-porosity soil or other material, including other types of waste. The backfill constitutes a cell and is covered with a layer of controlled-permeability soil for a cap. The excavation walls and floor are assumed to be of negligible permeability. The concept of such a disposal is shown in Figure 1. As long as the container remains intact and does not leak, no volatilization will occur. However, corrosion may eventually cause it to perforate, and PCB will vaporize.

The second type of disposal considered is that of a soil, such as a riverbed sludge, uniformly contaminated with PCB in the absorbed state. The sludge layer will be covered with a cap of controlled-permeability soil, as in the first case, and the walls and floor of the excava-

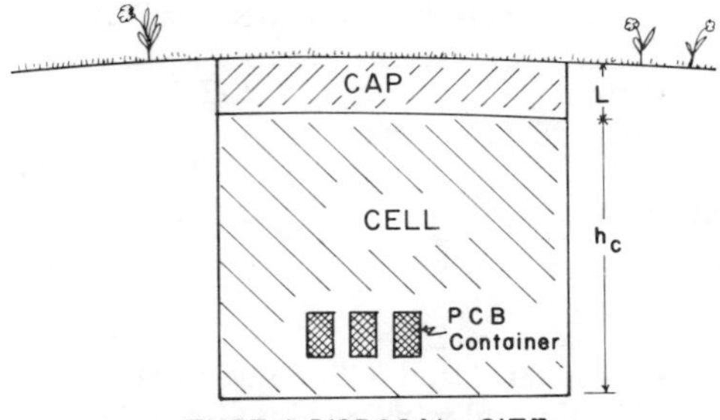

Figure 1. Conceptualization of a typical disposal of pure PCB.

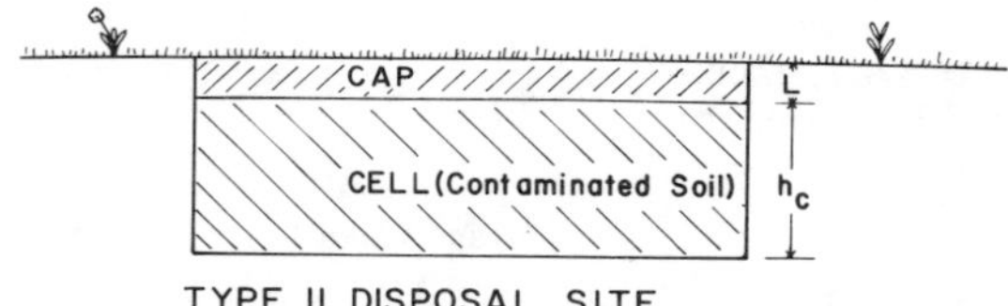

Figure 2. Conceptualization of the disposal of contaminated soil or riverbed sludge.

tion are again assumed to be essentially impermeable. Such a disposal is pictured in Figure 2.

The choice for a typical PCB was Aroclor 1248, which would be mostly a 4-Cl-substituted material with a molecular weight of about 292. The vapor pressure of this material would be expected to be about 9.2×10^{-4} mm Hg at 25°C (National Research Council, 1979).

In the second case (i.e., the PCB absorbed in riverbed sludge), the presence of PCB as a physically absorbed material is assumed to lower the equilibrium partial pressure somewhat, but little information exists to suggest a reasonable value for this decrease. In this study, the equilibrium partial pressure is assumed to be about one-half of the vapor pressure, or about 5×10^{-4} mm Hg.

To assess the importance of the presence of other wastes, assumptions are made about varying rates of biogas generation. The riverbed sludge is assumed to have fairly low biogas generation rates because only the organic component of the sludge would be responsible for the generation. However, the pure material might be in the presence of biodegradable organic wastes in varying amounts so that biogas generation rates could vary from essentially zero to nearly the maximum possible. The values of the various parameters and the ranges used in this study are shown in Table I for pure PCB disposal, in Table II for

Table I. Properties Assumed for PCB[a]

Molecular Weight	292
Vapor Pressure (25°C)	9.8×10^{-4} mm Hg (1.3×10^{-6} atm)
Diffusivity in Air	0.031 cm^2/s

[a]These properties are similar to Aroclor 1248.

Table II. Parameters Used in Simulation Study for "Pure" PCB Disposal with No Gas Generation

Constant Values			
Cell Porosity (ϵ_c)		0.2	
Cell Depth (h_c), m		3.0	
Cell Density (ρ_c), g/cm^3		2.0	
PCB Partial Pressure, atm		1.3×10^{-6}	
Varied Parameters			
	Base Case	**Other Values**	
Cap Depth (L), m	3.0	1.0	0.5
Cap Porosity (ϵ)	0.1	0.08	0.2

Table III. Parameters Used in Simulation Study for "Pure" PCB Disposal with Accompanying Biogas Generation

Constant Values			
Cell Porosity (ϵ_c)			0.2
Cell Depth (h_c), m			3.0
Cell Density (ρ_c), g/cm^3			2.0
PCB Partial Pressure, atm			1.3×10^{-6}
Varied Parameters			
	Base Case	**Other Values**	
Cap Depth (L), m	3.0	1.0	0.5
Cap Porosity (ϵ)	0.1	0.08	0.2
Gas Generation Rate (r_b), $cm^3/g \cdot s$	3×10^{-7}	2×10^{-7}	1×10^{-7}
Second Base Case (very low porosity cap)			
Cap Porosity (ϵ)	0.06		
Cap Depth (L), m	3.0	1.0	0.5

disposal in the presence of biogas generation, and in Table III for disposal of contaminated soil.

In the matter of the equilibria between the sludge and the PCB, the PCBs are known to be absored readily by soils that have a high organic content (Tofflemire et al., 1979; Tofflemire and Shen, 1979). However, adsorption isotherms are not available, and even if some adsorption data were available, it would be highly specific. Therefore, the concentration of PCB vapors in a disposal cell is quite speculative.

The barometric pressure fluctuations of a 2-week period in Fayetteville, Arkansas were assumed to be typical and were used for the simulation. The normal pressure in Fayetteville is about 1.4 in. (35 to 36 mm Hg) less than at sea level because of elevation.

For comparison, because PCB properties are quite variable, depending on the composition, two base-case simulations were run using the properties of monochloro biphenyl (MCB), which would represent an upper limit of PCB volatility and mobility.

The parameters used in the study are summarized in Tables I through IV.

Table IV. Parameters Used in Simulation Study of Contaminated Soil Disposal

Constant Values			
Cell Porosity (ϵ_c)		0.2	
Cell Depth (h_c), m		10.0	
Gas Generation Rate (P_c), $cm^3/g \cdot s$		5×10^{-8}	
PCB Partial Pressure, atm		5.3×10^{-7}	
Varied Parameters			
	Base Case	**Other Values**	
Cap Porosity (ϵ)	0.1	0.08	0.2
Cap Depth (L), m	1.0	2.0	0.5

SIMULATION RESULTS

The results of the simulations are shown in Tables V, VI, and VII for the cases of disposal with gas generation, without gas generation, and with contaminated sludge, respectively.

As expected, decreased cap porosity and increased cap depth reduces the level of PCB emissions. However, in the presence of biologically generated sweep gas, the properties of the cap have much less effect on the emission rates; when some gas generation exists, the emission will be nearly proportional to the gas generation rate.

The simulations also suggest some significant pressure buildup when biogas generation occurs beneath a relatively tight (low-porosity or increased-depth) cap. Some of the computed pressures are shown in Table VIII.

Table V. Simulation Results—
"Pure" PCB Disposal with Biogas Generation

Cap Parameters		Gas Rate	Flux (mg/m²·d)	
Porosity	Depth (m)	(cm³/g·s)	PCB	MCB
0.1	3.0	3×10^{-7}	2.42	95.0
		2×10^{-7}	1.6	
		1×10^{-7}	0.8	
0.1	3.0	3×10^{-7}	2.42	
	1.0		2.46	
	0.5		2.51	
0.1	3.0	3×10^{-7}	2.42	
0.2			2.65	
0.08			2.40	
0.06	3.0		2.31	
	1.0		2.36	
	0.5		2.39	

Table VI. Simulation Results—
"Pure" Disposal Without Biogas Generation

Cap Parameters		Flux (mg/m²·d)	
Porosity	Depth (m)	PCB	MCB
0.1	3.0	0.08	5.2
0.08		0.05	
0.2		0.33	
0.1	3.0	0.08	
	1.0	0.24	
	0.5	0.45	

Table VII. Simulation Results—
Disposal of Contaminated Sludge[a]

Cap Parameters		Flux	Cell Pressure
Porosity	Depth (m)	(mg/m²·d)	(atm)
0.1	1.0	0.53	1.2
0.08		0.30	2.0
0.2		0.91	1.0
0.1	1.0	0.53	1.2
	0.5	0.62	1.1
	2.0	0.48	1.4

[a]Gas generation rate assumed to be 5×10^{-8} cm³/g·s.

Table VIII. Simulation Results—
Cell Pressures Developed with Biogas Generation

Cap Parameters		Gas Generation Rate ($cm^3/g \cdot s$)	Maximum Cell Pressure (atm)
Porosity	Depth (m)		
0.1	3.0	3×10^{-7}	2.1
		2×10^{-7}	1.7
		1×10^{-7}	1.3
0.1	3.0	3×10^{-7}	2.1
	1.0		1.3
	0.5		1.2
0.1	3.0	3×10^{-7}	2.1
0.2			1.0
0.08			9.2
0.06	3.0	3×10^{-7}	110
	1.0		36.7
	0.5		19.0

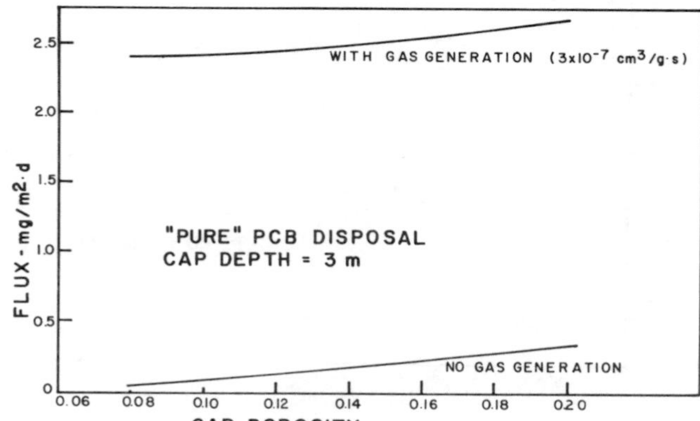

Figure 3. Effect of cap porosity on the emission rate of PCB from a type I or pure PCB disposal. This figure shows the results of both cases, with and without the accompanying generation of gases.

The effects of the variables are shown graphically in Figures 3 through 7.

CONCLUSIONS

Probably the most important finding of this study is that the emissions of PCB from landfills may be expected to be quite small. If the

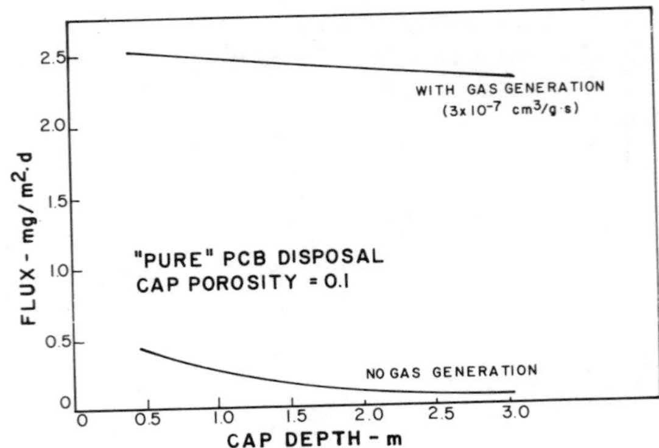

Figure 4. Effect of cap depth on the emission rate of PCB from a disposal pure PCB. This figure shows the results of both cases, with and without the accompanying generation of gases.

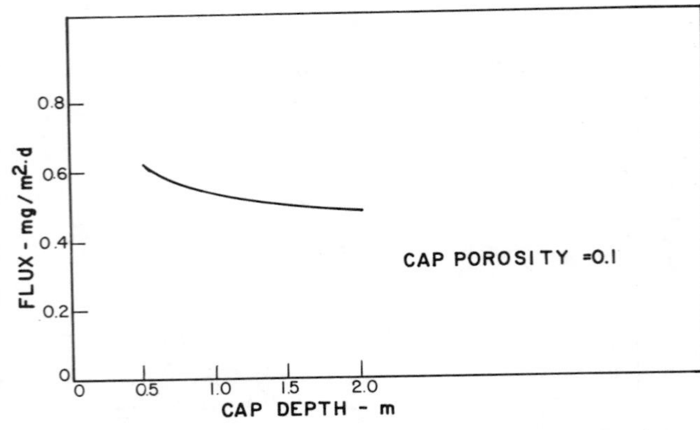

Figure 5. Effect of cap depth on the emission rate of PCB from a disposal of contaminated sludge or soil. The simulation parameters assume a low rate of accompanying gas generation.

environment can tolerate any PCB vapor emission at all, then any reasonable landfill that confines the material under a meter or more of soil will provide for adequately low emissions. Typical emissions from a 100- by 100-m landfill would be about 1 kg/year when there is no sweep gas source.

Although it may not be readily apparent that the computed emissions are small, the effects of these emissions can be conceived by considering possible airborne concentrations that might result in a particular circumstance.

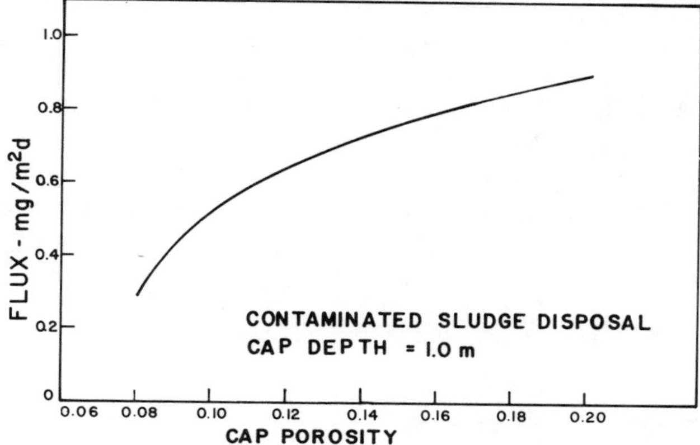

Figure 6. Effect of cap porosity on the emission of PCB from a disposal of contaminated sludge or soil. The simulation parameters assume a low rate of accompanying gas generation.

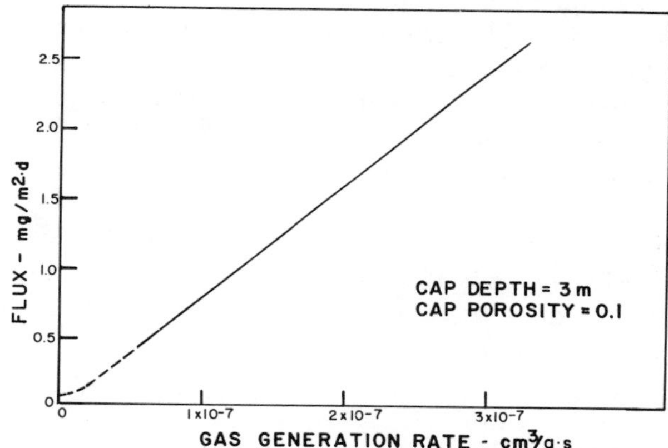

Figure 7. Effect of gas generation rate on the emission of PCB from a disposal site.

Given the gas generation base-case emission of about 2.4 $mg\,m^{-2}\,d^{-1}$, one can deduce that an airborne concentration of about 0.02 $\mu g/m^3$ would result at a location 100 m downwind in a typical 1- to 2-m s^{-1} wind. This concentration is about the lower limit of our ability to detect PCB in air. For comparison, the National Institute of Occupational Safety and Health (NIOSH) standard for airborne PCB specifies that it should be no more than 0.1 $\mu g/m^3$ for an 8-h exposure.

One aspect that does not often receive much attention is the consequences of biogas generation under a well-compacted cap or perhaps, a plastic film barrier. The simulation showed that excessive pressure can be generated under a low-porosity and deep cap. Although the actual pressures in Table VII would not be expected to occur, the simulation results clearly show that sufficient pressures can be generated to crack or otherwise compromise nearly any cap that might be placed over a landfill. Therefore, neither PCB nor any hazardous chemical should be deposited in a landfill with material that might biodegrade to form a gas unless some provision is made for venting to relieve the pressure.

SIMULATION LIMITATIONS

The model presented earlier has been confirmed by experimentation and field measurements to only a limited extent. Those measurements that have been made, however, suggest that the model is a reasonably accurate description. In application to an actual site, however, one might have considerable difficulty evaluating the various parameters that control the emission rate.

Some variables are not accounted for in the model, especially any adsorption of the volatile matter in the cap material. If this adsorption were significant, it could reduce, rather than increase, the initial emission rate.

Moisture in the soil can only be accounted for by the effect it would have on porosity. There is little question that an increased soil moisture content would reduce the emission rate, although decreased porosity may not be the only effect.

RECOMMENDATIONS

Materials of low volatility which are disposed of in properly constructed landfills will present little hazard from vapor emissions to the air. Care must be taken, however, when constructing any hazardous chemical landfill, to limit the amount of biodegradable material in the same site because gas produced can either greatly increase the emission rates by sweeping action or generate pressure that might rupture the cap and destroy its integrity.

Near-surface landfills may not be suitable repositories for relatively volatile, hazardous wastes. The possible limitations for landfill disposal of volatile materials require more study because emission hazards depend heavily on the properties of the hazardous materials and on the nature and properties of accompanying materials.

REFERENCES

National Research Council. 1979. "Polychlorinated Biphenyls," Report by the Committee on the Assessment of Polychlorinated Biphenyls in the Environment, National Academy of Science, Washington, DC.

Thibodeaux, L. J. 1981. "Estimating the Air Emissions of Chemicals from Hazardous Waste Landfills," *J. Hazardous Mater.* 4:235-244.

Thibodeaux, L. J., C. Springer, and L. M. Riley. 1981. "Models of Mechanisms for Hazardous Chemical Emissions from Landfills," Symposium on Toxic Substances Management Programs, 181st American Chemical Society National Meeting, Atlanta.

Tofflemire, T. J., and S. O. Quinn. 1979. "PCB in the Upper Hudson River: Mapping and Sediment Relations," New York State Department of Environmental Conservation, Technical Paper No. 56.

Tofflemire, T. J., and T. T. Shen. 1979. "Volatilization of PCB from Sediment and Water: Experimental and Field Data," in *Proceedings of the 11th Annual Industrial Waste Conference, University Park, PA.*

DISCUSSION

***Harry Freeman*, Office of Appropriate Technology, State of California:** I have a question about other volatiles in landfills. In California, we are getting more concerned about putting volatiles in landfills to the point that some recommendations are going to come out shortly, probably to start banning halogenated solvents and things. Do you have any comment to make about the effect of the ambient air volatility around landfills, or is there any way we can translate your findings and conclusions to substances with higher vapor pressures?

C. Springer: Yes, we can translate those conclusions. I told you, of course, that we haven't been working on this very long, and we don't have a big data base from which to work. But as a matter of fact, compounds with a high vapor pressure will get out; they will in fact diffuse through the cap. They may also diffuse through horizontal sand layers and so forth. Our model clearly shows that compounds of higher vapor pressure will, in fact, escape from the landfill. We did one simulation on benzene. For example, a 55-gal drum of benezene under a capped landfill, roughly equivalent to what I've been talking about, would have a lifetime of a month so so. Of course, all of this varies quite a bit with the kind of cap over the landfill and whether or not there is an allowance for any sweep gas and so forth.

***David Wilson*, Harwell:** I've been reading your papers in the literature for some time with considerable interest, and I have several questions I would like to take the opportunity to ask you. First of all, how far have your results been verified against actual field experience?

C. Springer: Well, we tried to verify the model with one set of field measurements, and it compared about as well as you might except. Our emission rates were within a factor of 1.5 of what was apparent. Now, we took this from some other measurements, not our own measurements. We had not had a chance to make our own measurements. We hope to in the future. One problem with this model, of course, is evaluating those parameters. How do you evaluate these parameters in an actual site? We can make intelligent guesses, but we don't know for sure. We also find that the only provision that our model has for the soil moisture content would be the effect it might have on the porosity. There may be some other effects for some chemicals; they would probably be rather minor, but there might be some effect. In verification of a model like this, a factor of 1.5 is probably not bad.

David Wilson: I would also like to ask you a couple of questions about the assumptions underlying your work, but I find it a little difficult to see just what you are doing. The first question is, are you assuming circulation of air through the wastes in the absence of gas generation? And the second question, for your high rates of gas generation you are saying that high rate of gas generation associated with biodegradable wastes is a bad thing. If PCB is disposed of with municipal refuse or indeed with contaminated soil, then we have already heard this morning that the PCB will be strongly adsorbed onto the waste matrix; have you taken this adsorption into account in your calculations?

C. Springer: The answer to the first question is simply no. And the answer to the second question is that adsorption which very likely would occur will not affect the ultimate steady-state emission rate. Once the soil through which the gas is passing has been saturated, no more PCB will be adsorbed, and it will go right on through. In a steady state, we feel there would be no effect; in the unsteady state, there would be a profound effect.

CHAPTER 18

MULTIMEDIA MODELING OF TRANSPORT AND TRANSFORMATION OF CONTAMINANTS

James W. Falco
Exposure Assessment Group
Office of Research and Development
U.S. Environmental Protection Agency
401 M Street, SW
Washington, DC 20460

Lee A. Mulkey
Environmental Research Laboratory
Office of Research and Development
U.S. Environmental Protection Agency
Athens, Georgia 30613

John Schaum
Exposure Assessment Group
Office of Research and Development
U.S. Environmental Protection Agency
401 M Street, SW
Washington, DC 20460

ABSTRACT

Passage of the Resource Conservation and Recovery Act (RCRA) and the Comprehensive Environmental Compensation and Liability Act has focused research efforts on environmental problems associated with the disposal of solid wastes. Methods for predicting human and environmental exposure to chemicals disposed of in landfills are being investigated. Multimedia models for predicting environmental concentrations are being developed to provide a means of estimating environmental concentrations of substances released into the environment from landfills.

The role of multimedia models in estimating exposure to pollutants released from landfills is defined in this chapter. Major issues addressed include optimization of temporal and spatial scales of application in each medium, and data requirements to calibrate and use the components of exposure assessment.

Major research gaps that must be addressed to improve exposure estimates are (1) lack of knowledge of chemical transformation pathways and rates within the landfill and groundwater; (2) validation and testing of existing multimedia models; and (3) linkage of these models with other analyses required to estimate exposure.

INTRODUCTION

Exposure assessments have been conducted as part of risk assessments for many problems. Recently, this area of research has aroused a great deal of interest because of its increased use in implementing regulations. As defined in proposed U.S. Environmental Protection Agency (EPA) guidelines, exposure is the contact between a subject of concern and a chemical or physical entity. The magnitude of an exposure is determined by measuring or estimating the amount of an agent available at exchange boundaries (i.e., lungs, gut and skin) during some specified time. Exposure assessment is the determination or estimation of the magnitude, frequency, duration and route of exposure.

Although the calculation of dose is not included in this definition of exposure assessment, it is often the reported result. Human health effects data are often reported in terms of dose-response curves. In such cases, dose or intake estimates are the logical endpoint for exposure assessment.

In some cases, ambient environmental concentrations as a function of time are the reported results of an exposure assessment. Some human health effects data are reported in terms of ambient concentration vs health effects, particularly when epidemiological studies are used in assessments.

Exposure assessments consider three aspects of pollutant characterization and contact with people: (1) estimation of release rates of toxic chemicals into the environment, (2) estimation of ambient environmental concentrations, and (3) estimation of dose rate. Estimating the quantities resulting from disposal of hazardous solid wastes is difficult.

In relation to evaluation of hazardous solid wastes, two types of problems are of immediate concern: contamination of the environment at high levels near a landfill and low-level contamination of groundwater aquifers at points distant from the disposal site. Near-field con-

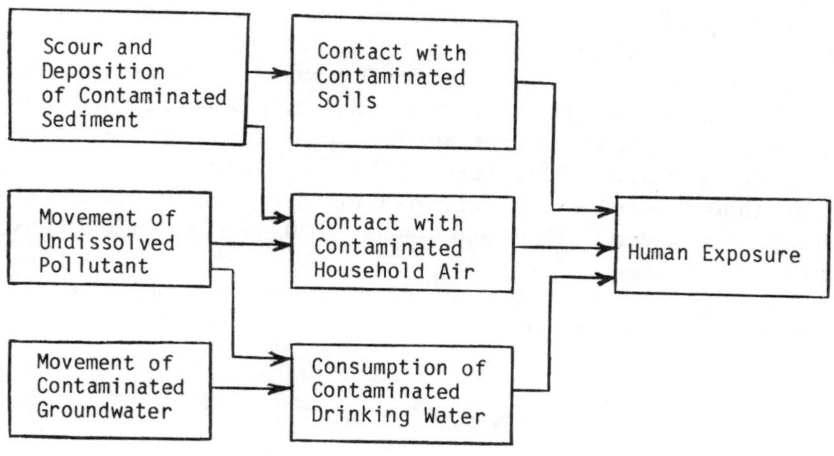

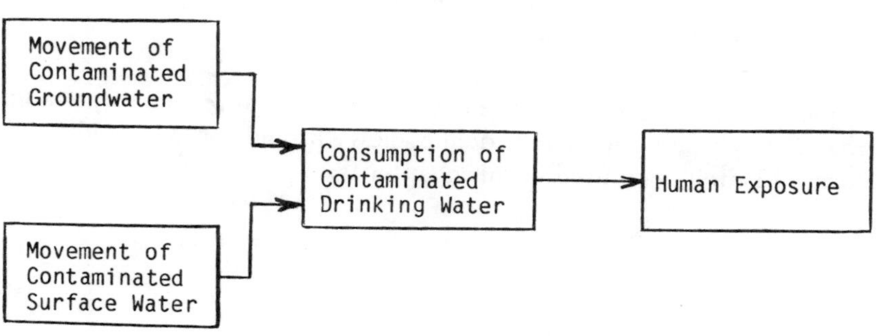

Figure 1. Opportunities for human exposure related to near- and far-field perspectives.

tamination (i.e., contamination of areas near landfills) have the greatest potential for multimedia exposure. Far-field contamination is mainly a drinking-water exposure problem. Figure 1 illustrates opportunities for human exposure related to near- and far-field perspectives.

PREDICTING GROUNDWATER CONCENTRATIONS

Estimating groundwater concentration profiles is a key element of near- and far-field evaluations. Mathematical models are being developed to predict the geographical extent and intensity of contamination. These models vary in the level of detail to which phenomena are defined, but all are based on momentum balances to predict groundwater movement and mass balances to predict pollutant movement. A review of models now available has been published by Bachmat et al. (1978), and a thorough review of the principles used in developing such models has been presented by Faust and Mercer (1980) and Mercer and Faust (1980).

To obtain estimates of groundwater concentration profiles, groundwater velocities and flowrates are estimated by solving:

$$\nabla \cdot \mathbf{K} \cdot \nabla h + R = S_s \frac{\partial h}{\partial t} \tag{1}$$

where h = hydraulic head
$\mathbf{K}$ = hydraulic conductivity tensor
R = summation of sources and sinks
S_s = specific storage

Equation 1 is the form of a momentum balance that results by assuming that Darcy's law adequately defines flow through porous media. The form of Darcy's law used in this chapter is:

$$q = \mathbf{K} \cdot \nabla h \tag{2}$$

where q = specific discharge

Appropriate boundary and initial conditions must be defined to obtain a usable solution to Equation 1. After Equation 1 has been solved for head as functions of time, depth and lateral position, Equation 2 can be used to determine specific discharge. Often, velocities are vertically averaged, and the two-dimensional forms of Equations 1 and 2 are solved.

The mass balance needed to estimate pollutant concentration is usually written as

$$\nabla \cdot \phi \mathbf{D} \cdot \nabla C - \nabla \cdot (q\,C) + R\,C_1 = \frac{\partial[(\phi C)(1 + k \cdot s)]}{\partial t} \tag{3}$$

where C = concentration
k = soil/water partition coefficient
$\mathbf{D}$ = dispersion tensor
ϕ = porosity
C_1 = concentration of pollutant in source or sink
R = volumetric source or sink flowrate
s = solid mass fraction of soil medium

Instantaneous equilibrium between sorbed and dissolved pollutant has been assumed in the derivation of Equation 3.

Three major problems relate to applying models based on these equations to groundwater problems. Solution of these equations requires computer computations in virtually all practical applications. Computation costs are very high when compared with equivalent problems for surface waters. The reason for such high cost is that both Equations 1 and 3 must be solved in three-dimensional form in many cases and at least in two-dimensional form for almost all other applications. In contrast, surface water quality models are usually one-dimensional and, at most, two-dimensional.

The second major problem is related to the cost of monitoring programs for obtaining data necessary for model calibration. In general, wells must be drilled to obtain data on the movement of groundwater and pollutants near landfills. The cost of drilling wells into deep aquifers is quite high. Even for shallow aquifers, costs are significant. Furthermore, because groundwater moves at relatively slow velocities, monitoring must be done over extended periods of time (years to decades). Again, large costs may be involved.

The third problem is specific to predicting groundwater concentration profiles in the immediate area of a landfill. Soil properties and elevation of groundwater tables vary from point to point, and local changes can be extreme. When predictions are made over large areas, this local difference averages out, and properties can be approximated by some mean value. For small areas, a compromise must be made between detail of physical description of an area and cost of computation and monitoring necessary to estimate parameters.

Although we have no immediate solution to these problems, we have proposed a development project that will attempt to address the first problem discussed. The parameters used in models to describe pollutant transport in groundwater vary over a finite range in the environment. Consequently, by varying values of parameters such as soil conductivity and porosity, the range of pollutant concentrations in groundwater for a specific release rate from a landfill can be estimated.

By running a groundwater pollutant model for several different values of environment- and chemical-specific parameters, a set of normalized pollutant concentration profiles can be generated. To evaluate a specific site, these profiles could be searched to select the profile that was produced by using parameter values that most closely approximate the site conditions and chemicals being investigated. This process would eliminate the need to rerun an expensive model for every new application.

A file of such profiles could be extremely valuable in emergency situations. Groundwater concentrations could be estimated in a few hours by searching the file and performing a few simple calculations. Because profiles depend on the sorption coefficients, maintaining a file of such partition coefficients or using structure-activity relationships to estimate such coefficients would be advantageous. A set of profiles is shown in Figure 2.

SURFACE-WATER CONTAMINATION

Outside the immediate vicinity of a solid waste disposal site, the major exposure route for a pollutant released from a landfill is exposure through drinking water. Drinking water contamination may occur through contamination of both ground- and surface-water supplies. Surface-water supplies can be contaminated directly or may receive contaminated groundwater.

Direct contamination of surface waters must be measured. Such measurements must span both wet and dry weather periods because release rates are a function of soil moisture and surface runoff patterns. After direct releases are defined, various mathematical models can be used to estimate ambient water concentrations.

Depending on the level of detail of the exposure assessment, different models may be selected. At the grossest level of detail is a simplified calculational approach developed by Falco et al. (in press). Information needed includes definition of degradation processes and rate coefficients, as well as equilibrium parameters, such as octanol–water partition coefficients that define partitioning of pollutants among air, water and sediment phases.

This model contains partial descriptions of two intermedia phenomena, volatilization of pollutants from water to air and sorption of pollu-

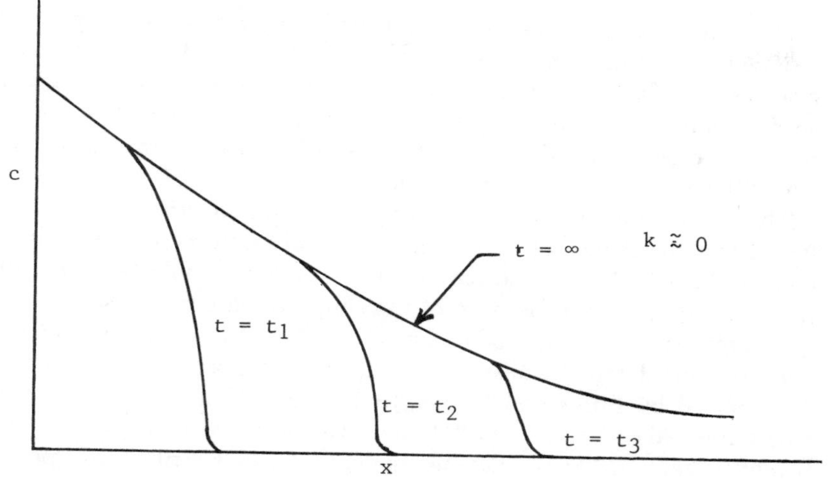

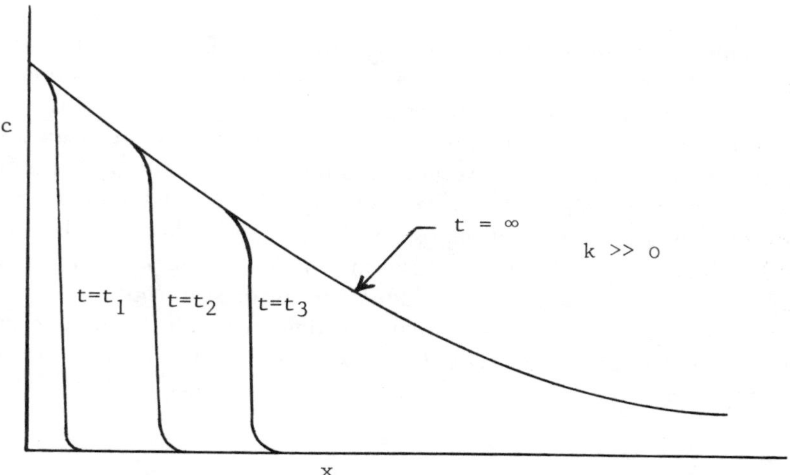

Figure 2. Illustrative curve of concentration vs distance for two values of partition coefficient.

tants onto suspended particulates. Settling of suspended particulates and sorbed pollutants is accounted for only in lakes and reservoirs. Movement of pollutants from air to water and from settled sediments to overlying waters is not estimated.

In addition to the information indicated earlier, pH and suspended sediment concentrations must also be provided. Sediment trap efficiencies must be estimated for lake and reservoirs. Because the model accepts only one average value for all parameters, steady-state conditions must be assumed. Large changes in environmental conditions cannot be accounted for in this model.

The model results in a steady-state concentration profile of a pollutant down the length of a river reach. Steady-state concentrations in lakes and reservoirs are also estimated. This model should be used only for screening purposes because of the simplistic approach used.

The exposure analysis model system (EXAMS) is a computer model that contains detailed transport and degradation process descriptions. As reported by Burns and Cline (1981), EXAMS allows variation in environmental exchange of pollutants between bed sediments and overlying waters. EXAMS is a flexible model that yields estimates of pollutant concentration profiles in rivers, lakes and reservoirs. Both velocity and volume flowrate are assumed to be constant in relation to time. EXAMS is probably adequate for most surface water quality evaluations in which emissions can be approximated by a series of point source discharges.

Finally, the most detailed model available is the Chemical Migration and Risk Assessment (CMRA) methodology developed by Onishi et al. (1980). It contains transport descriptions that allow velocity and flowrates as a function of time and distance. However, it contains no lake or reservoir component. This model provides estimates of pollutant movement, which varies temporally and spatially. A detailed description of sediment scour and deposition is provided. Although chemical and biological degradation processes are modeled in less detail than in EXAMS, process descriptions are adequate for most purposes.

We have presented three of many models available to predict surface-water concentrations. Others exists. However, these three models are representative of models that are used at various levels of detail.

ESTIMATION OF DRINKING WATER CONCENTRATIONS

Predicting surface water concentration profiles satisfies only one aspect of estimating the dose rate to populations of pollutants emitted from landfills. Because most drinking water supplied from surface waters is treated, pollutant levels in raw water drawn into a supply system may be altered before the water is consumed. A model developed by Hedden (1980) estimates the impact of water treatment on pollutant levels. Processes considered include settling of particulates, volatilization during aeration and hydrolysis as affected by pH adjustments.

The formation of halogenated hydrocarbons as a result of chlorination is discussed, but process estimates cannot be made without compound-specific data. Development of a procedure to predict the formation of halogenated hydrocarbons without such data is unlikely.

Groundwater supplies usually receive little or no treatment before being used as a drinking water supply. Consequently, drinking water pollutant concentrations should be approximately equal to groundwater pollutant concentrations in most cases. Assuming that concentrations are equal may lead to high estimates, particularly when groundwater is passed through activated carbon filters before consumption.

NEAR-FIELD EXPOSURE

Estimating the exposure of people to pollutants near a land-disposal site requires much information. Exposures may be triggered by storm events, which raise water tables, flood basements, scour contaminated sediments from the surface of the landfill, and subsequently deposit these sediments nearby. Of course, opportunities for exposure can be minimized by designing containment and treatment facilities. Concrete or clay caps over landfills, for example, eliminate the possibility of surface sediments being scoured at the landfill site and being deposited elsewhere. Pumping groundwater at the landfill site retards movement of contaminated groundwater throughout the site area. However, such measures are costly.

In dealing with near-field problems, modeling can play only a limited role in estimating release rates of chemicals and environmental concentrations after release. In evaluating such situations, data obtained by monitoring must provide much of the basis for evaluation. This situation arises from temporal and spatial variations associated with release rates and transport of pollutants away from points of release. Also, air exposure levels and levels of exposure to highly contaminated soils are difficult to estimate.

The most important role of modeling relating to near-field problems is scenario building. Scenarios are studied to prioritize existing problems and to predict the occurrence of new problems. Use of models for this purpose usually involves running the model over a range of possible release rates and sets of values of environmental parameters that are likely to occur in nature.

As an example of the type of estimates that can be made, a simplified calculation for predicting indoor air exposure will be presented. A possible route of contamination of indoor air is volatilization of pollutants from groundwater and subsequent vertical diffusion into the basements of structures. When the water table is low (i.e., the depth of the water table is much greater than the length of a structure), the concentration profile as shown in Figure 3 is relatively simple. Lines of constant concentration parallel lines of constant depth. The concentration at the soil surface is small compared with the concentration of vapor immediately

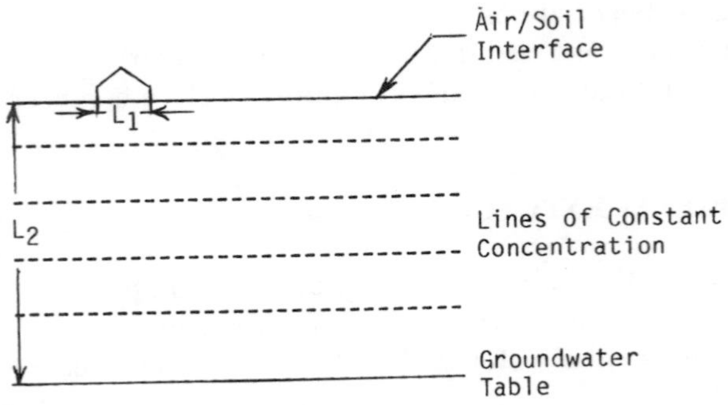

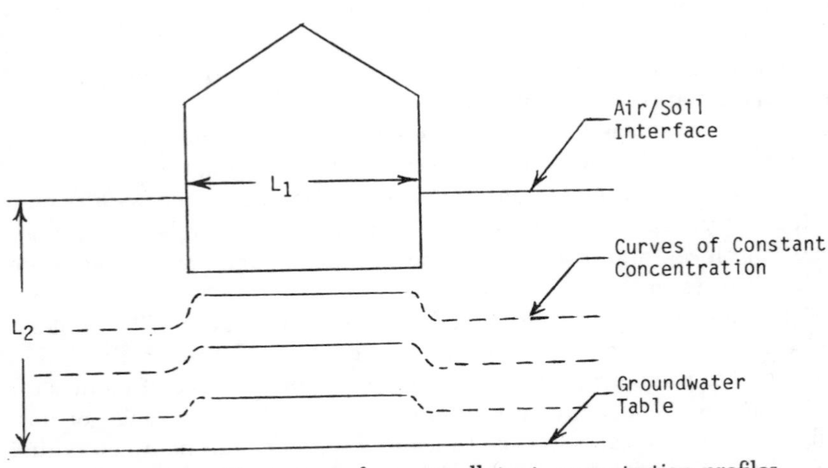

Figure 3. Illustration of vapor pollutant concentration profiles.

above the groundwater. The low concentration at the soil surface results from the large capacity of the atmosphere to transport.

If the water table is high (i.e., the depth of the water table is approximately equal to or less than the length of the structure), the concentration profiles are complex. Curves of constant concentration no longer coincide with lines of constant depth, and significantly more detailed models are required to estimate the concentration of pollutants in soil beneath a structure. Paved areas or other impermeable surfaces require more detailed analysis.

To illustrate the types of results that can be derived, a simplified model is presented that treats the case of a high water table. Performing a simple mass balance on pollutants of concern and assuming the dominant transport mechanism of vapors in soil is diffusion result in the differential equation,

$$(1 + ks)\frac{\partial c}{\partial t} = \left(D\frac{\partial^2 C}{\partial x^2} + \frac{\partial^2 C}{\partial y^2} + \frac{\partial^2 C}{\partial z^2}\right) \tag{4}$$

and boundary conditions,

$$\text{B.C. 1: } C(x, y, 0) = 0 \text{ if } x \leq x_1 \text{ or } x \geq x_2 \text{ or } y \leq y_1 \text{ or } y \geq y_2$$

$$\left.\frac{\partial C}{\partial z}\right| z_1 = k_1 (C - C_B) \quad y_1 < y < y_2 \text{ and } x_1 < x < x_2$$

$$\text{B.C. 2: } C(x, y, L_2) = C_{GW}$$

$$\text{B.C. 3: } \left|\frac{\partial C}{\partial x}\right._{x \to \infty} = 0$$

$$\text{B.C. 4: } \left|\frac{\partial C}{\partial x}\right._{x \to \infty} = 0$$

$$\text{B.C. 5: } \left|\frac{\partial C}{\partial y}\right._{y \to \infty} = 0$$

$$\text{B.C.6: } \left|\frac{\partial C}{\partial y}\right._{y \to \infty} = 0$$

where C = concentration of a gaseous pollutant at a function of position

C_B = concentration of a gaseous pollutant in the basement

C_{GW} = concentration of a gaseous pollutant in vapors immediately above the water table

D = dispersion coefficient
k = vapor/soil partition coefficient
k_1 = mass transfer coefficient appropriate for basement floor
L_2 = depth of water table
z_1 = depth of basement

If we assume that lateral dispersion of gases is small compared with vertical dispersion, Equation 4 can be reduced to a one-dimensional form. Furthermore, if groundwater elevation and pollutant concentrations change slowly compared with structure air exchange rates and transport rates of vapors vertically through the soil, steady-state concentration profiles may be assumed. These assumptions reduce Equation 4 and associated boundary conditions to

$$0 = \frac{D\partial^2 C}{\partial z^2} \qquad (5)$$

B.C. 1: $C(0) = 0 \quad x \leqq x_1$ or $x \geqq x_2$ or $y \leqq y_1$ or $y \geqq y_2$

B.C. 2: $D\frac{\partial C}{\partial z}\bigg|_{z_1} = k_2\ (C - C_B) \quad x_1 < x < x_2$ and $y_1 < y < y_2$

B.C. 3: $C(L_2) = C_{GW}$

The solutions to Equation 5 for these boundary conditions are

$$C = \frac{k_1\ (C_{GW} - C_B)\ (z - L_2)}{D + k_2(L_2 - z_1)} + C_{GW} \text{ for } x_1 < x < x_2 \text{ and } y_1 < y < y_2 \qquad (6)$$

and

$$C = \frac{z}{L_2} \cdot C_{GW} \text{ for } x \leqq x_1 \text{ or } x \geqq x_2 \text{ or } y \leqq y_1 \text{ or } y \geqq y_2 \qquad (7)$$

After the concentration of pollutant in the soil has been estimated, the next step is to estimate movement of the pollutant into the structure. For homes that contain basements, this step involves estimating transport through walls and basement flooring. Transport through these barriers depends on properties of construction materials. For homes that have no basements, but are constructed on a concrete slab or for which there is a crawl space, similar analysis can be performed.

Performing a mass balance on basement air and assuming that rates of transport from soil are proportional to the difference in pollutant soil vapor and basement air concentration yield the differential equation,

$$\dot{V}_B \frac{dC_B}{dt} = -\dot{V}_{AB}(C_B - C_A) + A_1k_1[C(z_1) - C_B] - \dot{V}_{BH}(C_B - C_H) + A_2k_2 \int_{z_1}^{0} \frac{C(z)dz}{z_1} - C_B \quad (8)$$

where
V_B = volume of basement
A_1 = area of basement floor
A_2 = area of basement walls below soil surface
C_A = concentration of gaseous pollutant in outdoor air
C_H = concentration of gaseous pollutant in the structure
k_2 = mass transfer coefficient appropriate for basement walls
$\dot{V}_B$ = volume of the basement
$\dot{V}_{AB}$ = volume rate of exchange between indoor and outdoor air
$\dot{V}_{BH}$ = volume rate of exchange between indoor and basement air

Similarly, a mass balance can be written for household air:

$$V_H d\frac{C_H}{dt} = \dot{V}_{HB}(C_B - C_H) - \dot{V}_{AH}(C_H - C_A) \quad (9)$$

where V_H = volume of house

If steady-state conditions are assumed for household and basement air, the time derivatives in Equations 8 and 9 are zero, and Equations 6 through 9 become a set of algebraic equations that can be solved to predict the concentration of pollutant in the basement and upper stories of a structure. If the concentration of pollutant in outdoor air is assumed to be negligible, the solutions of these equations are:

$$C_B = [DA_2k_2z_1 + k_1(L - z_1)A_2k_2z_1 + 2LDA_1k_1](\dot{V}_{AH} + \dot{V}_{BH})C_{GW} \{2L[(A_2k_2\dot{V}_{BH} + \dot{V}_{AB}\dot{V}_{BH} + \dot{V}_{AH}\dot{V}_{BH} + A_2k_2\dot{V}_{AH} + \dot{V}_{AB}\dot{V}_{AH}) [D + k_1(L - z_1) + (\dot{V}_{AH} + \dot{V}_{BH}) Dk_1A_1]\}^{-1} \quad (10)$$

$$C_H = \frac{\dot{V}_{BH} \cdot C_B}{\dot{V}_{AH} + \dot{V}_{BH}} \quad (11)$$

Nominal parameter values for estimating typical household air concentrations are:

$$
\begin{aligned}
A_1 &= 600\ \text{ft}^2 \\
A_2 &= 700\ \text{ft}^2 \\
z_1 &= 7\ \text{ft} \\
L &= 10 \text{ or } 7.5\ \text{ft} \\
\dot{V}_{AH} &= 6000\ \text{ft}^3/\text{h} \\
\dot{V}_{AB} &= 450\ \text{ft}^3/\text{h} \\
D &= 0.12\ \text{ft}^2/\text{h}
\end{aligned}
$$

The exchange rate between indoor and outdoor air used is the value reported by Bery and Ross (1981) for homes that have been weatherized to minimize heating requirements. Thus, numerical results shown in Table I are somewhat higher than would actually occur in most homes. The diffusion coefficient used is in the midrange of values reported for molecular diffusivities of gases in air, as reported by Perry et al. (1969).

Table I. Illustrative Results

	Case		
Parameter	I	II	III
K_1, ft/h	0.0002	0.24	0.24
L, ft	10	10	7
C_B/C_{GW}	0.00042	0.0014	0.091
C_H/C_{GW}	0.000067	0.00022	0.015

The numerical values reported in Table I are for three example conditions. In case I, the basement floor and wall are very impermeable and the water table is 3 ft below the basement floor surface.

In case II, the basement floor is soil and the water table is 3 ft below the basement. In case III, the basement floor is soil and the water table is 6 in. below the basement floor.

As the results clearly indicate, the concentrations of pollutant vapor in the sample structure do not exceed 2% of the concentration in vapors immediately above the water table. The concentrations in basements never exceed 10% of the concentration occurring in vapors immediately above the water table. Thus, except when a basement is about to be flooded, relatively small amounts of pollutant reach basement air. If the basement floods or if material from the landfill penetrates basement walls or flooring, the situation may change dramatically. When floods occur or when liquids from a landfill penetrate basement walls, there may be much higher vapor concentrations.

CONCLUSIONS

Some multimedia aspects of exposure assessment related to hazardous solid wastes have been summarized briefly. Much research is under way to address these and other aspects of exposure assessment related to disposal of hazardous solid wastes. The stage of development of such techniques is still quite primitive. Only limited testing of models has been conducted.

In the Exposure Assessment Group, we hope to devote much of our effort in this area to testing the accuracy and precision of these models and providing guidance as to their use. We intend to conduct sensitivity analyses of these models to determine their operating characteristics. Last, we hope to participate with other groups to find ways to reduce the cost of application of these models and to improve their predictive capabilities.

REFERENCES

Bachmat, Y., B. Andrews, D. Holta and S. Sebastian. 1978. "Utilization of Numerical Groundwater Models for Water Resource Management," EPA-600/8-78-102, U.S. EPA, R. S. Kerr Environmental Research Laboratory, Ada, OK.

Bery, D., and H. Ross. 1981. "Workshop on Indoor Air Quality Research Needs," U.S. EPA, Office of Environmental Engineering and Technology, Washington, DC.

Burns, L. A., and D. M. Cline. 1981. "EXAMS, An Exposure Analysis Modeling System," U.S. EPA, Athens, GA.

Falco, J. W., L. A. Mulkey, R. R. Swank, and R. E. Lipcsei. In press. "A Screening Procedure for Assessing the Transport and Degradation of Solid Waste Constituents in Subsurface and Surface Waters," Proceedings of the Annual Meeting of the Society of Environmental Toxicology and Chemistry.

Faust, C. R., and J. W. Mercer. 1980. "Groundwater Modeling: Numerical Models," *Groundwater* 18(4):395-407.

Hedden, K. F. 1980. "Water Treatment Model for Pollutant Exposure Assessment," *Environ. Sci. Health* A15(4):285-305.

Mercer, J. W., and C. R. Faust. 1980. "Groundwater Modeling: Mathematical Models," *Groundwater* 18(3):213-227.

Onishi, Y., A. Olsen, R. Ambrose and J. Falco. 1979. *Proceedings of the Symposium on Hydrologic Modeling,* ASAE Publication No. 4-80 (American Society of Agricultural Engineers).

Perry, R. H., C. H. Chelton, and S. D. Kirkpatrick. 1969. *Perry's Chemical Engineers' Handbook,* 4th ed. (New York: McGraw-Hill Book Company).

SECTION 4

ENVIRONMENTAL AND HEALTH CONSEQUENCES

CHAPTER 19

OVERVIEW

R. E. Franklin
Office of Health and Environmental Research
U.S. Department of Energy
Washington, DC 20545

Solid waste disposal, from the standpoint of health and environmental consequences, certainly is one of the most serious problems we face. The problems associated with solid waste disposal are brought on largely as a result of:

1. An expanding population: more people, more demand for goods and services result in more wastes being generated, and more pressure on land use.
2. Our casual treatment of wastes: there has been no incentive to conserve resources; disposal has been by the easiest and cheapest means without concern for the ultimate consequences; and we have been overconfident that time and nature will handle the problem.
3. Improved detection measures: we have more sophisticated instruments for detection of chemicals at ultralow concentrations, and sophisticated methods for detecting damages to chromosomes, cells and tissues caused by chemicals.

The general public has been made acutely aware of the solid waste issue over the past few years because of much-publicized "horror stories." However, the increased awareness is not necessarily accompanied by an increased understanding. Problems are readily picked up by the news media and publicized along with instant analyses by experts, pseudo experts, those who are directly affected and people who are simply concerned. More often than not, the experts cannot agree on how serious the problem is and what to do about it; thus, not only are the problems extremely complex, but complexity often leads to misunderstanding on the part of the general public. The health and environmental consequences of mishaps may be real, and they may be serious. Other times, the mental anxiety and worry created by misunderstanding the seriousness of the problem may be the primary consequence.

Therefore, in structuring this section, we were not sure how to do justice to the broad scope of the solid waste disposal issue. The variety of solid wastes, the many variations in disposal techniques and the far-ranging environmental settings present us with a vast and complex array of situations. Each waste has its own set of chemical characteristics, and each operation has its set of environmental and site-specific characteristics. There is one thing each situation has in common: something is to be thrown away. And therein lies the problem. As one of my colleagues used to say, "there is no more away."

That is certainly true for existing problems, because in those instances, wastes had been put where they did not remain away from people or that part of the environment we seek to protect. Our strategy in dealing with this subject was to sample several topics we considered important, such as:

- public health: what we know from previous experiences on case studies, what we need, what we should be doing and what the real problems are;
- biomedical test systems: what tests are available to estimate potential hazards and exposures; what the various test data mean, or how the data can be used;
- land resources: what the implications are in terms of future land use, how secure various disposal systems are, the compatibility of waste disposal with the use of land for agriculture and urban needs;
- water resources: effects of disposal schemes on water resources, both surface and groundwater; the implications to use of water resources for drinking water, agriculture and other uses;
- public perceptions: what the concerns are of the general public and the organized environmental community to solid waste problems; how the responses and attitudes of industry or generators of solid waste are perceived by the responsible state, federal and local governmental agencies and the scientific community; and
- risk analysis: how we use biomedical and ecological data to estimate risk, and what is the bottom line.

The authors in this section were given this guidance but were urged to use it only to stimulate their thinking. The result was six chapters that provide an interesting view of the solid waste issue.

Logan and Miller describe two basic strategies for land application of municipal organic wastes: (1) low application rates on land used for agricultural purposes, and (2) disposal of high concentrations at dedicated sites. In the first case, a host of trace metals can present problems and limit the rate at which sustained applications can be made if toxic accumulation in soil and food crops is to be avoided. The second case requires securing the waste in such a manner as to prevent contamination of groundwater resources. The authors discuss the pros and cons of these two strategies.

Aquatic ecosystem effects are discussed by Giddings. He stresses the need to examine the ecosystem as a whole in answering questions about the environmental impacts of solid waste disposal. As a backdrop for the discussion, he examines the logic pathway to answer a seemingly straightforward question posed by a sport fisherman regarding the effect of an industrial landfill on his favorite fishing hole. He examines the complex set of issues that must be considered in answering this question:

- the role of bioassay in assessment of toxicity;
- the extrapolation from controlled laboratory studies to natural conditions where animal behavior and the complexity of food webs play a role; and
- extrapolation from organisms to populations to ecosystems.

Florence and Sheldon present insights on public perceptions. They discuss the growing distrust of both the scientific community and the government, and people's attitude on siting of waste facilities. The chapter examines the results of a survey conducted to estimate public awareness of solid waste issues, their misunderstanding of the problem, their opinion of government's role and performance, and the responsibility for paying the bill for safe disposal. As with so many other complex technical issues that we encounter on a fairly regular basis, sources of public information and education limit our ability to provide intelligent responses.

One of the facts of life these days seems to be that we amass quantities of technical data in describing health and environmental problems. It also seems that the more information we gather, the more uncertain we are about what to do. Several attempts have been made to organize this information through a systematic analysis of risks. Fundamental problems remain, such as how we extrapolate from animal studies and short-term bioassay tests to man, and what the long-term consequences are of low-dose chronic exposures to biologically active compounds. Still, a systematic approach offers considerable promise in finding our way through the maze of data.

Gratt deals with this subject. He is conducting a risk analysis of the health and environmental consequences of oil shale technologies, and we asked him to share his approach and his experience in the activity.

CHAPTER 20

ASSESSING THE EFFECTS OF CONTAMINANTS ON THE STRUCTURE AND FUNCTION OF AQUATIC ECOSYSTEMS

Jeffrey M. Giddings

Environmental Sciences Division
Oak Ridge National Laboratory
Oak Ridge, Tennessee 37830

ABSTRACT

Ecosystems are the context in which all organisms live, reproduce, interact, and evolve. A balanced approach to environmental impact assessment includes consideration of effects on whole ecosystems as well as analyses of effects on organisms and populations. Potential effects of contaminants on whole ecosystems include changes in energy flow, nutrient flow, or trophic structure. Such changes are generally not predictable from measurements of responses of individual species, but must be determined by investigating whole ecosystems. Responses of aquatic ecosystems to organic and inorganic contaminants have been demonstrated in microcosms (laboratory models of ecosystems) simulating various aquatic environments.

INTRODUCTION

This chapter examines the range of potential interactions between chemical pollutants and aquatic ecosystems. The focal point of the discussion is the ecosystem, rather than any particular group of aquatic organisms, because the ecosystem—the biological community and its abiotic environment—is the context in which organisms live, reproduce, interact, and evolve. Because problems of pollution occur in whole ecosystems, not in bioassay jars, we need to understand how ecosystems work and how particular organisms fit into and interact with the rest of the system. From such an understanding, new methods emerge for addressing pollutant effects in and on ecosystems. These methods go beyond the acute bioassay to a more complete and realistic style of environmental assessment (O'Neill and Waide, 1981). The following discussion is an introduction to the modern science of ecosystem toxicology.

BIOASSAYS

As individuals and as a society, we are more concerned about some species than others. If a fisherman learns that an industrial landfill is to be constructed alongside his favorite trout stream, his first question is likely to be, "Will the trout be back next season?" To answer him, we examine the waste and the landfill site, and then predict eventual streamwater concentrations of heavy metals, trace organics, radionuclides, and other hazardous substances. We compare these estimated concentrations with certain bioassay endpoints or with standards based on bioassay results. We tell the angler, "The concentrations of pollutants that will be found in this stream had no effect on our test fish. Pollution from this landfill, if any, will be within acceptable limits." Often, but not always, our exposure predictions and bioassay results lead to accurate assessments.

In a conventional bioassay, laboratory-reared plants or animals are placed in beakers or aquaria in water containing various concentrations of the material being tested. Standard conditions (e.g., water quality, temperature, and feeding regime) are maintained so that results will be comparable with those of other tests. After an exposure of one or a few days, the survivors are counted, the results are analyzed statistically, and the concentration at which 50% mortality occurred (LC_{50}) is estimated.

When pollution is not episodic but continuous, as in the case of leachate from a landfill, acute (short-term) bioassays underestimate toxicity, and chronic (long-term) bioassays are necessary. Concentrations of some chemicals as low as 0.1% of the acute LC_{50} can reduce the fitness of animals exposed for their entire lifetime (Mount and Stephan, 1967; Parkhurst et al., 1981). Sensitivity to a toxicant varies with the age or life stage of an organism. The eggs and larvae of fish are more sensitive than adults to most chemicals; concentrations that do not affect the eggs and larvae are safe for the fish throughout its life (Pickering and Thatcher 1970). Thus an acute bioassay with eggs and larvae can be substituted for a chronic bioassay (Macek and Sleight, 1977).

Another way of estimating chronic toxicity is to measure physiological effects after short exposures. Here the objectives and methods of environmental toxicology merge with those of classical toxicology and pharmacology. Physiological endpoints have not been widely adopted in environmental toxicology, however, possibly because the ecological significance of physiological changes is uncertain (Stephan and Mount, 1973).

Many animals of concern to us cannot be maintained in the laboratory for study. For many others, the laboratory is so unlike the natural habitat that experimental results are irrelevant. Even with the most tractable species, the necessary artificialities of the bioassay preclude direct application of results to nature. Fish, for example, do not naturally live in beakers and eat trout chow; they live in spatially complex environments and eat a variety of foods. Furthermore, the physical and chemical features of the natural environment fluctuate daily and seasonally, and rarely, if ever, correspond to the standard bioassay conditions. The dependence of bioassay results on physical and chemical conditions is well recognized, and bioassay protocols specify standard conditions so that results will be repeatable and comparable among tests. This dependence also implies that responses of organisms in nature are likely to differ from those in bioassay and are frequently unpredictable or stochastic (O'Neill et al., 1982).

POPULATION INTERACTIONS

The survival of organisms and populations depends not only on tolerable environmental conditions but also on interactions with other organisms. Some chemicals can cause serious ecological disturbance through effects on interactions between predators and prey, parasites and hosts, or competitors. Few attempts have been made to measure such effects in the laboratory (Giddings, 1981). Kania and O'Hara (1974) exposed groups of mosquitofish (*Gambusia affinis*) to mercury for 24 h, then offered the exposed fish and equal numbers of unexposed fish to a predator, the largemouth bass (*Micropterus salmoides*). The untreated fish took refuge in a shallow area provided for them in the test aquarium; the exposed fish, which were hyperactive, wandered into the open water where they were caught by the bass. Similar effects were observed with fathead minnows (*Pimephales promelas*) exposed to cadmium (Sullivan et al., 1978). In both cases, vulnerability to predation increased at toxicant levels well below the lethal concentrations. Contaminants can also impair the ability of predators to capture food. For example, the predation efficiency of largemouth bass on mosquitofish declined when both species were exposed to low concentrations of ammonia (Woltering et al., 1978). At the highest prey densities, the mosquitofish reportedly harassed the bass.

The competitive balance among species sharing a limited resource such as an essential nutrient or a preferred food can be very delicate. Among planktonic algae, for example, competition for inorganic nutrients appears to play a critical role in determining which species are abundant and which are rare (O'Neill and Giddings, 1979; Tilman, 1977). Contaminants that affect competitors to varying degrees may lead to replacement of one species by another. Such an occurrence may mean the difference between an algal community that is a suitable food base for invertebrates and fish and a nuisance growth of unpalatable species. A series of experiments at the Woods Hole Oceanographic Institute (Fisher et al., 1974; Mosser et al., 1972) demonstrated that concentrations of polychlorinated biphenyls (PCB) as low as 0.1 μg/L can alter competition between two common marine species, the green alga *Dunaliella tertiolecta,* and the diatom *Thalassiosira pseudonana.* When these species were grown together in a laboratory culture, the diatom normally reached densities eight or nine times higher than those of the green alga. Low concentrations of PCB impaired the diatom's capacity to absorb nutrients from the water and permitted the green alga, which was unaffected, to reach higher population densities. The PCB concentration that produced this effect was at least two orders of magnitude below the concentration that inhibited pure cultures of the diatom.

A shift in focus from individual organisms and populations to interactions among populations thus reveals that effects of environmental contaminants may be manifested as alterations in feeding behavior, predator-prey relationships, and interspecific competition. The significance of such effects is twofold: (1) they often occur at low toxicant concentrations, and (2) they are usually unpredictable from bioassay results alone. By broadening our perspective still further, we see that each interaction is embedded in an intricate network that involves every organism and population in the ecosystem. Such a network constitutes a distinct unit of ecological organization—it is not simply an assemblage of organisms, just as an organism is more than an assemblage of tissues. The fundamental biological processes—growth, development, transformation of matter and energy, and reproduction—are expressed in ecosystems in a manner unique to that level of organization (Warren and Liss, 1977). The following section will examine the properties of whole ecosystems and consider how these properties might be affected by exposure to pollutants.

ECOSYSTEMS

We can describe the network of interactions that constitute an ecosystem by tracing the flows of energy and nutrients through the system. The pattern of sequential energy transfers among organisms, known as the food web, is a distinctive feature of every ecosystem. Figure 1 depicts a generalized aquatic food web. The four rectangles in the figure represent trophic levels or stages in the transfer of energy. Primary producers (plants) capture solar energy and incorporate it by photo-

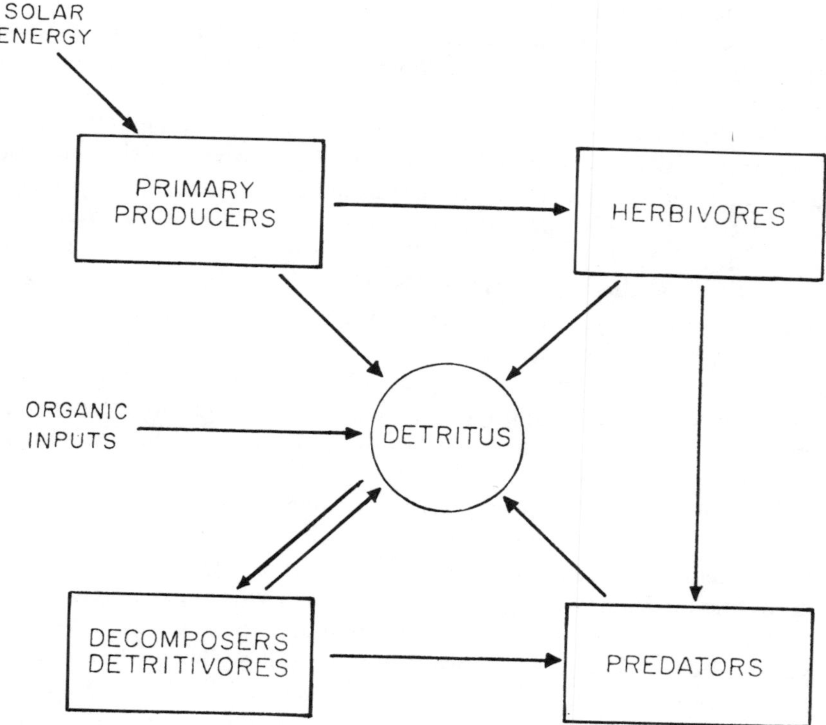

Figure 1. Generalized aquatic food web. Arrows indicate pathways of energy flow.

synthesis into chemical bonds. In lakes, floating microscopic algae (phytoplankton) are the dominant primary producers; in ponds, rooted vascular plants (macrophytes) fill this role; and in streams, primary production is carried on by algae attached to the substrate (periphyton). In each system, a fraction of the primary production passes to herbivores and then to predators. Small crustaceans (zooplankton) are the main herbivores in lakes; snails, crayfish, insects, and certain fish are herbivores of importance in ponds and streams. Predators also vary among ecosystems; they include zooplankton in lakes, insects in ponds and streams, and fish in nearly all aquatic ecosystems.

Not all plant production is eaten by herbivores, of course; unconsumed plant tissue dies and becomes detritus. Excretions and the dead bodies of animals also contribute to the detritus, which in turn supports another diverse and highly productive food web. The detrital food web begins with bacteria and fungi that colonize dead particles and eventually decompose them. Many small animals including protozoans, nematodes, burrowing worms and insect larvae feed on the colonized particles, stripping off the decomposers as the detritus passes quickly through their gut (Berrie, 1976). These animals are an important food source for larger organisms, including bottom-feeding fish.

The patterns of energy flow vary among different types of aquatic ecosystems. Because phytoplankton contain relatively little cellulose and other indigestible material, a substantial fraction of the primary production in a lake is passed to herbivores. Macrophytes are generally less edible than algae, and as much as 90% of the primary production in a pond may become detritus (Wetzel, 1975). In many rivers and streams, primary production is relatively unimportant, and detritus from the surrounding watershed is the major source of organic matter sustaining the ecosystem. Eventually, nearly all of the energy entering an ecosystem is either exported or is released by respiration; a relatively small percentage may remain within the ecosystem as a net accumulation of living biomass or detritus.

Most biological hardware (e.g., cell membranes, enzyme systems, and genetic material) contains nitrogen, phosphorus, sulfur, and other nutrient elements. Nutrients must be available for energy flow to continue, and some of the energy processed by the food web is expended in converting nutrients from inorganic to organic form and back again. At various points along the pathways of energy flow, nutrients are incorporated into or released from biological structures. Energy and nutrient flows are interdependent; any disturbance in the pattern of nutrient flow will be reflected in the flow of energy and vice versa.

Changes in energy and nutrient flow may have far-reaching consequences for organisms and populations in an ecosystem. If this appears to be a radical suggestion, it is because the context, not the concept, is unfamiliar. It is clear that altering the flow of water through an aquatic environment (e.g., by building a dam, channelling a river, or draining a marsh) is liable to cause serious ecological changes. Is it unreasonable to suppose, then, that altering the flow of energy or nutrients can also have significant effects? Patterns of energy and nutrient flow are hard to comprehend and harder still to measure, but that should not deter us from incorporating them into the framework of environmental assessment. In fact, methods for measuring contaminant effects on whole ecosystems are beginning to emerge, and evidence of the sensitivity of ecosystem properties to stress is slowly accumulating (Giddings, 1981; Sanders et al., 1980).

Disruptions in energy and nutrient flow can be detected in a variety of measurable properties. Total ecosystem primary production (P) and respiration (R) are integrated measures of energy flow. The ratio P:R is particularly responsive to contaminant effects; normally around 1 in most aquatic ecosystems, P:R falls below 1 when ecosystems are disturbed (Giddings, 1981; Giddings and Eddlemon, 1978). Effects on nutrient flow may appear as changes in nutrient concentrations in water or sediments or as altered rates of specific nutrient cycling processes, such as uptake of inorganic nutrients by plants, excretion by animals or transformations by microorganisms. Changing patterns of energy and nutrient flow may also be seen as changes in trophic structure, standing crop, or species diversity in an ecosystem.

The science of measuring or predicting effects of contaminants on ecosystem properties is in its infancy; theories are more plentiful than data. However, some observations of ecosystem-level effects have been made in laboratory model ecosystems or microcosms (Giddings, 1981). The following summaries of three microcosm studies illustrate many of the effects just described.

Microcosm Study 1

The first example comes from an experiment at Oak Ridge National Laboratory in which 80-L pond microcosms containing sediment, water, and a complex community of organisms from a natural pond were treated with an aqueous extract of a coal-derived oil (Giddings, in press). The treatment caused an immediate decline in P and a smaller decline in R. Both measures of energy flow reached minimum values within two weeks and remained at low levels for two weeks longer. The

P:R ratio was below 1 for 50 d, resulting in a gradual depletion of the ecosystem's reserve of organic matter (detritus). Nutrient uptake by plants ceased for several weeks, but nutrient regeneration continued, leading to an accumulation of inorganic nitrogen in the water and the sediment. The trophic structure of the ecosystems changed; herbivores (zooplankton and snails) decreased, whereas bacteria and detritivores (ostracods and oligochaetes) became more abundant. The microcosms began to recover 4–6 weeks after treatment, as surviving macrophytes resumed growth and herbivore populations returned. Two months after treatment, P, R, and nutrient concentrations were back to pretreatment levels, although significant changes in community composition were still evident.

Microcosm Study 2

The second example of experimentation at the ecosystem level involves a series of outdoor channel microcosms constructed at the Savannah River Plant in South Carolina by the U.S. Environmental Protection Agency (Giesy et al., 1979). Each channel was 92 m long, 0.6 m wide and 0.2 m deep, with a continuous supply of well water and a biological community typical of shallow, slow-moving streams. For one year, cadmium was added to the inflowing water at concentrations equal to or less than the federal criterion for public water supplies. The Cd inputs caused a variety of changes in the populations inhabiting the channels; some species decreased, whereas others, freed from competition or predation, increased in abundance. Total ecosystem production and respiration declined, as did the P:R ratio. Less detritus was formed, and export of organic particles from the channels decreased. In nature, the organic matter exported from one section of a stream is an important energy source for downstream ecosystems. The Cd treatments also interfered with nutrient flow; uptake and retention of dissolved inorganic nitrogen were lower in the treated channels than in the controls, implying less efficient use of the available nitrogen supply by the community.

Microcosm Study 3

Whereas the first two examples concerned microcosms that were modeled after specific ecosystem types (ponds and streams, respectively), the third example is a study of simpler, more generalized model ecosystems. Cadmium was again used as the stressing agent. Hendrix et al. (1981) applied Cd to a series of microcosms consisting of 7-L polyethylene containers of artificial nutrient media inoculated with bacteria, phytoplankton, and zooplankton from a pond. Microcosms constructed in this way have no counterparts in nature, but they do exhibit universal ecosystem properties (Gorden et al. 1969) and are therefore suitable for investigating the fundamental responses of ecosystems to stress (Ferens and Beyers, 1976; Leffler, 1978). In the experiments with Cd, both static (no water replacement) and flow-through (semicontinuous water replacement) microcosms were used, nutrient regimes were varied, and the Cd applications included single acute doses, repeated doses, and continuous exposure. Cadmium caused decreases in herbivores and, consequently, increases in the standing crop of plants, decomposers, and detritivores. The net effect was a change in the pattern of energy flow from a food web dominated by herbivory to one dominated by detritivory. Production, respiration, and nitrogen retention were also affected by Cd treatments, but the magnitudes and directions of changes depended on the configuration of the microcosms. For example, the P:R ratio declined in the static microcosms but was unaffected in the flow-through microcosms. The responses of the flow-through microcosms to a pulse of Cd late in the experiment were influenced by the ecosystem's previous history of Cd exposure. For example, when Cd was added to previously unexposed microcosms, P and R increased, but when Cd was added to microcosms that had also been exposed earlier, P and R decreased. Effects of Cd also varied between high- and low-nutrient microcosms. The interactions between the toxicant and the ecosystem were therefore partly determined by the nature of the ecosystem and the pattern of exposure.

In each microcosm study, the stressed ecosystems showed signs of recovery after the contaminant had dissipated. To "kill" an ecosystem, to render it devoid of life, is extremely difficult. More often, nature rebounds from the insult, and an ecosystem of some sort persists. Contaminant effects on ecosystems are therefore not absolute, like death of an organism or extinction of a population, but a matter of degree. Environmental toxicologists may aim at predictions of the probable extent of ecosystem change (O'Neill et al., 1982), but the decision as to how much change is acceptable depends on cultural values and is properly left to society at large.

CONCLUSION

The concerns of our imaginary trout fisherman cannot be fully addressed through standard, single-species bioassays alone. Organisms and populations do not simply inhabit an environment, but rather are interconnected components of a higher organizational unit, the ecosystem. Contaminants that affect an organism's ability to locate and capture food, avoid predators, or compete for limited resources are ultimately as lethal as those that kill the organism outright. Furthermore, contaminants may affect an organism indirectly by altering the structure or function of the ecosystem of which the organism is a part. The implication for environmental toxicology is obvious: the ecosystem as a whole is the proper context for hazard assessment. Measuring effects of toxic substances on a few isolated species of economic or cultural importance is not enough; we must also ascertain potential effects in and on entire ecosystems, outside of which our favored species cannot survive.

ACKNOWLEDGMENT

Research sponsored by the Office of Health and Environmental Research, U.S. Department of Energy, under contract W-7405-eng-26 with Union Carbide Corporation.

REFERENCES

Berrie, A. D. 1976. "Detritus, Micro-organisms and Animals in Fresh Water," in *The Role of Terrestrial and Aquatic Organisms in Decomposition Proccesses,* J. M. Anderson and A. Macfadyen, Eds. (Oxford: Blackwell Scientific Publications), pp. 323-328.

Ferens, M. C., and R. J. Beyers. 1972. "Studies of a Simple Laboratory Microecosystem: Effects of Stress," *Ecology* 53:709-713.

Fisher, N. S., E. J. Carpenter, C. C. Remsen, and C. F. Wurster. 1974. "Effects of PCB on Interspecific Competition in Natural and Gnotobiotic Phytoplankton Communities in Continuous and Batch Cultures," *Microbial Ecol.* 1:39-50.

Giddings, J. M. 1981. "Laboratory Tests of Chemical Effects on Aquatic Population Interactions and Ecosystem Properties," in *Methods for Ecological Toxicology: A Critical Review of Laboratory Multispecies Tests,* A. S. Hammons, Ed. (Ann Arbor, MI: Ann Arbor Science Publishers, Inc.), pp. 23-91.

Giddings, J. M. In press. "Effects of the Water-Soluble Fraction of a Coal-Derived Oil on Pond Microcosms," *Arch. Environ. Contam. Toxicol.*

Giddings, J. M., and G. K. Eddlemon. 1978. "Photosynthesis/Respiration Ratios in Aquatic Microcosms under Arsenic Stress," *Water Air Soil Poll.* 9:207-212.

Giesy, J. P., Jr., H. J. Kania, J. W. Bowling, R. L. Knight, S. Mashburn, and S. Clarkin. 1979. "Fate and Biological Effects of Cadmium Introduced into Channel Microcosms," EPA-600/3-79-039, U.S. EPA, Washington, DC.

Gorden, R. W., R. J. Beyers, E. P. Odum, and R. G. Eagon. 1969. "Studies of a Simple Laboratory Microecosystem: Bacterial Activities in a Heterotrophic Succession," *Ecology* 50:86-100.

Hendrix, P. F., C. L. Langner, E. P. Odum, and C. L. Thomas. 1981. "Microcosms as Test Systems for the Ecological Effects of Toxic Substances: An Appraisal with Cadmium," EPA-600/S3-81-036, U.S. EPA, Washington, DC.

Kania, H. J., and J. O'Hara. 1974. "Behavioral Alterations in a Simple Predator-Prey System Due to Sublethal Exposure to Mercury," *Trans. Am. Fish. Soc.* 103:134-136.

Leffler, J. W. 1978. "Ecosystem Responses to Stress in Aquatic Microcosms," in *Energy and Environmental Stress in Aquatic Systems,* J. H. Thorp and J. W. Gibbons, Eds. (Oak Ridge, TN: U.S. Department of Energy Technical Information Center), pp. 102-110.

Macek, K. J., and B. H. Sleight, III. 1977. "Utility of Toxicity Tests with Embryos and Fry of Fish in Evaluating Hazards Associated with the Chronic Toxicity of Chemicals to Fishes," in *Aquatic Toxicology and Hazard Evaluation,* ASTM STP 634, F. L. Mayer and J. L. Hamelink, Eds. (Philadelphia: American Society for Testing and Materials), pp. 137-146.

Mosser, J. L., N. S. Fisher, and C. F. Wurster. 1972. "Polychlorinated Biphenyls Alter Species Composition in Mixed Cultures of Algae," *Science* 176: 533-535.

Mount, D. I., and C. E. Stephan. 1967. "A Method for Establishing Acceptable Toxicant Limits for Fish—Malathion and the Butoxyethanol Ester of 2,4-D," *Trans. Am. Fish. Soc.* 95:185-193.

O'Neill, R. V., and J. M. Giddings. 1979. "Population Interactions and Ecosystem Function: Phytoplankton Competition and Community Production," in *Simulation Analysis of Ecosystems,* G. S. Innis and R. V. O'Neill, Eds. (Fairland, MD: International Cooperative Publishing House), pp. 103-123.

O'Neill, R. V., and J. B. Waide. 1981. "Ecosystem Theory and the Unexpected: Implications for Environmental Toxicology," in *Management of Toxic Substances in Our Ecosystem: Taming the Medusa,* B. Cornaby, Ed. (Ann Arbor, MI: Ann Arbor Science Publishers, Inc.).

O'Neill, R. V., R. H. Gardner, L. W. Barnthouse, G. W. Suter, S. G. Hildebrand, and C. W. Gehrs. 1982. "Ecosystem Risk Assessment," *J. Soc. Environ. Toxicol. Chem.* 1:167-177.

Parkhurst, B. R., J. L. Forte, G. P. Wright, M. P. Farrell, and R. H. Strand. 1981. "Reproducibility of a Life Cycle Toxicity Test with *Daphnia magna," Bull. Environ. Contam. Toxicol.* 26:1-8.

Pickering, Q. H., and T. O. Thatcher. 1970. "The Chronic Toxicity of Linear Alkylate Sulfonate (LAS) to *Pimephales promelas* Raffinesque," *J. Water Poll. Control Fed.* 42:243-254.

Sanders, F. S., S. M. Adams, L. W. Barnthouse, J. M. Giddings, E. E. Huber, K. D. Kumar, D. Lee, B. Murphy, G. W. Suter, and W. Van Winkle. 1980. "Strategies for Ecological Effects Assessment at DOE Energy Activity Sites," ORNL/TM-6783 (Oak Ridge, TN: Oak Ridge National Laboratory).

Stephan, C. E., and D. I. Mount. 1973. "Use of Toxicity Tests with Fish in Water Pollution Control," *Biological Methods for the Assessment of Water Quality*, ASTM STP 528, J. Cairns, Jr. and K. L. Dickson, Eds. (Philadelphia: American Society for Testing and Materials), pp. 164-177.

Sullivan, J. F., G. J. Atchison, D. J. Kolar, and A. W. McIntosh. 1978. "Changes in the Predator-Prey Behavior of Fathead Minnows (*Pimephales promelas*) and Largemouth Bass (*Micropterus salmoides*) Caused by Cadmium," *J. Fish. Res. Board Can.* 35:446-451.

Tilman, D. 1977. "Resource Competition Between Planktonic Algae: An Experimental and Theoretical Approach," *Ecology* 58:338-348.

Warren, C. E., and W. J. Liss. 1977. "Design and Evaluation of Laboratory Ecological System Studies," EPA-600/3-77-022 (Washington, D.C.: U.S. Environmental Protection Agency).

Wetzel, R. G. 1975. *Limnology*. (Philadelphia: W. B. Saunders Company).

Woltering, D. M., J. L. Hedtke, and L. J. Weber. 1978. "Predator-Prey Interactions of Fishes under the Influence of Ammonia," *Trans. Am. Fish. Soc.* 107:500-504.

DISCUSSION

***John J. Suloway*, Charles T. Main, Inc.:** Do you think that you could conduct the suggested studies and then correlate them to slightly cheaper tests such as subchronic tests or bioassays so that you could apply a factor to get the same results after you have done some initial research?

J. Giddings: Hopefully. The first step, of course, is conducting the studies, and we are just beginning on that. The methods have not been standardized for any of the things I talked about. We are now at the point of trying to accumulate some basic data, and then we can take a look at some of those correlations. About the only thing that emerges so far is the effects on the population interactions. I think it is too early to say anything about ecosystem-level effects, but effects on population interactions tend to occur at levels well below the LC_{50} for the respective organisms. I do not think we can make any generalization beyond that yet, and the next step is to gather the needed data.

John Suloway: That is why I suggested using the subchronic tests where you would be looking at something other than the acute data. Also, now you can speculate: do you think that funding will be available to do microcosm tests when we heard suggestions that microcosm studies are so oversimplified they really do not simulate what is happening in a full-scale ecosystem?

J. Giddings: You raised several points. Some microcosms are so simplified that they do not behave like any real ecosystem. Others, such as the pond microcosms that we studied, seem to behave quite a bit like natural ponds.

But in terms of where the money is going to come from, I don't know. The EPA [U.S. Environmental Protection Agency] has supported a good deal of microcosm research so far; whether they will continue, I would rather not have to guess.

Harry Freeman, **Office of Appropriate Technology, State of California:** You mentioned the study of the Savannah River Plant where they looked at the cadmium. I found that kind of alarming at the levels you were talking about. Did they look at any other metals in that same level of input, the seemingly safe levels?

J. Giddings: They did a study of mercury before the one of cadmium and that was aimed at looking at uptake and distribution of the mercury. As far as I know, they were not looking at effects. I think cadmium is the only one that they have actually looked at for effects on the system.

Bart Simmons, **California Department of Health Services:** Are the behavioral effects due to a toxic effect on the generation exposed or more likely to teratogenic effects?

J. Giddings: I do not think any of the studies that I talked about dealt with looking at subsequent generations. They were all direct and usually short-term effects. For instance, that mercury experiment was a 24-h exposure of adult mosquitofish.

Ed Portner, **Johns Hopkins Applied Physics Laboratory:** It would seem that in a complex system, like an estuary, the test you set before us is almost impossible. Would you see any simplification possible that would make an ecosystem study feasible with the decision-makers who are focusing primarily on the total effect on the top predators? In other words, if they were not really concerned about the changes in the community structure at the lower trophic levels but primarily were interested in all the impacts on the top predators?

J. Giddings: I would suggest that they can't fully address the top predators if they have not looked at the changes in the whole system below those predators, and there are studies under way examining estuaries which are, I agree, much more complex than the ponds or those channels I showed you. The longer-term problem of how we address ecosystem-level problems in complex ecosystems is one that I do not know if we will ever be able to make any real progress in until we have studied ecosystems as ecosystems—more than we have so far. Very few studies have looked at a whole ecosystem. Most of these have been in these model ecosystems and whether they simulate real ecosystems enough to draw valid conclusions is still a very open question. I would say in the near term, the only answer is that some additional basic research is needed on whole ecosystems.

Harry Freeman: You mentioned how site-specific each of the ecological areas were. Are they all so site-specific that you cannot generalize from one to the other at all? Is there any continuity? Can it be translated from one study to the other study or are all ponds so different that nothing can be said?

J. Giddings: How much you translate, of course, depends on the degree of difference between the system you have studied and the system you are trying to extrapolate to. I am sure you caught the common thread in the three examples of ecosystem-level experiments that I showed, like the changes from a herbivorous food web structure to one that is based on the detritus, also a decrease in the ratio of production to respiration; there are other common elements as we do more of these experiments that we may be able to be a little more certain about. The approach that my group takes in addressing that question is to do the whole ecosystem experiment in such a way that we can distinguish between the direct effects of the toxicant and the indirect effects that are caused by interaction within the system. Then, if we happen to know enough about another system to know what interactions are occurring in the other system, we can guess which of those things we saw in our experiment might happen in the other system and which ones probably would not. Again, that depends on a knowledge of ecosystem structure and function that we are only beginning to put together.

CHAPTER 21

RESPONSIBLE LONG-TERM USE OF AGRICULTURAL AND URBAN LAND FOR SOLID WASTE DISPOSAL

Terry J. Logan and Robert H. Miller

Agronomy Department
The Ohio State University
Columbus, Ohio 43210

ABSTRACT

Because the nation's agricultural soils constitute one of its most valuable resources, long-term productivity of such soils must be maintained or enhanced. At the same time, proper and responsible use of soils for disposal of certain solid waste materials can recycle valuable nutrients, improve the physical properties of soils, and provide a cost-effective method for waste disposal with financial benefits to the urban sector. Likewise, some solid wastes, primarily composted organic materials and fly ash, aid in reclamation of land disturbed or impaired by surface mining. These reclamation efforts may increase the land's agricultural productivity beyond its premining capability. Organic solid wastes can substitute, in the urban sector, for commercial products in landscaping (e.g., of highways, parks and homes).

The long-term potential hazards of solid waste disposal on agricultural and urban land are from those waste constituents that will accumulate in soil and are not readily recycled through plants, primarily heavy metals and stable toxic organics. For domestic sewage sludges, garbage, and agricultural processing wastes, annual application rates that satisfy the nitrogen or phosphorus requirements of agricultural or horticultural crops (1–10 t/ha) will prevent accumulation of these materials, and regulations as presently formulated for cadmium and polychlorinated biphenyls (PCB) can control total accumulation in a particular soil. Disposal of more contaminated solid wastes on land may require the use of dedicated sites with controlled drainage and more extensive monitoring. The waste disposal history of all sites should be maintained as guidance for long-term land use change.

INTRODUCTION

The historical use of land, primarily agricultural land, for solid waste disposal has involved those waste materials whose characteristics were such that the soil-enriching benefits they provided through their nutrients and organic matter were considered to be more important than any health hazard they may have posed. Livestock manures, human "night soil," and the garbage heap helped man maintain the fertility of his soils before the advent of chemical fertilizers. By the latter half of this century, however, land was also being used for disposal of various industrial wastes and excessive applications of municipal wastes and manure from confined feedlot operations. Disposal, not utilization, was the primary goal of many of these operations, and environmental and resource management objectives often meant containing or reducing the contamination of the disposal site ecosystem. Although many industrial, mining, and processing solid wastes must continue to be handled as the potentially harmful products they are, many other solid waste materials either are of such low-level toxicity or offer sufficient potential benefits to the land that they are considered to be worthy candidates for long-term land disposal. Included in this category are organic, nutrient-containing materials such as manures, sewage sludges, and various food-processing wastes; low-nutrient organic materials such as some paper mill wastes and some composts; and inorganic waste products such as fly ash, water treatment lime sludge, and various slag materials.

The discussion excludes livestock manures because well-accepted mechanisms already exist for their disposal and use in the agricultural production system, and focuses instead on the land disposal of municipal organic wastes, primarily digested sewage sludges.

GENERAL PROPERTIES OF COMMON SOLID WASTES

Table I summarizes the general beneficial and potentially harmful properties of solid waste materials that are commonly applied to land.

Table I. Relative Beneficial and Harmful Properties of Selected Solid Waste Materials

Solid Waste	Properties[a]									
	Beneficial					Harmful				
	Organic Matter	Nutrients			Lime Equiva-lent	Heavy Metals	Toxic Organics	Pathogens	Salt	Acidity
		N	P	K						
Digested Sewage Sludge	M	H	H	L	M	V	L	M	M	M
Composted Sewage Sludge	H	M	M	L	M	L	L	L	L	L
Processing Wastes	H	M	M	M	V	L	L	L	V	V
Industrial Wastes, Organic	H	V	L	L	L	L	H	L	L	M
Industrial Wastes, Inorganic	L	V	V	V	V	V	L	L	H	V
Fly Ash	L	L	V	L	H	L	L	L	H	L
Water Treatment Lime Sludge	L	L	L	L	H	L	L	L	M	L
Basic Slags	L	L	V	V	H	M	L	L	M	L

[a]H, M, L = high, medium, and low; V = variable.

The major benefits are organic matter and nutrients (particularly nitrogen and phosphorus), whereas the major long-term hazards are heavy metal and toxic organic accumulations. Some materials, particularly water treatment lime sludge, offer beneficial liming effects, and many of the wastes provide varying amounts of trace nutrients. Many of the properties given in Table I are rate variable. Digested sewage sludges contribute significant amounts of organic matter only at high application rates, and the total amounts of heavy metals added to soil are more important in heavy metal uptake by plants than the concentrations of those metals in sludges. Although salt accumulation can be a problem for long-term land application of some solid wastes in climatic regions with low net percolation rates, it has not been a problem in more humid regions.

LONG-TERM IMPACTS OF SOLID WASTE DISPOSAL ON LAND

Effects on Soil Properties

The most apparent impact of any long-term application of solid waste to land is the accumulation of those waste constituents that are conservative and not readily leached or taken up in large quantities by harvestable biomass. Such constituents would include the heavy metals, phosphate, and elements such as iron and aluminum that exist in high concentrations in some solid wastes but are low compared to natural levels in the soil to which they are added. Highly resistant organic compounds such as the halogenated hydrocarbons will accumulate in soil by adsorption into the lipid fraction of soil organic matter. Other compounds, such as paraquat, are tightly bound in the soil by electrostatic forces, whereas others containing ring structures are chemically incorporated into soil humus by covalent bonds (Bartha, 1980). Bartha (1980) has cautioned, however, that although these compounds are inactivated in the soil, they may be released in the future as biologically active molecules by microbial decomposition of the organic matter, and their accumulations in soil from solid wastes must be considered as potential hazards to the ecosystem.

Although the chemical effects of solid wastes such as sewage sludges on soils have received considerable attention (CAST, 1980), solid wastes can also affect soil physical properties. Large applications of organic wastes, over time, can increase the organic matter content of soils;

this increase can lead to changes in moisture-holding capacity, bulk density, aggregate stability, and other properties. These beneficial effects on soil physical properties are major objectives in the use of large amounts of organic solid wastes on depleted or drastically disturbed soils (Sopper et al., 1980). Chang et al. (1977) found that, in general, fly ash improved or had little negative effect on agronomic soils. Except for highly toxic materials, most solid wastes do not appear to significantly affect soil microbiological processes, and no long-term accumulations of pathogenic organisms in soils result from solid wastes. In regions of low rainfall, salt accumulations from long-term waste applications without appropriate leaching management must be considered a potential problem.

Effects on Agronomic, Silvicultural and Horticultural Plants

The beneficial and deleterious effects of heavy metals and other solid waste constituents on the growth and elemental composition of major agronomic crops have been extensively researched, and the report by the Council for Agricultural Science and Technology (CAST, 1980) summarizes these effects for cadmium and zinc. Heavy metal effects on food crops involve both phytotoxicity (e.g., Zn, Cu, and Ni) and food-chain contamination (e.g., Cd and Pb). Although food-chain contamination is not a concern with silvicultural and horticultural species, phytotoxicity by heavy metals in solid wastes must be considered. Of equal consideration in long-term solid waste disposal is the integrity of land use on a particular parcel of land. The forest land used for disposal of heavy metal–containing solid waste today could be converted to food crop production tomorrow. This issue will be discussed later in the chapter.

Phytotoxicity to silvicultural and horticultural species by heavy metals is not as well documented as for agronomic crops, and effects on nursery stock and seedlings are likely to be more prevalent than they are on established plants. Effects of salt accumulations on silvicultural and horticultural species are well documented (Stewart and Meek, 1977; USDA, 1954), and the range of salt tolerance among species is wide. Livestock waste and commercial sewage sludge products such as Milorganite have been used successfully in the horticultural industry for years, and no long-term problems of using some solid wastes on these species have been indicated. Likewise, sewage sludge has been used with success on forested sites (Smith and Evans, 1977).

Effects on Water Resources

The long-term effects of solid waste disposal on our nation's water quality will involve those waste constituents or transformation products that are toxic, stable, and either (1) of low affinity for soil particles and therefore leachable or (2) of high affinity for soil particles but labile. Heavy metals and many persistent pesticides that have strong affinities for soil particles are effectively retained in the disposal zone of the soil and do not leach to groundwater. Their greatest threat to long-term water quality is through land runoff in dissolved and particulate (sediment-bound) phases and accumulation and recycling of the pollutant in the aquatic system. This threat, however, can be greatly minimized by land disposal management, including site selection, waste incorporation, low application rate, timeliness of application, and runoff diversion.

A great threat to long-term water quality is the leaching of solid waste products to groundwater. Nitrate leaching is a common problem when nitrogen-containing wastes are applied at rates in excess of the assimilative capacity of the vegetation on the site and net percolation is sufficient for leaching. This problem is readily controllable by limiting waste application rates, as discussed later. Of more concern, perhaps, is the slow, low-level movement of some organic compounds to groundwater. The number of reports of contamination of groundwater supplies by organics is increasing (EPA, 1976). Concentrations of many of these compounds in waste materials are very low, their analytical costs are high, and our knowledge of their retention and stability in soil and groundwater (Roberts et al., 1980) is limited.

TWO STRATEGIES FOR LAND APPLICATIONS OF SOLID WASTES: EXTENSIVE (UTILIZATION) AND INTENSIVE (DEDICATED SITES)

Solid Waste Utilization at Low Application Rates on Agronomic Crops

Nutrient-containing solid wastes are applied annually at rates that can supply the nitrogen and phosphorus requirements of agronomic crops. For digested sewage sludges, about 2–20 t/ha is required; Logan and Miller (1980) calculated the effects of low application rates on accumulations of cadmium in soil. Table II shows that annual applications of 2–10 t/ha/y can supply 15–75% of the nitrogen requirements of a corn crop, and phosphorus in amounts greatly in excess of agronomic crop requirements. In Ohio, we have been recommending that farmers use sludge at annual rates of 2–10 t/ha, depending on individual crop requirements for nitrogen and phosphorus, and Forster et al. (1981), in a 1980 survey of 56 sludge land-spreading operations in Ohio, reported a median annual application of 8.5 t/ha. Table II also shows that, for a sludge of median cadmium content (15 mg/kg), annual cadmium applications would be well below the proposed U.S. Environmental Protection Agency (EPA) guidelines, which are to be reduced from 2 to 0.5 kg/ha by 1987. The median cadmium sludge would also allow repeated sludge applications for 30–300 years, depending on annual sludge application rate and soil pH. With a sludge high in cadmium (100 mg/kg), a sludge application rate of 10 t/ha would not exceed the annual cadmium loading criteria until after 1987, but the lifetime loading on a particular site would be reduced to as low as five years.

Application rates as low as 2–10 t/ha require that sufficient agricultural land be available annually in the area. Figure 1 shows that, at these rates, 0.5–2.6% of Ohio's cropland would be required to apply all of the state's sewage sludge. The national picture can be seen in Table III, which shows that, at an annual application rate of 11.2 t/ha, 0.09–1.3% of the cropland would be required in various regions to apply all sludge produced. The greatest land requirement is in the Northeast, but this figure may be optimistically misleading because agricultural land is often not close enough to urban centers for sludge to be transported and spread economically. A high percentage of land, devoted to sensitive crops (e.g., vegetables and tobacco), would probably not be used for sludge disposal. Table III also indicates that the greatest land-spreading activity is in the corn belt, lake, and Pacific states. This activity may reflect the interest and progress in land application of sewage sludge in these areas as well as the problems with cropland availability in areas such as the Northeast and Appalachia.

The many environmental and economic benefits of low rates of application of solid wastes explain the increasing popularity of this strategy. If land is available within reasonable hauling distance of the source, low application rates provide the following benefits:

1. environmental: slower accumulation of heavy metals; lower threat of nitrate contamination of groundwater; reduced potential of surface runoff of waste materials; reduced potential for odor problems;
2. economic: reduced soil, plant, and water monitoring costs; some opportunity to recover disposal costs by farmer payments for nutri-

Table II. Effect of Low Sewage Sludge Application Rates (2-10 dry t/ha/y) on Corn Nutrient Requirements and Cadmium Accumulations[a]

Sludge Application Rate (t/ha/y)	Percent of Nutrient Requirement of Corn Crop Provided by Sludge		Annual Cd Loading (kg/ha/y)		Lifetime Years of Sludge Application Based on Cd Accumulation[e]			
					Low-Cd Sludge		High-Cd Sludge	
	Nitrogen[b]	Phosphorus[c]	Low-Cd[d,f] Sludge	High-Cd[d,f] Sludge	pH < 6.5	pH > 6.5	pH < 6.5	pH > 6.5
2	15	167	0.03	0.2	166	332	25	50
4	30	333	0.06	0.4	83	166	13	25
6	45	500	0.09	0.6	56	112	8	17
8	60	667	0.12	0.8	42	84	6	13
10	75	833	0.15	1.0	33	67	5	10

[a]From Logan, T. J., and R. H. Miller. 1980. "Ohio's Program for Application of Municipal Sewage Sludge on Farmland," in *Proceedings of the National Conference on Municipal and Industrial Sludge Utilization and Disposal* (Silver Spring, MD: Information Transfer Inc.), pp. 71-75.

[b]Assumes 19 kg/t available nitrogen in the sludge (100% of the NH_3-N and 30% of the organic-N), and a requirement of 250 kg/ha of nitrogen for a 10,000-kg/ha corn crop.

[c]Assumes a corn requirement of 30 kg/ha of phosphorus, and a sludge phosphorus content of 25 kg/t.

[d]The low-Cd sludge contains 15 mg Cd/kg, a median value for Midwest municipal sludges. The high-Cd sludge contains 100 mg Cd/kg.

[e]EPA regulations allow a total Cd accumulation of 5 kg/ha for a soil with medium CEC (5-15 meq/100-g soil) when background pH is <6.5 and 10 kg/ha for soils with pH >6.5 (*Federal Register*, September 13, 1979).

[f]EPA proposed annual Cd application rates are: present to June 30, 1984, 2.0 kg/ha/y; July 1, 1984 to December 31, 1986, 1.25 kg/ha/y; after January 1, 1987, 0.5 kg/ha/y (*Federal Register* September 13, 1979).

Table III. Estimate of Total Cropland Acreage Needed for Annual Utilization of All Sewage Sludge in the United States[a]

Region	Land Area (10^3 ha)			Sludge Production (10^3 dry t/y)	Cropland Needed		Percent of Sludge Currently Used on Land
	Cropland	Pasture and Rangeland	Forest		Hectares	Percent of Total	
Northeast	7,021	2,416	24,637	980.8	87,488	1.25	5.4
Lake States	17,730	2,434	17,091	402.9	35,972	0.20	33.6
Corn Belt	36,158	10,151	9,982	813.8	72,657	0.25	39.8
Northern Plains	38,300	32,766	777	68.9	6,156	0.20	9.0
Appalachia	9,140	7,449	24,978	291.7	26,042	0.28	6.6
Southeast	7,102	6,973	26,016	280.3	25,029	0.35	16.6
Delta States	8,580	5,334	16,641	113.7	10,149	0.12	10.1
Southern Plains	17,005	51,590	5,744	268.2	23,944	0.14	26.9
Mountain	16,993	79,391	8,917	174.6	15,593	0.09	28.6
Pacific	9,386	24,734	13,079	486.9	43,473	0.46	37.7
Hawaii, Puerto Rico, Alaska	238	4,767	769	19.7			0.2

[a]From USDA. 1978. "Improving Soils with Organic Wastes," U.S. Department of Agriculture report to Congress in response to Section 1461 of the Food and Agriculture Act of 1977, U.S. Government Printing Office, Washington, DC.

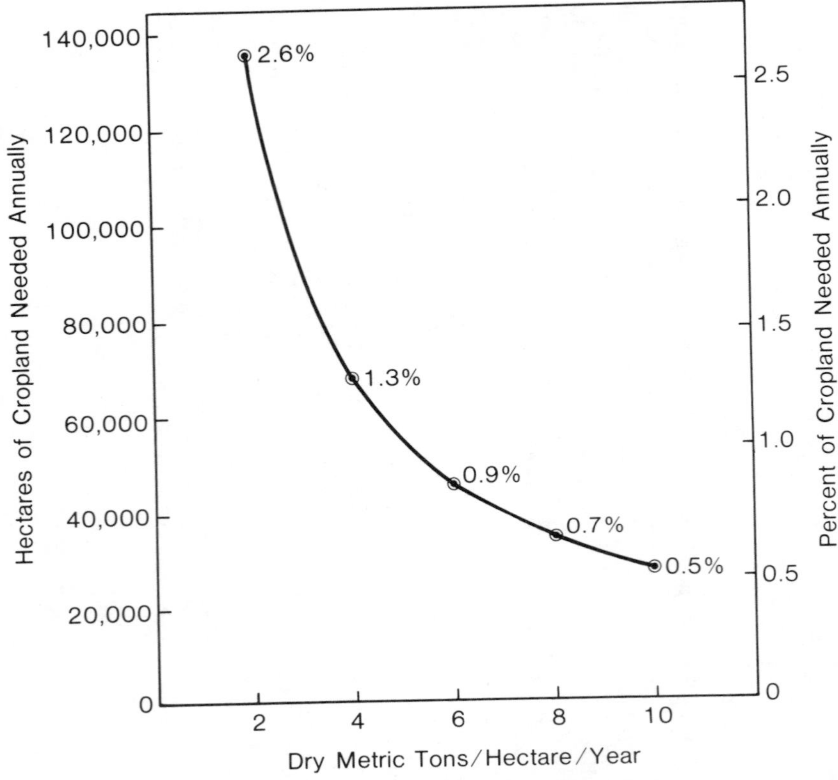

Figure 1. Ohio cropland needed annually to spread all of the state's sewage sludge at application rates of 2–10 t/ha [from Logan, T. J., and R. H. Miller. 1980. "Ohio's Program for Application of Municipal Sewage Sludge on Farmland," in *Proceedings of the National Conference on Municipal and Industrial Sludge Utilization and Disposal* (Silver Spring, MD: Information Transfer, Inc.)].

ents in the waste; reduced costs for land acquisition or rent; and possible savings in disposal equipment costs;

3. management: greater flexibility in time of waste disposal; single-pass application, which reduces compaction compared to multiple passes.

Continued long-term application of solid waste on private land (land not owned or controlled by the waste producer), however, requires a commitment for cooperation between the waste producer and the local user community. In Ohio this involves input from the waste producer, the user group, and all technical agencies involved in regulating or making recommendations on waste disposal. Although this input is most important in the initial stages of a program, a continuing effort is required to ensure long-term success.

The greatest potential problem of the low application, "extensive" strategy of waste disposal is the lack of land-use control on the application sites. Land used today for grain crops that accumulate low levels of heavy metals could be used in the future for accumulator crops such as tobacco or leafy vegetables. To guard against this, complete records of waste disposal, including annual and cumulative applications and waste analysis, should be a required component of any waste disposal program and part of the legal deed of that property. Records of this type are already commonly used in land application of sewage sludge, but requirements vary widely from state to state and even from county to county.

Solid Waste Disposal at High Rates on Dedicated Sites

The dedicated site or "intensive" strategy of solid waste disposal involves application of the waste to land at rates that are usually higher than can be used by the existing vegetation or that might contaminate food-chain crops. The advantages of the dedicated site strategy are:

1. prevents contamination of the food-producing land resource, except for the small area of the dedicated site itself;
2. provides the solid waste manager with total control of the disposal site;
3. allows concentration of waste management resources at a single site;
4. may reduce costs for waste transport and application;
5. is easier to monitor than many sites extensively; and
6. requires less cooperation with the local community.

These advantages are attractive, especially in regions where land is not readily available for more widespread disposal and where the diversity of food crops may increase the fear of food-chain contamination by waste disposal. However, major problems with this strategy also exist. The potential for surface and groundwater contamination is greater with repeated high applications to the same site, and accumulation of heavy metals or salts could reduce the effective lifetime of the site. Likewise, a high degree of management would be required for dedicated site disposal, including the high costs of extensive plant, soil, and water monitoring. Finally, the dedicated site strategy would probably prohibit any long-term benefits to the nation's soils from nutrient and organic matter recycling.

SUMMARY

Although the long-term effects of disposal of some solid wastes on land, particularly wastes containing stable and toxic organic compounds, are uncertain, research indicates that some opportunity exists for the disposal of solid wastes without the threat of long-term impairment of land and water resources and endangerment of food-producing capacity. Environmental and human health problems are greatly reduced when wastes are applied to the land at low rates, but this extensive strategy involves a high degree of cooperation between the waste producer and the land owner or user and would require rigorous documentation of waste disposal to a specific site to aid in planning for any future land use changes.

On the other hand, an intensive strategy of solid waste disposal on dedicated sites, perhaps appropriate for the most hazardous materials, provides protection of the present and future food-producing potential of our land, but at significant costs. Although the dedicated site strategy may be appealing to the waste disposal manager or the regulatory official because of the greater control that it offers over the waste disposal process, we should not forgo the opportunity to use the nutrients and organic matter provided by more environmentally benign materials such as domestic sewage sludges low in heavy metals.

REFERENCES

Bartha, R. 1980. "Pesticide Residues on Humus," *Am. Soc. Microbiol. News* 46:356-360.

CAST. 1980. "Effects of Sewage Sludge on the Cadmium and Zinc Content of Crops," Report No. 83, Council for Agricultural Science and Technology, Ames, IA.

Chang, A. C., L. J. Lund, A. L. Page, and J. E. Warneke. 1977. "Physical Properties of Fly Ash–Amended Soils," *J. Environ. Qual.* 6:267-270.

EPA. 1976. "Waste Disposal Practices and Their Effects on Ground Water." U.S. EPA, Office of Water Supply and Office of Solid Waste Management Programs, Washington, DC.

Forster, D. L., T. J. Logan, R. H. Miller, and R. K. White. 1981. "State of the Art in Sludge Landspreading: 1980 Perspective," Ohio Agricultural Research and Development Center Research Bulletin No. 1134, Wooster, OH.

Logan, T. J., and R. H. Miller. 1980. "Ohio's Program for the Application of Municipal Sewage Sludge on Farmland," in *Proceedings of the National Conference on Municipal and Industrial Sludge Utilization and Disposal* (Silver Springs, MD: Information Transfer, Inc.), pp. 71-75.

Roberts, P. V., P. L. McCarty, M. Reinhard and J. Schreiner. 1980. "Organic-Contaminant Behavior During Ground Water Recharge." *J. Water Poll. Control Fed.* 52:161-172.

Smith, W. H., and J. O. Evans. 1977. "Special Opportunities and Problems in Using Forest Soils for Organic Waste Application," in *Soils for Management of Organic Wastes and Waste Waters,* L. F. Elliott and F. J. Stevenson, Eds. (Madison, WI: American Society of Agronomy), pp. 429-456.

Sopper, W. E., S. N. Kerr, and D. T. Murray. 1980. "Mined Land Reclamation in Pennsylvania with Municipal Sludge," in *Proceedings of the National Conference on Municipal and Industrial Sludge Utilization and Disposal* (Silver Spring, MD: Information Transfer, Inc.), pp. 90-96.

Stewart, B. A., and B. D. Meek. 1977. "Soluble Salt Considerations with Waste Applications," in *Soils for Management of Organic Wastes and Waste Waters,* L. F. Elliott and F. J. Stevenson, Eds. (Madison, WI: American Society of Agronomy), pp. 219-234.

USDA. 1954. "Diagnosis and Improvement of Saline and Alkali Soils," Agriculture Handbook No. 60, L. A. Richards, Ed., U.S. Department of Agriculture, Washington, DC.

USDA. 1978. "Improving Soils with Organic Wastes," U.S. Department of Agriculture report to Congress in response to Section 1461 of the Food and Agriculture Act of 1977, U.S. Government Printing Office, Washington, DC.

DISCUSSION

***Harry Freeman*, Office of Appropriate Technology, State of California:** You mentioned 0.5% of all the agricultural land in Ohio. Are you talking about using that to dispose of all the sludge that is generated in Ohio?

T. Logan: Yes. Right.

Harry Freeman: Did you look at how that would be affected if composting came in in a big way? Would the numbers go up or down, or how would that affect you?

T. Logan: Well, we view composting as a completely different product. We do not really see compost and sewage sludge as being used routinely on agricultural land. First of all, in order to supply nutrients for crop production, the farmer would have to apply a lot more wastes. The main benefit, as I see the compost, is in the organic matter. Now, if the soil is badly worn out and it needs organic matter, or if we are talking about strip mine reclamation, or the case of sod producers that routinely scrape off a couple of inches of soil when they remove the sod, we see those as outlets for compost. In fact, we recently started a market survey for composted products in Ohio. What we find is that there appears to be no problem in getting rid of the compost or even selling it, but I think the compost will be going into routes other than traditional agriculture. There is a lot of strip-mine land in Ohio that needs organic matter.

***J. Hansen*, University of Aalborg, Denmark:** Isn't impairment after 200–300 years a long-term impairment? After such time, land would not be available for food crop production even when applying low-cadmium sludge. Why not combine extensive usage and nonfood crop production?

T. Logan: Well, I think that there are instances where that certainly may be the strategy that we want to go with. But what I am saying is that there is a large area in the United States, in the corn belt for example, where I do not think that is necessary. Considering the clean sludges and the rates we are talking about, I do not see a long-term impairment problem.

J. Hansen: Just a followup. You did not answer my second question: whether you could combine the extensive use in nonfood chain crops? Because if you look at the acreage you need, about 0.5–1.5% is available as marginal land already.

T. Logan: Yes. It is being done on such things as forested land, strip-mined land. Of course, when you get into forested land, you have an application problem. It is a lot more difficult to spread sludge between trees than on a nice open area.

***John Cherry*, University of Waterloo:** Chromium is one of the most mobile of the heavier transition metals in groundwater; I did not notice chromium on your list. Do you have any comments on chromium?

T. Logan: Again, our experience in soils is that chromium is not that mobile, especially in terms of plant uptake. You can have much higher levels of chromium addition to soil without seeing the uptake that you would, say, with cadmium or zinc. I think Rufus pointed that out.

***Rufus L. Chaney*, USDA:** Would you comment, Terry, about whether you think that the cumulative loading limits that EPA, Food and Drug, and several other agencies have indicated in the United States would have any impairment in the worst-case analysis.

T. Logan: No, and again this is just a personal opinion, but looking at those proposed limits and being somewhat involved in how they were developed, I think there is a tremendous degree of conservatism built in. I don't think the public is aware of that. Even at the accumulative levels of 20 kg/ha, which is the highest you can get under the high pH and high CEC, I think there is still a high degree of safety.

CHAPTER 22

WHO IS RESPONSIBLE FOR THIS MESS?—PUBLIC PERCEPTIONS OF THE WASTE MANAGEMENT ISSUE

Lois A. Florence and Joanne Sheldon

Environmental Action Foundation
724 Dupont Circle Building
Washington, DC 20036

ABSTRACT

In the wake of Love Canal and reports of other tragic environmental problems, public concerns and fear in many communities have risen to near-hysterical levels. This concern has led to public distrust of the government agencies and industries involved in waste management.

To assess public attitudes toward these agencies and industries and to measure public knowledge of waste management issues, a national survey was conducted. A questionnaire was mailed to 1000 environmentalists who belong to groups that work on hazardous waste. The general public was surveyed through telephone interviews. Preliminary results of this survey are presented, and literature and other surveys available on the subject are reviewed.

THE PROBLEM

In 1980 the U.S. Environmental Protection Agency (EPA) estimated that the production of hazardous waste in this country, already at 57×10^6 t, was increasing 3.5% annually (EPA, 1980a). Unbelievably, as much as 90% of that total has been disposed of by practices that do not meet current federal regulations (EPA, 1978).

Mismanagement of hazardous waste has caused many problems, including contamination of groundwater, adverse health effects, crippling of municipal treatment plants, soil contamination, crop damage, and livestock loss (EPA, 1980b). Damage caused by improper hazardous waste disposal has been reported in every state. Indeed, almost every community seems recently to have discovered its own illegal dump site. Living next door to a dump site and hearing of illegal dumps across the country have made the public realize just how pervasive the problem of waste disposal has become.

Yet, as recently as five years ago, the phrase "hazardous waste" was not part of most people's vocabulary. In 1976 Congress passed the Resource Conservation and Recovery Act (RCRA; PL 94-580) to ensure that our country's wastes were properly managed. Subtitle C of RCRA addresses hazardous wastes specifically, imposing controls over the management of those wastes from generation to disposal.

Even as the law was being passed and regulations were being developed, thousands of abandoned dumps containing hazardous substances that had been improperly disposed of were being discovered with increasing frequency. Yet, hazardous wastes still did not become an issue of national public concern until almost two years after the law was passed, when a suburban neighborhood of Buffalo, New York made headlines.

The tragedy at Love Canal near Niagara Falls, New York has become the most widely reported incident of improper disposal of chemical wastes. It made national news in August 1978 when the New York State Commissioner of Health declared Love Canal an imminent health hazard. He recommended that children under the age of two and pregnant women be evacuated from a neighborhood literally built on top of a former disposal site for chemicals (Brown, 1980). The hazard was created when leaking drums, buried 25 years earlier, began to corrode and their contents percolated through the soil into yards and basements. One writer described Love Canal as a name that "will live in infamy as a synonym for uncontrolled and improper disposition of toxic wastes" (Fine, 1980).

The most recent legislative action to address these tragic problems was passed by Congress in late 1980. The Comprehensive Environmental Response, Compensation and Liability Act (PL 96-510), usually referred to as Superfund, established a $1.6 billion trust fund to provide emergency cleanup by the U.S. government for chemical spills and hazardous waste dumps that threaten human health and the environment.

Although Love Canal received more national publicity than any other incident of improper disposal of chemical wastes, it is not necessarily the worst case known to many individual Americans. Too many communities nationwide are discovering an excess of chemicals in their own towns—contaminating drinking water, poisoning air, damaging wildlife, and threatening human health.

A direct result of these discoveries of abandoned sites throughout the country has been public opposition to the construction of new, properly designed facilities for handling hazardous waste. Although RCRA regulations should prevent any new facility from threatening human health or the environment, most people remain steadfastly opposed to the siting of a facility in their communities. They fear that hazardous wastes cannot or will not be handled properly. This fear stems from past failures of industry and government to manage wastes safely (Centaur Associates, Inc., 1979).

New facilities for properly handling hazardous wastes are needed. RCRA regulations require disposal at approved facilities, and marginal sites now in operation may be forced to close because they cannot upgrade to meet federal standards. As a by-product of many manufacturing processes, hazardous wastes will continue to be produced, requiring that safe disposal sites for the substances be found.

The best way to allay the fears of the public is education. The public should be informed about hazardous wastes and about any proposed facility to increase public awareness of the issues involved.

THE STUDY

To determine public perceptions of hazardous waste issues, the Environmental Action Foundation (EAF), a national educational institute in Washington, DC is conducting a national survey. The major objectives of this survey are to examine:

1. public awareness of hazardous waste issues, including any problems caused by hazardous wastes in the Washington, DC area;
2. level of public concern about hazardous waste and how it affects them;
3. public awareness of the government's role in hazardous waste management and opinions about the effectiveness of governmental agencies in dealing with hazardous waste; and
4. accessibility and usefulness of information about hazardous waste and the degree of trust in the sources of this information.

Two target populations from the public were identified: nearby residents and environmentalists. Nearby residents are defined as adults who live within 20 miles of an abandoned hazardous waste site and who have a telephone. Nearby residents were chosen for study because they are more likely to be aware of hazardous waste issues and, therefore, are in a better position to make informed comments. Environmentalists are adult members of Environmental Action, Inc. (EA), Sierra Club, or Friends of the Earth (FOE). These three national groups were chosen because of their work on hazardous waste management issues, including study of abandoned sites and passage of Superfund.

To gather a random sample of nearby residents, a listing of each state's top-priority abandoned site for cleanup under Superfund was obtained (Finlayson, 1981). Five sites were randomly selected, each one representing a different EPA region. A listing of all existing telephone exchanges in the communities near each site was compiled, and phone numbers were chosen through a simple random selection process known as random digit dialing (Waksburg, 1978). A total of 125 telephone interviews was completed with approximately an equal number of responses from each location. About 70% of all respondents contacted agreed to be interviewed.

A sample of environmentalists was selected randomly from the membership lists of EA, Sierra Club, and FOE. A total of 500 questionnaires was mailed to EA members. At this writing 62% have been returned, with 225 of these responses included in this report.

The survey was carefully designed and executed to avoid biasing the results. A conscientious effort was made to design clear questions that are neutral in tone. Interviewers were carefully instructed on the principles of interviewing to ensure that they minimally affected responses. All appropriate precautionary measures were taken to guarantee that the samples selected were indeed random and representative of the populations as they are defined. Each environmentalist and each nearby resident had an equal chance of being selected.

The findings reported are strictly preliminary. They indicate possible trends in public opinion on hazardous waste management issues. At this stage of the study, it cannot be inferred that the opinions expressed necessarily reflect the views of the general public. As more information is gathered from returned questionnaires and from telephone interviews, more conclusive results will be reported.

FINDINGS

Concern

To determine public perception of the importance of current national issues, especially in relation to environmental issues, the EAF study asked respondents how concerned they are about each of six national problems. Overall, the most pressing problem named by nearby residents is crime, about which 98% of the respondents are "very concerned" or "somewhat concerned." Environmentalists are more concerned about hazardous waste than any of the other issues mentioned, with 100% reporting that they are "very concerned" or "somewhat concerned" about the issue. For both groups concern over the disposal of hazardous wastes is higher than the concern expressed for any other environmental issue.

A report on national public opinion released in 1980 indicated that similar findings have been reported in past environmental surveys. The study, completed by Resources for the Future (RFF) for the President's Council on Environmental Quality (CEQ) found that 64% of those polled expressed "a great deal" of concern about the disposal of industrial chemical wastes that are hazardous (RFF, 1980).

In early 1980 Cambridge Reports conducted an opinion survey for the Chemical Manufacturers Association questioning politically active individuals, chemical industry neighbors, and influential opinion leaders. The survey showed that, of all the health and safety issues involving chemical companies, waste disposal was the issue of greatest concern for the three groups interviewed. The respondents also felt that the level of industry's concern about the waste disposal problem was significantly lower than their personal concerns (Cambridge Reports, 1980). These findings indicate that hazardous waste is a subject of grave concern to most Americans.

Not only are most respondents surveyed by EAF very concerned about hazardous waste issues on the national level, they are also worried about how hazardous waste in their own communities may be affecting them. Of those surveyed, 56% believe that hazardous waste has created a problem in their area. The vast majority stated that the most serious problem is within 5 miles of their home.

The problem most often cited by nearby residents is hazardous waste disposal (41%). Many stated that barrels of hazardous waste are being dumped nearby, and many knew of apartment complexes built on abandoned hazardous waste sites. Many respondents (29%) fear that hazardous wastes are leaking and contaminating their drinking water source. Water pollution is the local hazardous waste problem most frequently mentioned by environmentalists (44%).

Most nearby residents and environmentalists report having personal concerns about the problems they mentioned (32 and 48% respectively). Many indicated that they would feel concerned personally if they lived closer to the problem. The most frequently expressed concerns of both groups center around the manner in which the problem is being dealt with, the threat that hazardous waste poses to human health, and drinking water contamination.

Knowledge

The EAF survey asked a series of questions designed to determine how knowledgeable the public is about some environmental topics that have been relatively well publicized, such as Love Canal and Agent Orange. Of the nearby residents, 25% correctly described what occurred at Love Canal. In a 1980 study of the general public conducted by RFF, 22% described what happened at Love Canal correctly (RFF, 1980). Of the environmentalists questioned in the EAF study, 68% gave a correct description.

An opinion survey sponsored by the Chemical Manufacturers Association found that recall of specific news stories about chemicals or the chemical industry (including hazardous waste management problems) is relatively low. The news item most often mentioned was Love Canal.

The EAF study also asked respondents to define Agent Orange. Of the respondents living near sites, 38% gave a correct definition, compared with 68% of the environmentalists. In a study conducted by Duke University to assess North Carolina residents' perceptions of hazardous substances (Richardson, 1981), the incident of polychorinated biphenyl (PCB) disposal along a roadside in the state was used to measure knowledge. Most (90%) of the respondents said they had heard of the incident, and 62% correctly identified the substance involved.

The Duke study is the only piece of research reviewed here that found most of the general public to be knowledgeable about hazardous waste. The remaining studies have indicated a relatively low level of public knowledge about hazardous waste issues. The RFF survey points out that a low level of public knowledge is not unique to environmental issues but is consistent with levels found in other policy areas. Simply because the public has a relatively low level of knowledge in a policy area does not invalidate opinion research results on the issues involved. "Just as people have opinions about inflation even though they may not know the current rate of inflation within several percentage points, so their views on waste dumps for hazardous chemicals should be taken seriously even though they may not be able to say what happened at Love Canal" (RFF, 1980). Although a respondent may confess to being unfamiliar about an issue as widely publicized as Love Canal or Agent Orange, he or she will certainly have plenty of informed comments to make about the hazardous waste facility a couple of miles down the road.

Environmentalists in the EAF study were also asked to define midnight dumping and Superfund. (These two questions were asked only of environmentalists and not of nearby residents.) The questions were designed to determine how successful the environmental groups have been in their efforts to inform their members on hazardous waste issues. Midnight dumping was correctly defined by 59% of the environmentalists, and 29% could describe Superfund. A partly correct definition of Superfund was given by 8%. The discovery that so few environmentalists know about Superfund, as compared with other hazardous waste issues, suggests that public awareness of the legislative component of hazardous waste issues is lacking.

Agency Performance

Federal Government

To assess the visibility of the federal agency responsible for dealing with hazardous wastes (EPA), respondents were asked by EAF to name it. Most nearby residents (90%) and environmentalists (62%) could not name EPA. The Duke study found that 76% of North Carolinians surveyed could not name the federal agency responsible. Respondents were then asked if they were satisfied with the way the agency was dealing with hazardous waste in their area. Of the environmentalists that correctly named EPA, 26% replied that they were satisfied, whereas 47% were "not very satisfied" or "not satisfied at all." When asked to explain their opinions, most (25%) of the environmentalists stated that EPA's flaws stemmed from problems within EPA itself. Some think that the agency is understaffed or that the amount of work to be done exceeds the organization's capacity.

In addition, a sizable number of respondents (20%) complained that EPA is too slow and is essentially inactive; 11% think EPA's problems are caused by insufficient funds, whereas another 11% think EPA tends to yield too much to business and political interests. Only 19% said something positive about EPA, with the most common response being that EPA has been successful in some of its efforts in dealing with hazardous wastes.

State Government

The same agency-directed questions were asked about state agencies responsible for dealing with hazardous waste. The Duke study found that 92% of the respondents could not name the state agency responsible. The EAF study found that only 7% of nearby residents could name a state agency involved in the management of wastes, while 30% of the environmentalists could name such an agency. Environmentalists tend to have more favorable attitudes toward their state government agencies, with 60% responding that they were satisfied with how the agency is dealing with hazardous wastes in their area. Many (40%) commended their state agency, with the most common positive comment being that the agency is successful in at least some of its efforts. The most common criticism is that the state government does not do enough one way or another in dealing with hazardous wastes. Although those who are aware of the government's role in hazardous waste management have strong opinions on the subject, most of the public does not know what the government's function is in dealing with hazardous wastes.

Information

Even though many people are unfamiliar with the government's role in managing hazardous waste and with nationally publicized hazardous waste issues, 90% of nearby residents and 85% of environmentalists report having heard "a fair amount" or "a great deal" about hazardous waste in the past couple of years. Television seems to be the most popular source of information on hazardous waste. Of both the nearby residents in the EAF study and North Carolinians in the Duke study, 49% got most of their information from television. These findings suggest that the best way to reach the public with information on hazardous waste may be through the television medium. Not surprisingly, most environmentalists get their information from reports and pamphlets from environmental organizations.

The second most popular source for all three audiences is newspapers. Magazines are a distant third, with 5% of the nearby residents, 6% of the North Carolinians and 10% of the environmentalists reporting it as their primary source. Yet many national magazines have had considerable coverage of hazardous waste incidents. A cover story in *Time* (Magnusan, 1980) drew criticism from the chemical industry as irresponsible and sensational when it pictured a skeleton submerged in a murky solution labeled toxic chemical waste (McCarville, 1980). This incident may have been particularly disturbing to the chemical industry because a study conducted for Chemical Manufacturers Association found that *Time* was one of the two periodicals most frequently read by the public (Cambridge Reports, 1980).

Coverage of problems related to improper hazardous waste disposal has even invaded the comic pages. A recent example shows a young woman introducing her date to her parents with, "Eddie makes BIG money. He's in illegal toxic waste dumping" (Berry, 1981).

A mere 6% of nearby residents have tried to get specific information about hazardous waste, whereas 22% of the environmentalists surveyed have made such an effort. What is more revealing of the general public's need for information is the finding that 25% of those living near sites and 44% of the environmentalists have at some point wanted to get such information.

Trust

The EAF survey probed respondents on their level of trust in five different groups, asking how confident they are that the information provided by each is reliable. Most nearby residents feel "somewhat confident" in the information provided by state government, scientists, environmental organizations, and the federal government. Most report that they are "not very confident" in the information provided by industry.

Not surprisingly, most environmentalists are "very confident" in the information provided by environmental organizations. Most are "somewhat confident" in that provided by state government and scientists, are "not very confident" in that provided by the federal government, and are "not very confident at all" in the material provided by industry.

The 1980 RFF survey found that 67% of the public had "a great deal" or "some" confidence (13 and 54%, respectively) that the federal government will be able to protect the environment. Other interesting results reported by RFF (1980) include that

> Far fewer people believe that the government is responsive to the public's views. Less than a majority of respondents felt that the federal government provides "citizens like you" with "a great deal" or "some" opportunity to express their views on environmental issues. Still fewer thought that the views of individuals are heard and only one in three believes that their views would be paid a "great deal" or "some" attention. Respondents felt that the government pays attention to the views of national environmental groups, however. In this case almost three out of four people believe that the groups are listened to "a great deal" or "some" (RRF 1980, pp. 38-41).

Who Should Pay?

Disposal of industrial waste in a manner that protects human health and the environment is costly. The EAF study asked respondents who should pay for such safe disposal. Two-thirds of the nearby residents said that industry should pay the entire cost for disposal of hazardous wastes. An identical number of environmentalists agreed that industry should bear the entire cost. Respondents from both groups remarked that the consumer would ultimately pay the cost through increased product prices. The validity of this feeling is supported by a poll conducted for Union Carbide, which found that nine out of ten Americans believe that the consumer ultimately pays the cost of regulation (Lennett, 1980).

CONCLUSIONS

Some interesting conclusions may be drawn from the results of the surveys reviewed in this report. Findings indicate that a relatively small percentage of the population is knowledgeable about hazardous waste issues that have received national attention. At first, hazardous waste management may appear to be a low-profile issue. Yet, public concern about hazardous waste is nearly as high as that for issues of undisputed importance such as inflation and crime. True, it is startling to learn that only 37% of nearby residents can give a correct or partly correct description of the incident at Love Canal. However, more than two-thirds of these respondents gave at least a partly correct definition of hazardous waste. Well over half of both populations surveyed are aware of a problem in their area stemming from hazardous waste management. Two-thirds of nearby residents and 87% of environmentalists expressed personal concern about the effects of hazardous waste on their community and on their health. These findings indicate a trend toward increasing public awareness of the severity of problems associated with hazardous waste management in individual communities.

Although most of the public is at least aware of hazardous waste, efforts must be made to educate the public in this area. More than one-third of respondents surveyed said they have wanted to get specific information about hazardous waste at some point. More than three-fourths of them feel it would be difficult to get such information on their own. As we have seen, the concern definitely exists. The desire and need for information also exist. The only way that the general public can make informed decisions about hazardous waste is to provide them with the information they need and have indicated they want.

REFERENCES

Berry, J. 1981. "Berry's World," *Washington Post* (Sept. 5).

Brown, M. 1980. *Laying Waste: The Poisoning of America by Toxic Chemicals* (New York: Washington Square Press).

Cambridge Reports. 1980. "Opinion Survey: Overview and Summary," prepared for the Chemical Manufacturers Association.

Centaur Associates, Inc. "Siting of Hazardous Waste Management Facilities and Public Opposition," SW-809, U.S. EPA.

EPA. 1978. "Hazardous Waste. Proposed Guidelines and Proposal on Identification and Listing," *Federal Register* 43(243):58946-49028.

EPA. 1980a. "Hazardous Waste Information," SW-737.

EPA. 1980b. "Damages and Threats Caused by Hazardous Materials Sites."

Fine, J. 1980. "Toxic Waste Dangers," *Water Spectrum* (Winter), pp. 23-30.

Finlayson, R. A., Ed. 1981. "Dumps of Defunct Firms and Unidentifiable Generators Submitted as Top Superfund Sites," *Haz. Waste News* 3(29): 227-229.

Lennett, D. J. 1980. "Handling Hazardous Wastes, an Unsolved Problem," *Environment* 22(8):6-15.

Magnusan, E. 1980. "The Poisoning of America," *Time* 116(12):58-69.

McCarville, J. 1980. "Toxic Wastes and Public Trust," *Waste Age* (September), pp. 28-29.

RFF. 1980. "Public Opinion on Environmental Issues, Results of a National Public Opinion Survey," conducted by Resources for the Future for CEQ, DOE, EPA and USDA, Washington, DC.

Richardson, C. J. 1981. "Hazardous Substances in North Carolina: An Analysis of Public Perceptions," School of Forestry and Environmental Studies, Duke University, Durham, NC.

Waksburg, J. 1978. "Sampling Methods for Random Digit Dialing," *J. Am. Stat. Assoc.* 73(361):40-46.

DISCUSSION

***Helga Gerstner,* ORNL:** How do you use reports that you generate such as this one? Is it provided to industry, the Environmental Protection Agency, the Council on Environment Quality, and others that are involved in this issue, or is there some connection where it would be of benefit in general?

L. Florence: I certainly hope so. This is the first kind of project that our Waste and Toxic Substances Group within the EAF has taken on. We have already been approached by two different magazines that cover the waste management field to write an article relative to the results of our survey. We are also writing specific reports for the Sierra Club, Friends of the Earth, and Environmental Action. At the same time, we are very much interested in sharing this with governmental and industry people who can use it.

***David R. Cope*, International Energy Agency, London:** I wonder if, in your survey or any of the other surveys you reported, questions were asked of the public on their knowledge of the source of the wastes? In other words, which industrial product is giving rise to a particular waste, and what are the attitudes as to whether there should be a continuation of production of those products giving the wastes? Also, was there any attempt to get any idea of the extra costs the public would be willing to pay in terms of product costs in order to have high standards of waste disposal facilities?

L. Florence: In our survey, we did ask who people thought generally produced the most hazardous wastes, and the answer was industry, quite closely followed by government. We did not ask any of those other questions, but there were surveys that did, most notably the Richardson survey from Duke University, which gave people a series of trade-off questions. If they had $10 of their tax money, would they want it to go to other social issues or problems or toward hazardous waste management? I cannot give you the exact figures, but an overwhelming majority said that they would like to see more than that $10 going to management of the waste disposal problem. I think that could be unique to North Carolina in light of the fact that this finding said that 90% of the people were aware of their PCB incident relative to dumping along the roadside; so people know it's a real problem there. Also asked was what kind of disposal facility or what type of facility to handle the wastes would be preferred. Would it be incineration, land disposal, or recycling of wastes? There again I don't have the exact figures, but I can tell you that land disposal was the last method preferred.

CHAPTER 23

RISK ANALYSIS: SOLID WASTES FROM OIL SHALE AS AN EXAMPLE OF THE USE OF HEALTH EFFECTS AND ECOLOGICAL DATA FOR RISK ESTIMATION

Lawrence B. Gratt

IWG Corporation
975 Hornblend Street
San Diego, California 92109

ABSTRACT

Risk analysis is a predictive process that provides information to aid in formulating and managing a program of environmental research. Health risks are determined by an interactive process that uses (1) exposure resulting from source terms, as characterized by the probability of exposure to pollutants and accidents, based on a reference industry fuel cycle; (2) health effects, as characterized by the conditional probability of effect given the exposure; and (3) the population at risk. The health risk is defined as the expectation of the population health effects for all exposures under consideration.

The oil shale industry will be an immense solids-handling industry moving 1.9×10^6 t of raw shale to produce 1×10^6 bbl/d of oil. Solid waste generated by the oil shale industry will increase air and water pollution. Air pollutants from solid waste piles are primarily fugitive dust. Water pollutants derived from solid wastes consist of leachates percolated from spent shale.

Health effects, both public and occupational, for oil shale can be analyzed for both threshold and nonthreshold effects. Airborne pollutants, especially dust and particulates, are a major concern for the occupational workforce. The potential for serious water pollution exists, especially from leachates from the spent shale.

Ecosystem risks result from the loss of habitat associated with the enormous amount of solid waste generated by a commercial-size oil shale industry. Crude models for indicator species using available data indicate that the mule deer population may decline by one-fifth, and trout populations in the region may be drastically curtailed under extreme conditions.

INTRODUCTION

The objective of the oil shale risk analysis (OSRA) project is to analyze potential risks to human health and the environmental effects associated with an oil shale industry capable of supplying a significant fraction of the U.S. energy demand. The OSRA results are reported in a health and environmental effects document (HEED). This document quantitatively describes the knowledge and uncertainty about potential health and ecosystem effects of an oil shale industry and provides information to help in formulating and managing a program of environmental research.

A commercial U.S. oil shale industry will be an immense solids-handling industry, which moves 1.9×10^6 t of shale to produce 1×10^6 bbl/d of shale oil. The industry will also have an enormous disposal problem, with about 1.7×10^6 t/d of spent shale (corresponding to a volume of about 1.3×10^6 m^3). The solid wastes are indicative of a source of air emissions and water effluents of concern to human health and the ecosystem. The disturbed surface area used for storing and disposing of raw and spent shale represents large disturbances in a semiarid environment of significant recreational value. This chapter illustrates the use of health effects and ecological data in risk estimation.

METHODOLOGY

The overall OSRA estimate of risk associated with an oil shale industry is the result of an interactive process of (1) exposure resulting from source terms (defined as pollutants, fugitive emissions, and accidents), (2) health effects of those exposures, and (3) population at risk. Exposure to source terms is established on the basis of probability of exposure and analogous industry data for the oil shale fuel cycle. The health effects are characterized by the conditional probability of effect, given the exposure. The risk is defined as the expectation of population effects (e.g., health effects) under the exposures considered.

The oil shale reference industry is based on a fuel cycle for oil shale, from in-place resource extraction to the end use of the products, and a scenario for producing 1×10^6 bbl/d of shale oil. The scenario for the 1×10^6 bbl/d industry was based on 15 sites in the Piceance Creek and Uintah Basins and a mix of the current oil shale production technologies, including the modified in situ (MIS) and true in situ (TIS) technologies. This scenario, presented in Table I, is used to develop site-specific pollutant source terms, including the magnitude of the solid wastes generated at each site. The solid wastes produced at these sites are sources of air emissions and water effluents. The primary effect of the solid waste will be the disturbed surface area used for storage and disposal. A rough estimate of land needed to store spent shale after 30 years of operation is presented in Table I for the oil shale scenario.

Table I. Risk Analysis Scenario with Solid Waste Land Disturbance for 1×10^6 bbl/d Shale Oil Production

Tract (Group)	Process[a]	Production Level (bbl/d)	Land Disturbance[b] (km^2/30 y)
Surface Induced			
Long Ridge (Union Oil Co.)	Union B	150,000	36
Superior (Superior Oil)	Superior	46,000	7.8
Dow (Colony, DCO)	TOSCO	47,000	12
Sandwash, Utah	TOSCO	50,000	12
U-a, U-b (Combined White River Project)	Union B	100,000	24
Paraho, Utah	Paraho	30,000	7.5
Parachute (Chevron)	Paraho	100,000	24
E. Parachute Creek (Mobil)	TOSCO	100,000	24
Parachute Creek (Getty)	TOSCO	50,000	12
Naval Oil Shale Reserve	Paraho	100,000	24
Multiminerals	Superior	50,000	8.4
C-a (Rio Blanco)	TOSCO	100,000	24
Modified in situ: C-b (OXY/Tenneco)	OXY (MIS)/ TOSCO	100,000	9.6
True in situ			
Equity	BX pure/steam	10,000	0
Geokinetics	LOFRECO, in situ	10,000	0

[a]Process assigned based on available characterization data for analysis purposes; not necessarily a currently planned process.

[b]Estimates based on spent shale piles 30 m high, 50% backfill into mines, and 1360 kg/m^3 (85 lb/ft^3) compaction of spent shale.

By far, the major solid waste concern is the spent shale because only about 50% can be disposed of by backfilling the mines. The remaining will be stored in piles or used to fill in canyons, followed by revegetation. The source terms are transported and transformed in the biosphere, resulting in exposure to man and the ecosystem.

The projected populations at risk, for the occupational workforce (59,000) and public (685,000) within the 18,000-mi^2 oil shale "extraction region," are combined with the human health effects from the pollutants of the oil shale industry to predict the health risks. The evaluation of ecosystem risk resulting from an environmental perturbation involves predicting the ecosystem response. Species diversity has been chosen as an initial measure of ecosystem risk. However, more data are needed before a comprehensive, community diversity approach can be used to estimate ecosystem risk related to the oil shale scenario. The initial efforts in this approach are based on estimating the risk to selected indicator species of the Colorado River drainage from obvious major disturbances caused by development of the oil shale resource.

HEALTH EFFECTS

Human populations working in and living near the oil shale industry may be exposed to many pollutants: sulfur oxides, nitrogen oxides, particulates, ozone, carbon monoxide, dozens of trace elements, and a myriad of hydrocarbons, including polycyclic aromatic hydrocarbons and nitrogen-containing aza-azarenes. Many of these pollutants have been associated with adverse health effects, either through toxicologic investigations (in vitro and in vivo studies) or through epidemiologic studies in the occupational or public setting.

Because of the complexity of the exposure, the methodology for the OSRA is to initially analyze the health risk, pollutant by pollutant. Thus, each pollutant is considered independent of the rest. An approach based on using sulfur oxides as a surrogate for all air pollutants is also used.

Excluding carcinogenic effects and the surrogate analysis, the health effects related to the oil shale industry pollutants will be assumed to have thresholds. A threshold is defined as the exposure level below which no adverse health effect is observed. If exposures are determined to be below the chosen threshold, then no effects will be observed and the quantifiable risk becomes zero. Health effects should not be confused with normal homeostasis and physiologic response to exposure. Although this homeostasis varies within the population, the threshold can be fairly well established within a certain range. Thresholds were chosen to protect the most sensitive subgroups. If a threshold is lower than normal for the subgroup, this lower threshold will be used. If exposures are below this lower threshold, the probability of effects occurring in the general population is extremely low.

Thresholds for this analysis are chosen from published reviews of the related toxicology and epidemiology. Where thresholds are very difficult to define, occupational threshold limit values (TLV; ACGIH, 1980) were used with a safety factor of 3. This factor allows for a 24-h exposure and yields conservative thresholds.

Carcinogenic effects will be considered to have no threshold. This assumption reflects the absence of data to the contrary and the need to be conservative with this effect. Although thresholds may differ for subgroups of the population (Rall, 1978), "examination of published dose-response data for chemical carcinogenesis in laboratory animals provides no clear indication of a threshold for any carcinogen" (Lepkowski, 1978). The carcinogen dose-effect relationship for benzo[*a*]-pyrene is based on results from a diesel risk analysis (Cuddihy et al., 1980).

Occupational exposures in the developing oil shale industry may sometimes be above established TLV. The ultimate risk of exposure to particulates depends on the efficiency of safety methods concerned with creating minimal exposure of individuals to the oil shale dust, spent shale dust, and gaseous emissions. Although the health effects resulting from exposure to source terms from the oil shale industry are understood in general terms, much work needs to be done. In particular, worker exposure needs to be quantified in terms of duration and quantity of exposure. Also, a better estimate of the actual number of workers who will be exposed is necessary. Further research is needed to develop dose-effect relationships associated with the oil shale industry exposures.

PUBLIC HEALTH RISK

The initial estimate of the public risk of persons living in the oil shale region from all pollutants of a 1×10^6 bbl/d oil shale industry is summarized in Tables II and III. This summary includes estimates for non-threshold health effects in terms of the potential for yearly excess cancers and excess deaths resulting from the airborne pollutants analyzed. The analysis was based on continuous exposure in the region, with an assumed average lifetime of 70 years. The uncertainty ranges indicate that the results may be from zero (no health effect) to as much as 13 times greater than the estimate (for benzo[*a*]pyrene as a polycyclic nuclear aromatic hydrocarbon health effect indicator). This risk estimate is 3.5×10^{-8} of the overall U.S. annual lung cancer rate of about 3.5% (Cutler, 1975). The excess cancers shown in Table II may not necessarily result in death.

Table II. Oil Shale Risk Analysis—Nonthreshold Public Health Risks from Airborne Pollutants

Controlled Emission	Type of Cancer	Annual Rate of Excess Cancers for a Total Population of 685,000 at Risk (uncertainty range)
Arsenic	Respiratory	4.5×10^{-3} (0–1.4×10^{-2})
Benzo[*a*]pyrene	Respiratory	5.5×10^{-3} (0–7.0×10^{-2})
Cadmium	Lung	4.3×10^{-5} (0–1.6×10^{-4})
Chromium	Lung	1.6×10^{-4} (0–6.4×10^{-4})
Nickel	Respiratory	7.1×10^{-5} (0–3.0×10^{-4})

The increased overall mortality with total pollution exposure was estimated by using sulfur oxides as a surrogate for all air pollutants combined. This is a crude measure of mortality effects based on correlation epidemiologic studies. The results are shown in Table III.

Table III. Oil Shale Risk Analysis—Public Health Risk from Airborne Pollutions Based on Sulfur Oxide Surrogate

Controlled Emission	Annual Rate of Premature Deaths for a Total Population of 685,000 at Risk (uncertainty range)
Sulfur Oxide Surrogate for All Pollutants	73 (0–330)

Analysis of the potential health effects from airborne pollutants indicates that all exposures are below the threshold levels except for hydrogen sulfide, sulfur oxides, and particulates. The number of persons exposed above the conservatively chosen thresholds (i.e., most sensitive member of population) is quite small.

The potential for health effects from waterborne pollutants performed for extreme discharge and leachates reaching public water supplies indicates exposures above the chosen health effects thresholds for sodium, fluoride, mercury, and selenium. The ability to sustain a chronic exposure at these levels is doubtful and will require further analysis of potential water problems. Under extreme conditions, arsenic exposure may present an annual rate of excess skin cancer of 9.4×10^{-3}, with an uncertainty range of 0–0.1 for the 685,000 persons at risk. This excess cancer rate is about 1×10^{-5} of the general U.S. population annual skin cancer rate, as estimated from El Paso cancer registry data (Helm, 1979).

ECOSYSTEM

The ecosystem risk methodology will be based on the concept of community diversity when appropriate data are available. Initial measures of risk are based on first-order estimates for selected indicator species with the terrestrial and aquatic communities.

Mule deer (*Odocoileus hemionus*) have been initially selected as an indicator species to demonstrate the effects of oil shale development on the terrestrial community. Mule deer are probably the terrestrial species evoking the most public interest; thus, more biological data are known about its life history than for any other terrestrial species in the basin. Also, deer hunting provides substantial economic benefit to the region. Population changes of mule deer affect community diversity and are a component of the terrestrial community at risk.

An oil shale industry in the Piceance Creek Basin will disrupt the migration and winter habitat of mule deer by destroying vegetation, erecting migration barriers, and increasing mortality from road kills and poaching. The population decline resulting from the development is difficult to predict because intensive deer management or revegetation success could keep mule deer population decline to a minimum. Even after the oil shale industry is in place, the resulting changes in mule deer populations will be difficult to assess, because natural mortality can cause severe population fluctuations that would effectively mask development-induced effects (Bartman, 1980).

The largest effect on the terrestrial community may come from habitat destruction or alteration from surface disturbance. The magnitude of the surface disturbance will depend on the location, size, and type of mining and processing options. Most of the surface disturbance will be from solid waste disposal. Solid waste disposed of on the surface was estimated on the basis of the oil shale scenario. If 50% of the material is backfilled into mines and the waste is piled 30 m high, the land disturbance from solid wastes in the baseline scenario will be about

230 km^2 (57,000 ac). If another 40 km^2 (10,000 ac) were used for other surface facilities and utility rights-of-way, and an additional 81 km^2 (20,000 ac) were used for associated urban development (DOI, 1973), then the total land disturbance for the 1×10^6 bbl/d baseline oil shale development scenario is 350 km^2 (87,000 acres). Clearly, the amount of habitat disturbance from the scenario also depends on the success of revegetation efforts. Success of vegetation, however, is usually determined from postvegetation productivity estimates and does not necessarily replace pinyon-juniper and sagebrush communities in structure or nutrient array. Thus, the revegetation process and results are associated with additional ecosystem risks. Unless toxic element uptake and salinity, pH, and boron tolerance problems are solved, revegetated communities on spent shale piles will not resemble the natural communities that were replaced. Revegetation of disturbed land other than shale piles is possible, given that revegetated areas will require several years of management before stable, viable communities are established which resemble the existing, natural communities.

Mule deer population decline resulting solely from habitat destruction of oil shale development in the Piceance Creek Basin can be estimated if several assumptions are considered valid:

1. All winter habitat is occupied to carrying capacity by mule deer.
2. There is an even distribution and constant density of deer in a given habitat throughout the Piceance Creek Basin.
3. Revegetated shale piles do not support a significant number of mule deer because suitable forage and cover are lacking.
4. Intensive habitat management will not successfully increase the carrying capacity of nondisturbed vegetative communities.
5. Only habitat loss is considered to affect mule deer populations. Migration barriers, noise, air and water pollution, and road kills are assumed to not cause a significant net decrease in mule deer population when compared with the effects of habitat loss.

Using these assumptions, one can estimate mule deer population decline as a function of land disturbance (Figure 1). The curve is based on an estimated average winter baseline population of 28,000 deer in the Piceance Creek Basin and a total habitat size of 1700 km^2 (425,000 ac). If 350 km^2 (87,000 ac) is disturbed under the baseline scenario, then the relationship given in Figure 1 indicates that a loss of about 5700 deer (20%) can be expected. Although natural fluctuations in the mule deer population are as high as $\pm 40\%$, at any given year the deer herd is predicted to have about 20% fewer individuals because of land disturbance related to oil shale development.

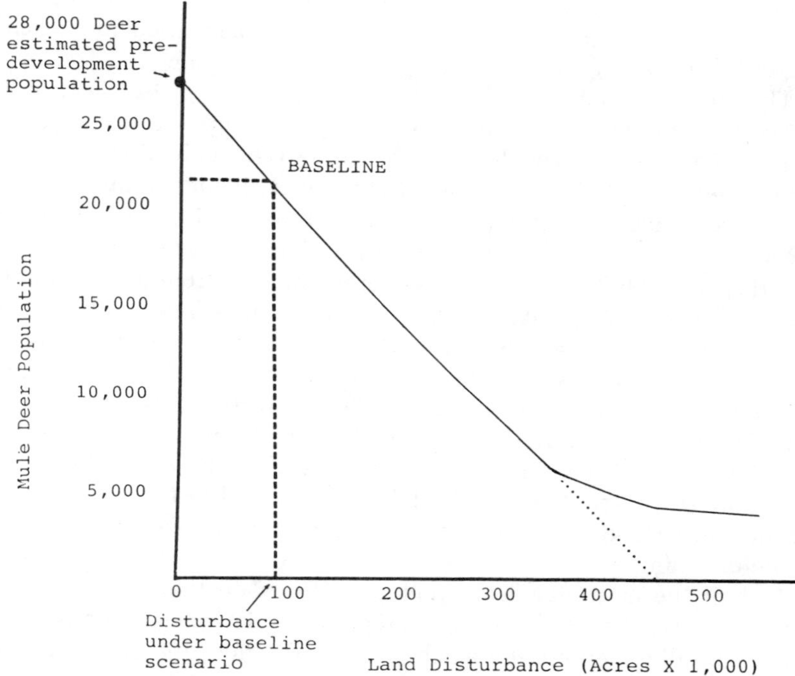

Figure 1. Estimated mule deer population decline because of land disturbance.

The relationship of mule deer decline and land disturbance has a significant uncertainty because of the as yet unknown use of reclaimed areas and the potential carrying capacity of undisturbed lands. The oil shale industry may have no net effect on the mule deer herd if undisturbed lands can be effectively managed for big game. Deer population numbers on the winter range are extremely sensitive to the natural elements, a sensitivity that may overshadow any effects from the oil shale industry.

Ecosystem risk in aquatic communities, as in terrestrial communities, is a function of changes in natality and mortality of populations. However, turnover rates in aquatic communities are usually much higher than those in terrestrial communities, and natural fluctuations in populations can be extreme. Changes in diversity are a more meaningful measure of ecosystem risk than absolute natality and mortality in aquatic communities, although risk to a specific species can be determined from probability of effluent discharge resulting in a population change.

Aquatic community diversity is a strong function of water quality. Water quality changes can result from direct effluent discharge, leaching of solid wastes, erosion and air pollutant deposition.

Threatened or endangered fish species would serve as excellent indicator species of ecosystem risk if suitable life history and toxicological information were known. However, because so few individuals of these diminishing populations exist, suitable numbers for laboratory or field study are difficult to obtain. Toxicological tests inevitably involve fish mortality, which is unacceptable when studying threatened or endangered species. On the basis of these considerations, trout are a suitable indicator species for aquatic ecosystem risk. Furthermore, adult trout are carnivorous and at the top of the aquatic ecosystem food chain. Trout also have a substantial history in the Piceance and Uintah Basins, and estimates of trout abundance can be found from 1969 to the present. The most recent study indicates that brook, brown, and rainbow trout are most common in the upper reaches of the Piceance Creek and its tributaries (Goettl and Eddy, 1978). Brook and brown trout are reproducing naturally in the Piceance Creek, but rainbow trout appear to be declining.

One of the most significant pollutants found in the water effluents from oil shale sites was ammonia. High ammonia concentrations cause trout mortality, depress egg hatchability, and reduce the growth of fry. The 96-h LC_{50} value (concentration resulting in 50% lethality) of ammonia with trout is 3300 mg/L. The ammonia concentrations from the extreme analysis indicate acute trout mortality. Linear extrapolation from the single LC_{50} value allows a first-order estimate of trout mortality from discharge of ammonia indicating population declines of 20%. This estimate is only for discharge ammonia, although ammonia has been indicated as the most potent fish toxicant in oil shale process waters. Although the available data are inconclusive, additional trout mortality and morbidity may also occur from heavy metal burdens.

The risk to the fisheries of the Piceance Creek, White River, and Colorado River depends largely on the frequency of accidental leaks and discharge of effluent. Effectiveness of leachate containment over an indefinite maintenance period will also determine the long-term water quality of the drainages. The response of the native fish to both acute and chronic water quality changes, as yet unknown, is a key factor of aquatic ecosystem risk.

The estimate of ecosystem risk has a basic uncertainty because of the inherent variability of any biological system. Moreover, the relative lack of knowledge of biological systems, compared with our understanding of physical systems, increases the apparent biological variability over that which is naturally present. Thus, even the best methods available for censusing animals can be highly inaccurate. Predictions based on even the best information can be rendered useless, for example, by changes in weather.

PERSPECTIVE ON RESULTS

The use of health effects and ecological data for risk estimation of oil shale pollutants has been presented. The analysis tends to be simple, and use of the results requires a proper understanding of the assumptions, estimates, and uncertainties. The analysis provides information to aid in formulating and managing a program of environmental research focused on providing information required to reduce uncertainties in critical areas. The 1981 OSRA has established a set of assumptions and a structure for the use of data and models to predict the potential human health and environmental risks associated with an oil shale industry. The results can serve as a structure for defining needed refinements of models and required data to allow timely evaluations as the industry develops. The results can also help define key control technology requirements for the emerging oil shale industry.

ACKNOWLEDGMENT

Research supported by the Health and Environmental Risk Analysis Program, Office of Energy Research, U.S. Department of Energy, Contract EV 10706.

REFERENCES

ACGIH. 1980. *Documentation of the Threshold Limit Values,* 4th ed. (Cincinnati, OH: American Conference of Governmental Industrial Hygienists).

Bartman, R. 1980. Wildlife Biologist, Colorado Division of Wildlife, Fort Collins, CO. Personal communication.

Cuddihy, R. H., F. A. Seiler, W. C. Griffith, B. R. Scott, and R. O. McClellan. 1980. "Potential Health and Environmental Effects of Diesel Light Duty Vehicles," U.S. DOE, Washington, DC.

Cutler, S. J. 1975. "Third National Cancer Survey: Incidence Data," *Nat. Cancer Inst. Monog.* 41:22.

DOI. 1973. "Final Environmental Statement for the Prototype Oil Shale Program, Vol. 1, Regional Impacts of Oil Shale Development," U.S. Department of the Interior, U.S. Government Printing Office, Washington, DC.

Goettl, J. P., and J. Eddy. "Environmental Effects of Oil Shale Mining and Processing. Part 1: Fishes of Piceance Creek, Colorado, Prior to Oil Shale Processing," Colorado Division of Wildlife, Fisheries Research Center, Fort Collins, CO.

Helm, F. 1979. *Cancer Dermatology* (Philadelphia: Lea and Febiger).

Lepkowski, W. 1978. "Extrapolation of Carcinogenesis Data," *Environ. Health Persp.* 22:173-181.

Rall, D. P. 1978. "Thresholds?" *Environ. Health Persp.* 22:163-165.

DISCUSSION

***Jeff Giddings,* ORNL:** You said that one of the purposes of the HEED document was to determine what research was needed. What were the conclusions regarding ecological research on oil shale?

L. Gratt: The conclusions were that the work on rehabilitation and revegetation needs to continue. The whole issue of how water will be managed in the regions of the Colorado River basin has to be examined because, with regard to any aquatic systems, we found that basically the quantity of water was the key issue. Right now, we really do not know how they are going to manage water at each of these individual sites. It was recommended that there be further research in virtually all the areas. Now we haven't done anything to rank what is more important relative to each research area. The aquatic systems tended to suffer very badly on the kind of releases we have anticipated there. We are currently refining the analysis. We have now identified what elements and compounds are of key concern. The next time we go through the analysis we will add the removal mechanisms and so on.

***Ishwar Murarka,* EPRI:** You indicated excess deaths in the population of 685,000 ranging from 0 to 1000. What would be the more realistic number in that range? Zero means there would be no risk involved by developing this large industry, and obviously 1000 out of 685,000 would be considered to be a risk. What would be a realistic lower limit other than zero?

L. Gratt: Roy Albert indicated that these models are subject to question, and we are taking data from a high end and extrapolating it to a low end. That model is very controversial; it is based on an East Coast urban environment where they are looking at sulfur oxides as a surrogate for all pollutants. The environment there is entirely different from the environment in the oil shale region, but our best estimate currently is 73 with a range from 0 to 330. As we refine the various components of the model, this can change. I agree there is a significant probability the answer might be zero.

***Ralph Franklin,* DOE:** Let me add a comment to that or give a different type of answer and stress that it is not Larry's job to decide what is acceptable. It is his job to simply lay the facts out for us and help us organize it. The decision regarding what is acceptable gets into a much more involved public and scientific debate. So we are using the risk assessment at this point as a sensitivity analysis of all of the problems.

Ishwar Murarka: I agree with that; I am not asking him to make a decision as to what is acceptable. But we do not have to go through any calculation whatsoever to start with a statement there is absolutely no risk. I could not help but note that in every case the lower bound was zero.

L. Gratt: The reason for that is the same as the argument Roy Albert gave. We are taking data that represent high-dose data over a short time and trying to predict what happens at a very low, chronic level. There are arguments saying there are no effects; for example, there really are thresholds, and there are people saying that there are not. We are trying to cover the uncertainty range so that they can be compared to other certainty ranges where we do have numbers that are not zero.

***Milt Schloss*, Bureau of Land Management:** Just a comment, Larry, regarding water availability. There is an oil shale act that reserves the water for developmental purposes. This water has a federal reserve water right. For the commercial development of oil shale the companies will have to go to the state to obtain a water right, but for developmental purposes the water needed will have a federal reserved water right. This is on public land.

Ralph Franklin: Is it a reasonable assumption that the deer population is going to decline in proportion to the extent the browse area is affected and forage available?

***Rufus Chaney*, USDA:** I would like to comment on that. I have seen revegetation studies on oil shales performed at Colorado State that indicate with a good revegetation practice man would be more productive than he is now, and there might be more mule deer and more diversity as a result of the changed land use than there was before with the natural land shapes and so on.

L. Gratt: I would like to comment on that because I have heard that same argument. They can certainly build blinds for the deer and everything else like this. However, the people know that this is the area where the deer are, and thus they go and harvest the deer at appropriate times.

Rufus Chaney: Is there a regular review of the risk analysis research? If so, by whom?

Ralph Franklin: We have a program manager for this here; if we could get the microphone over to Paul Cho, perhaps Paul would like to comment on how these kinds of programs and their analyses are reviewed on a regular basis.

***Paul Cho*, DOE:** We do have a comprehensive quality assurance type of review process, beginning from day one, when Larry started his risk analyses. The major review process is through the Oil-Shale Task Force that is multi-agency made up of interested personnel from federal, state and industrial organizations. In addition, we also have available to us some groups that are responsible for looking at draft reports and formal review processes.

Ralph Franklin: Excuse me, Paul, but you have annual meetings in which you invite corresponding people from EPA and other parts of the federal government to critique this. In addition to this, you have a National Academy of Sciences project that is, in fact, evaluating and critiquing these.

Paul Cho: Thank you for adding this. Of course, such a process is more or less open-ended in a sense that the risk analysis is an iterative process. For example, right now we have experimental groups as well as groups from industry like the American Petroleum Institute looking at a draft review. By next month we should be able to publish the first year's product.

***Bob Meglen,* University of Colorado:** Larry, I think that Ralph alluded to something here. He mentioned sensitivity measurements. Is there some way to get an estimate of the most important parameters, or things you think ought to be measured more carefully or more thoroughly? Can you do a perturbation study where you would put an artificial impact of, say, 50% error on one of the components in your measurement and see how large an effect that has on the overall analysis? This would determine which areas within this particular study have the highest uncertainty.

L. Gratt: I would like to answer yes to that, and we have done part of that. When we look at the risks, those ranges that are reported are based on a whole series of both quantitative and subjective numbers. The associated errors are in the individual components relative to that risk estimate. In general, dose effect is the largest area. It has associated with it a one- to two-order-of-magnitude error that is used in the risk analysis. We have done a whole series of sensitivity analyses looking at some questions like: What if there was all open pit mining as opposed to mostly underground mining? What if they went to all modified in situ process vs surface processing?

***Michael Dahlberg,* DOE:** I would like to comment on the deer population. Unless the deer population is at a carrying capacity already, reducing the habitat should not reduce the deer population. The population is determined mainly by the harvest. You can increase the population by reducing the harvest or vice versa. I would also like to inquire about the endangered fish species in the area. Is that a potential problem?

L. Gratt: Yes, the fish are a problem. There are at least three kinds of trout as well as some nongame fish. The fishing in the basin streams is rather limited. However, there are some people who claim they do quite a bit of fishing there at certain times of the year. It is a significant problem because it is of public visibility. As we saw in an early presentation, catching a big fish is one of the things of concern. Under our extreme analysis, if the plants release the constituents that we're talking about, the ammonia is strong enough to eventually wipe out the trout, based on LC_{50} values. We are currently redoing these analyses because they appear to be too extreme a case. The probability of it occurring is very low, and we want to get a more realistic case.

***J. Brown,* CEGB:** I would just like to make a comment. Lois Florence talked about public education in this field; we have just heard an analysis of risk assessment which indicates that it will be difficult enough to brief the legal people, let alone the general public. And I think you are going to have a great inquiry on this if you get to that stage.

SECTION 5

NEEDS AND CHALLENGES IN MONITORING

CHAPTER 24
OVERVIEW

Ishwar P. Murarka

Electric Power Research Institute
Palo Alto, CA 94303

Historically, wastes have been disposed of into the air and water and onto the land. However, during the past two decades, concerns have grown that the no-harm loading limits of the air and water are being rapidly approached. As a result, solid wastes are increasingly being generated and disposed of on the land. Of course, land disposal of wastes is not an entirely new development. However, the concern that groundwaters may be contaminated is a more recent development, one that can directly affect the methods used for land disposal of wastes. When we speak of concerns related to land disposal, we must clarify that concern exists only when waste has been improperly disposed of.

Because of the diversity of the environment and the many varieties of solid wastes, it is highly improbable that we will be able to understand completely all possible concerns related to land disposal of solid waste. What we can do is to employ the best available scientific methods to establish reasonable possibilities for any specific disposal question. Even after using the best available scientific or other reasoned methods, we would still need to monitor or obtain data specific to the sites in question.

Monitor is a somewhat vague term because the purpose of the monitoring usually dictates the kind and type of measurements to be made. Therefore, it is essential that the purposes be clearly stated and understood in the design and implementation of monitoring. The papers in this section cover monitoring needs and experiences in the context of groundwater investigations near waste disposal sites.

Nelson and Rice provide some insights into the subsurface hydrologic monitoring required to evaluate contaminant migration. Obviously, their discussion has taken the more comprehensive aspect of the end-use that data on contaminant distributions in the subsurface will provide inputs to assessing the overall environmental consequences.

Saar and Braids address the question of determining which specific chemical indicators to monitor to meet the purposes of monitoring. Implicit in their initial considerations are the costs and ease of obtaining the relevant data. The authors also identified the all-important problems that are encountered in chemical sampling of the groundwaters. Their thoughts are well expressed in the two questions they frame: "How can an investigator distinguish chemical variations arising from sample handling and analysis from actual variations at the well?" and "Will both types of variations make it forbiddingly difficult to measure the range of a leachate plume?" The authors draw on their practical experiences and identify a relatively small number of inorganic chemical elements that can, in most hydrogeological settings, define a plume near landfill that contains municipal wastes.

Two papers discuss electric-utility waste disposal sites that have been monitored for the purpose of observing the nature and extent of groundwater contamination that may have been caused by the disposal practices. These full-scale waste disposal facilities tend to be large and represent a variety of hydrogeological wastes and management practices.

Santhanam and Kleinschmidt discuss the methodologies used for monitoring solid waste disposal sites for coal-fired electric utilities. They emphasize the importance to monitoring, planning, and implementation of knowledge of potential human health and environmental impacts. In the context of the electric utility disposal monitoring, this EPA-sponsored study proposes a broad approach to comprehensive monitoring and data analysis. The authors provide a good overview of efforts to develop and implement a monitoring system for an electric-utility disposal facility.

Hupe and Golden summarized monitoring experiences at an FGD sludge disposal site. They identify the difficulties involved in monitoring the site, both in terms of suitable location of well clusters and in changes in the land use surrounding the disposal site. They conclude that sufficient background and upgradient monitoring are essential to avoid false conclusions.

Mason addresses the problems of gathering data at abandoned waste sites and providing insights he gained while working at Love Canal and other abandoned sites. Obviously, the monitoring purpose of such an application is somewhat different at the beginning of the field investigation. The first purpose usually is to identify the nature and extent of the problem. The author concludes that a key aspect of field studies at abandoned sites is that one should retain the best available scientists to design and carry out the investigations.

The final chapter in this section, by Meglen and Erickson, addresses questions related to data analysis. They conclude that it is always necessary to identify patterns in the data and then make sound interpretations. Without proper care, one can misread monitoring results and arrive at costly, erroneous conclusions. Not only is a proper data-analysis scheme needed to analyze the data, it is also helpful in the selection and design of the monitoring system itself. It is easy to accumulate quantities of data but quite difficult to obtain the relevant data unless thorough planning has been done at the beginning of the monitoring program. Together, these chapters define the difficulties and need for scientific cautions in the monitoring of disposal facilities.

CHAPTER 25

SUBSURFACE HYDROLOGIC MONITORING TO EVALUATE CONTAMINANT MIGRATION: REQUIREMENTS AND SOLUTIONS

R. William Nelson and Wendy A. Rice

Pacific Northwest Laboratory
Richland, Washington 99352

INTRODUCTION

Providing adequate subsurface monitoring for solid waste treatment and disposal facilities involves various complexities, many of which result from the diversity of the waste forms and the geologic and hydrologic settings encountered. An adequate, cost-effective evaluation of groundwater quality under such diverse conditions requires coordination of many technical disciplines and analytic methods.

A successful program results from detailed planning, capable management, and coordination of technical capabilities in subsurface hydrology, geochemistry, analytical chemistry, and subsurface environmental consequences analysis. Accordingly, appropriate planning, coordination, and continuing reevaluation of the monitoring function are central components of a successful subsurface monitoring program. Inadequate planning or the lack of continuing reevaluation of objectives in the early stages of monitoring tends to encourage increased sampling; later, the system may become an exercise in number generation and storage.

This emphasis on planning and continuing reevaluation of monitoring objectives is not meant to detract from technical considerations of monitoring (Kazmann, 1981; Pichens et al., 1981). Such factors include constructing and installing hydrologic sampling devices, obtaining realistic samples with the field-sampling structures, obtaining in-place measurements to supplement and enhance later laboratory analysis, preserving samples, providing accurate analytical measurements, checking measured results for consistency, and reporting and recording data results for historical records. These and other technical aspects must be incorporated into the planning, organizing, and continued updating of subsurface monitoring efforts.

This chapter (1) considers the relationship of monitoring and the more detailed subsurface environmental performance assessments; (2) shows how subsurface monitoring is best planned and organized within the framework of the overall environmental performance assessment objectives; and (3) explains the requirements and background needed for environmental evaluations of solid waste treatment facilities and disposal sites.

SUBSURFACE MONITORING AND OVERALL ENVIRONMENTAL PERFORMANCE ASSESSMENTS

The role of subsurface monitoring is shifting from an emphasis on groundwater quality measurements to a more important, overall supportive role within the framework of the more detailed subsurface environmental performance assessment. This changing function of subsurface monitoring provides a new scope of monitoring objectives.

Monitoring subsurface water quality, although important, is no longer the final analysis required. Previously, where little or no subsurface contamination was expected or where contamination already existed, monitoring alone satisfied public and control agency requirements. Today, however, the public and control agencies usually require an overall subsurface environmental performance evaluation or safety assessment before allowing construction of a solid waste treatment or disposal facility. Under these conditions, the monitoring system becomes a long-term effort that supports, studies, and gradually verifies the expected performance of the subsurface system, as evaluated by the safety assessment.

The subordinate, but cross-checking, role of monitoring is necessary when determining potential groundwater contamination problems, because such problems develop slowly and because the corrective measures also involve very long times. In fact, if problems are not corrected early in their development, remedies may not be economically feasible. The predominant purpose of the overall environmental performance evaluation is to determine, as accurately as possible, potential environmental consequences of a proposed waste treatment or disposal facility on the biosphere before the facility is constructed. The public and control agencies increasingly require such evaluations before construction of treatment or disposal facilities is approved. At the same time, greater emphasis is being placed on subsurface monitoring to test the expected performance determined during the environmental consequence analysis.

This represents a shift from the traditional monitoring network, which is designed to simply observe subsurface water quality, to the current, more elaborate, monitoring system needed to test contaminant predictive methods. Accordingly, the newer monitoring system must satisfy requirements more closely allied to the overall subsurface environmental performance assessments. From such requirements comes the need for monitoring capabilities that deal with problems that result from the diversity of subsurface hydrologic settings, where solid waste facilities and waste disposal operations are located.

REQUIREMENTS FOR ENVIRONMENTAL EVALUATION OF WASTE TREATMENT FACILITIES AND DISPOSAL SITES

The major objective in an environmental safety evaluation of solid waste facilities is to determine the potential extent of contamination that will ultimately reach the biosphere. Three questions must be answered in the specific site assessment (Nelson, 1976; Welby, 1981):

1. Where in the ground are potential contaminants located with passing time?
2. When (or at what times) will the contaminant reach the biosphere?
3. How much of the contaminant will reach the biosphere?

Location of the contaminant is important because a contaminant that is isolated from the biosphere may represent little immediate hazard, even in large quantities. However, small amounts of contaminant arriving at critical locations over a short time could cause severe hazard. The question of location is simplified by concentrating on those places where the contaminants will outflow to the biosphere.

The time needed for a contaminant to reach a critical location is also a vital factor. If arrival is imminent, corrective actions will be required immediately. Because longer arrival times are often involved in groundwater systems, long-range assessment is required.

Of the three items to be considered in the environmental assessment, the quantity of contaminant present is perhaps most important. Small amounts may be a nuisance, whereas larger quantities may constitute serious hazards. Thus, the amount and concentration of the contaminant must be known to effectively evaluate consequences to the environment.

These three interrelated factors—location of contaminant arrival at the outflow boundary, arrival time of contaminant at the boundary, and quantity of contaminant reaching the outflow boundary—are expressed by two relationships or distributions: the location and arrival-time distribution, and the location and outflow quantity distribution. These distributions are conveniently displayed as two simple summary graphs. Using these summary graphs, we can quantitatively show how much contaminated fluid will interface with the biosphere, where that contaminant will contact the environment, and the time at which the contaminant will reach the outflow location. These facts are needed by decision-makers to allow, disallow, or require corrective action, based on the projected environmental consequences.

Specific Evaluation Requirements

These arrival distributions satisfy five requirements that serve as a guide for evaluating any subsurface contamination problem—in this case, the solid-waste burial sites:

1. determine those locations where contaminated groundwater may contact the biosphere now or in the future;
2. provide the contaminated fluid location and arrival-time distribution for each outflow boundary;
3. provide the contaminated fluid location and outflow quantity distribution for each outflow boundary;
4. provide the location and arrival time and outflow quantity distribution for each chemical or biological constituent of environmental importance in the contaminated fluid; and
5. determine the amount and concentration of each contaminant constituent that will, with time, interface with the environment (this is accomplished by using the distributions from items 2 and 4).

Satisfying these five requirements assures that the environmental consequences of any subsurface contamination problem can be evaluated. Therefore, these requirements provide the technical basis for making decisions affecting overall management of groundwater quality. These five concise requirements provide a specific means to the scientist or engineer for setting goals and objectives and for communicating with political decision-makers and the public. Because decision-makers usually evaluate, authorize, and finance environmental assessments, a working communications link with them is essential. This communications link is facilitated by an understanding and use of the arrival distributions. The arrival distributions are detailed elsewhere (Nelson, 1978a–d,1981), so only an overview of the distributions is presented here.

Overview of Subsurface Contaminant Arrival Distributions

A rather simple groundwater contamination problem is used to introduce the arrival-distribution concepts and to illustrate their usefulness in evaluating solid-waste burial sites. In the two-dimensional example (Figure 1), precipitation percolates through the buried waste, dissolves contaminants during seepage to the water table, and gradually flows toward the river. Environmental consequences of the contaminated groundwater entering the river and reaching the biosphere can be determined by using the arrival distributions.

Location and Arrival-Time Distribution

The location and arrival time distribution shown as the center graph in Figure 2 is the contamination arrival time as a function of the outflow

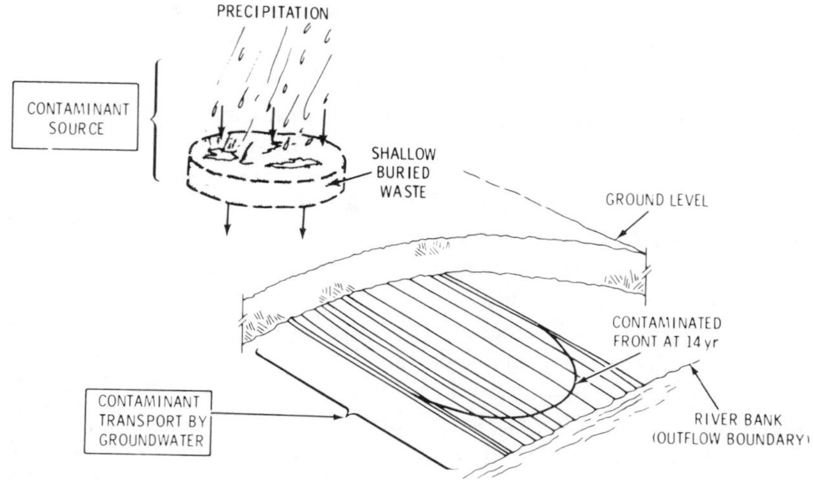

Figure 1. Schematic of the groundwater contamination problem.

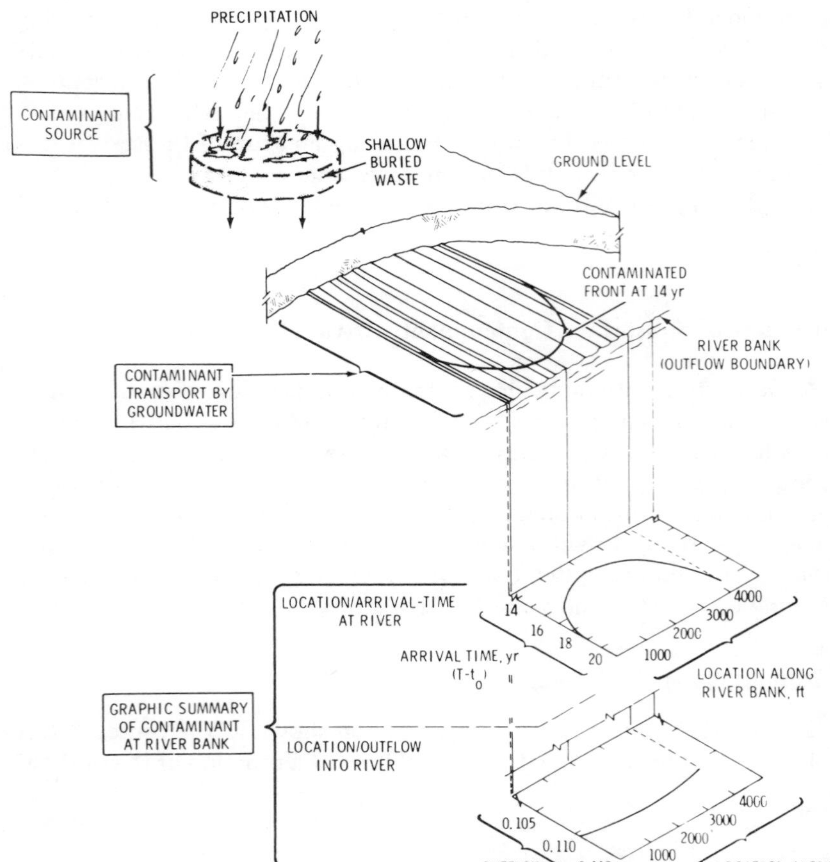

Figure 2. Basic summary relationships used to identify potential subsurface contamination.

location along the riverbank. The location and arrival-time curve quantitatively interrelates the *where* and *when* aspects, which are two of the three final factors required to assess the environmental consequences of subsurface pollution.

The location at which leachate from the buried waste will enter the river is provided at any elapsed time $(T - t_o)$ by the location and arrival time curve. The outflow location of the first contamination leaving the buried waste at $t_o = 0$ and the arrival time for any of the later departures (t_o) are provided by the curve. The single curve provides this information when the flow is steady, as in our example. A transient system, however, has as many arrival curves as departure times. Each arrival curve is considered and used in the same way as the one curve from a steady-flow system.

Location and Outflow Quantity Distribution

The second summary distribution needed, the location and outflow-quantity relationship, is illustrated as the bottom graph of Figure 2. Effects on flowrates into the river caused by differences in flow path lengths and movement rates are reflected in the location and outflow curve.

The single curve in Figure 2, representing the location and outflow distributions, results from the steady state of the sample flow system. In transient systems, several curves (each representing a different arrival time at the river) would make up the outflow quantity distribution. Each transient curve would be used in the same way as the single curve for the simpler case illustrated here.

The location and outflow quantity distribution summarizes the many interactions in the subsurface system that influence the *where* and *how much* aspects of the system evaluated. This summary distribution also provides a quantitative interrelationship between contamination location and quantity on a single-response basis.

Using Summary Results

The two arrival summaries can be used to determine the quantities of contaminants that may interface with the biosphere. In our example, the precipitation enters the ground at a constant rate. (For illustrative purposes, leachate concentration is 5 ppm for only the first 300 d, or 0.85 year.) After that time, no contamination is in the water seeping away from the disposal site. The specific location and arrival-time curve for the first contaminated rainfall passing through the waste at $t_o = 0.85$ year is plotted in Figure 3. These plotted results are obtained by using the t_o values in the arrival time curve from Figure 2. In the shaded areas of Figure 3, the concentration (5 ppm) entering the river is the same as that which originated in the buried waste.

Figure 3 displays factors needed to answer various questions. For example, it enables us to calculate the amount of contaminant a person fishing at a particular time and location along the riverbank would receive from the groundwater entering the river. By changing the fisherman's location with time, the worst-possible dose could be computed.

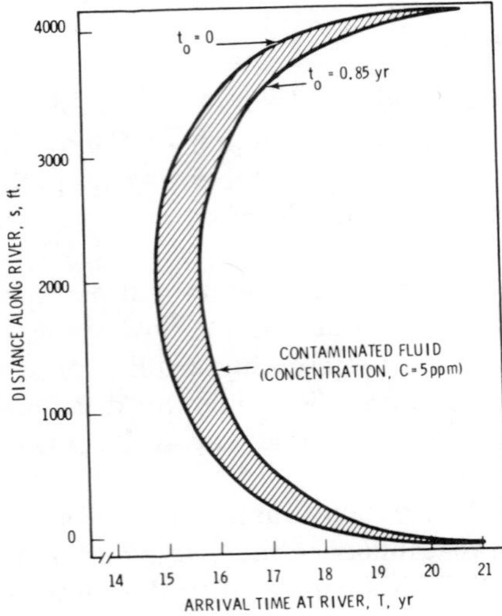

Figure 3. Use of location and arrival time distributions to determine the location of contaminant outflow along the river.

The location and arrival time and the location and outflow quantity distributions are used together to determine the second group of consequences from contaminants interfacing with the biosphere. Specifically, at a given arrival time, T, the product of the water outflow rates (from the bottom graph in Figure 2) and the concentrations from Figure 3 are integrated along the riverbank between the limits of contaminant outflow to obtain the total outflow at arrival time, T. Other arrival times are chosen, and corresponding integrals are evaluated to completely describe the total outflow rate or quantity of contaminant entering the river as time passes. Results of such evaluations are plotted in Figure 4 to give the variation of the contaminant outflow rate, W, with the river arrival time.

The flowrate of contaminants into the river in Figure 4 is the second of the two final results needed to completely evaluate the environmental consequences of the waste leachate. It shows directly how much of the contaminant will enter the river with time.

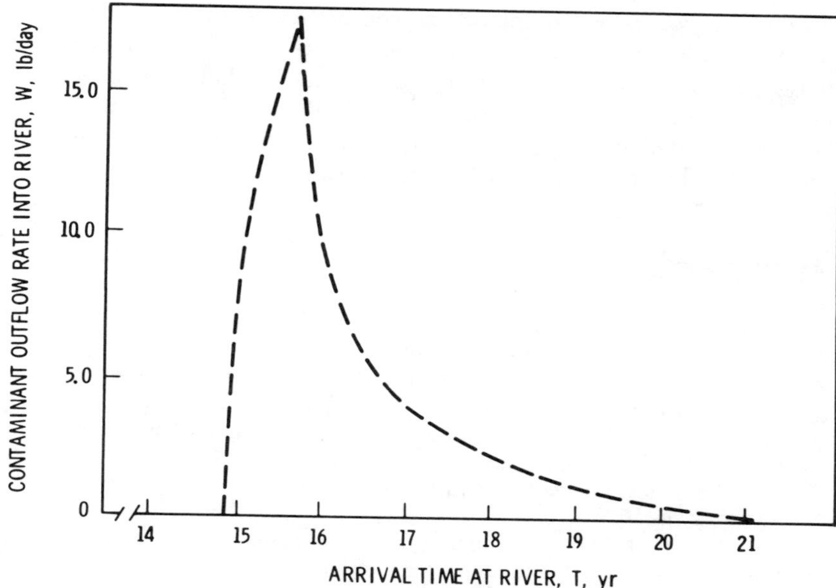

Figure 4. Contamination entering the river.

RESULTS

The above discussion illustrates that the results needed to determine the environmental consequences of subsurface pollution were obtained by combined application of location and arrival time and location and outflow quantity distributions. The method provides quantitative results about the quantity of the contaminant to interface with the biosphere, the location of contaminant contact, and the time when the subsurface flow system will no longer isolate the contaminant. These facts are needed by decision makers to choose alternatives and to allow, disallow, or require corrective action based on the environmental consequences. Thus, summary results shown in Figures 3 and 4 provide the primary information necessary to evaluate the environmental consequences resulting from solid-waste disposal.

EXPANDED ROLE OF MONITORING AS PART OF SUBSURFACE ENVIRONMENTAL ASSESSMENTS

The five requirements for subsurface performance assessment that we have discussed are significant because they provide the analysis needed to estimate the long-term consequences of essentially any potential subsurface pollution problems. The underlying assumption is that the potential for subsurface contamination should be evaluated before disposal facilities are approved or built. In addition, anticipated results provided by the arrival distributions must be tested, verified, and updated through appropriate monitoring.

REFERENCES

Kazmann, R. G. 1981. "An Introduction to Ground-Water Monitoring," *Ground Water Monit. Rev.* 1(1):28-31.

Nelson, R. W. 1976. "Evaluating the Environmental Consequences of Ground-Water Contamination," BCSR-6/UC-11, BCS, Richland, WA.

Nelson, R. W. 1978a. "Evaluating the Environmental Consequences of Ground-Water Contamination, 1, An Overview of Contaminant Arrival Distributions as General Evaluation Requirements," *Water Resources Res.* 14(3):409-415.

Nelson, R. W. 1978b. "Evaluating the Environmental Consequences of Ground-Water Contamination, 2, Obtaining Location/Arrival Time and Location/Outflow Quantity Distribution for Steady Flow Systems," *Water Resources Res.* 14(3):416-428.

Nelson, R. W. 1978c. "Evaluating the Environmental Consequences of Ground-Water Contamination, 3, Obtaining Contaminant Arrival Distributions for Steady Flow in Heterogeneous Systems," *Water Resources Res.* 14(3):429-440.

Nelson, R. W. 1978d. "Evaluating the Environmental Consequences of Ground-Water Contamination, 4, Obtaining and Utilizing Contaminant Arrival Distributions in Transient Flow Systems," *Water Resources Res.* 14(3):441-450.

Nelson, R. W. 1981. "Use of Geohydrologic Response Functions in the Assessment of Deep Nuclear Waste Repositories," PNL-3817, Pacific Northwest Laboratory, Richland, WA.

Pichens, J. F. 1981. "A Multi-level Device for Groundwater Sampling," *Ground Water Monit. Rev.* 1(1):48-51.

Welby, C. W. 1981. "A Graphical Approach to Prediction of Groundwater Pollutant Movement," *Bal. Assoc. Engo. Geol.* 18(3):237-243.

CHAPTER 26

CHEMICAL INDICATORS OF LEACHATE CONTAMINATION IN GROUNDWATER NEAR MUNICIPAL LANDFILLS

Robert A. Saar and Olin C. Braids
Geraghty & Miller, Inc.
6800 Jericho Turnpike
Syosset, New York 11791

ABSTRACT

To investigate the impact of a municipal landfill on groundwater, water samples from nearby observation wells can be tested for leachate indicators. Chemical results from such investigations may be difficult to interpret because leachate production varies with time and place, soils attenuate some chemicals more than others, and problems of sampling and analysis may introduce errors into the data set.

We have found that the following easily measured constituents are most helpful in distinguishing groundwater contaminated by leachate from that which is not: specific conductance, ammonium, chloride, bicarbonate, iron, potassium, and sulfate (as an inverse indicator). The reliability of data can be assessed in part by determining if contradictions exist in a data set through the use of interrelated analyses. Replicates and spiked samples can help to monitor sampling and laboratory performance.

INTRODUCTION

During the past decade, we have learned that municipal landfills are not the end of the line for many wastes. Rainfall percolates through mountains of garbage and forms a potent brew of inorganic and organic chemicals, called leachate, that ultimately travels toward the water table. These chemicals may contaminate the groundwater for miles down-gradient of a landfill.

The resulting leachate plume may degrade the quality of well water in a large area. It is important, therefore, to determine how large such a plume is and, if it has not reached a steady-state size, how fast it is growing or shrinking. Chemical tests on water drawn from supply or monitoring wells comprise the main method for assessing a plume. Many organic and inorganic species may be quantified, but interpretation of the analytical results is often difficult. Several types of problems plague all chemical sampling:

1. A sample may be contaminated as it is withdrawn from a well.
2. A sample that is improperly preserved or analyzed too late may not show the concentrations that were originally present even if no contamination occurred.
3. A laboratory may improperly analyze the sample.

Superimposed on these problems is the fact that groundwater chemistry at a well changes constantly with variations in weather, hydrogeology, and landfill contents and operation. How can an investigator distinguish chemical variations arising from sample handling and analysis from actual variations at the well? Will both types of variations make it forbiddingly difficult to measure the range of a leachate plume?

A review of the literature and our experience during the design of leachate monitoring programs have shown us that several of the many chemicals leaching from a landfill are particularly useful for delineating a plume. We have also found that several of the chemical indicators can be used together to check the internal consistency of a data set. If more than a few data points exist, one or more apparent anomalies are almost certain to occur; thus, the check for internal consistency can be extremely useful. It helps to answer the questions, "Should some data be deemphasized or ignored during the overall plume assessment? If so, which data are they?"

The following sections describe the chemical indicators found to be most useful when examining leachate plume data. Also described are ways to increase the reliability of field data should contradictions in a data set prove difficult to explain.

CHEMICAL INDICATORS OF CONTAMINATION

Several common analyses provide valuable information that can define a leachate plume from any municipal landfill in most hydrogeological settings. These analyses include those for specific conductance, ammonium (NH_4^+), chloride (Cl^-), bicarbonate (HCO_3^-), iron (Fe^{2+} and Fe^{3+}), potassium (K^+), and sulfate (SO_4^{2-}).

Specific Conductance

This general analysis is probably performed most often. The specific conductance measurement is sensitive to all dissolved ionic species; it is especially useful because it is easily done in the field. A delay of several weeks for laboratory analyses is avoided, and the field results can be used immediately to prepare leachate-plume maps or to adjust the sampling program or procedures.

Ammonium

The reducing conditions that usually arise under a landfill and the relatively large amounts of nitrogen in the wastes can lead to substantial amounts of ammonium ion. This constituent is an excellent indicator of leachate because it often increases in concentration more dramatically than do the concentrations of other species. Table I shows that ammonium ion concentration often increases above background levels by an order of magnitude more than does the concentration of total dissolved solids.

Chloride

Chloride nearly always leaches from landfills in large quantities. Background levels are easy to measure, but the chief virtues of this ion as an indicator are that it is not chemically or biologically converted into other forms of chlorine, and it moves at approximately the same speed as water because it is not substantially adsorbed onto soil surfaces. Chloride ions will generally move as fast as any chemical and, hence, serve as a forerunner of the plume.

Bicarbonate

The presence of this ion in landfill leachate results largely from intensive biological activity, which produces carbon dioxide (CO_2) gas. The

Table I. Sulfate and Ammonium in Background and Leachate-Contaminated Groundwater

Study	Background Samples (mg/L)			Contaminated Samples (mg/L)			Reference
	TDS	Sulfate	Ammonium	TDS	Sulfate	Ammonium	
Stadtwald, Frankfurt, 1967–1973	241	55	0.5	5634	170	111	Golwer et al., 1976
Grosskrotzenburg, Frankfurt, 1964–1967	426	87	0.01	3465	556	101	Golwer et al., 1976
Babylon, NY, 1972–1974	219	60	5.3	1220	19	7.8	Kimmel and Braids, 1980
Babylon, NY, 1981	160	23	0.3	1218	<1	70	Geraghty & Miller, 1981
Islip, NY, 1972–1974	223	74	0	1560	57	12	Kimmel and Braids, 1980
Banta Road and Tibbs Avenue, Indianapolis, 1975	341	40	0.46	2000	8	100	Pettijohn, 1977
West Raymond Street, Indianapolis, 1974	432	55	0.7	8600	9.7	210	Pettijohn, 1977
96th St. and Zionsville Rd., Indianapolis, 1974	342	3.8	0.54	4770	17	680	Pettijohn, 1977
Geneseo, IL, 1977	140	38		1190	12		Johnson and Cartwright, 1980

gas dissolves in water to form bicarbonate (HCO_3^-) and hydrogen ions. Partial pressures of CO_2 gas under a landfill can be several orders of magnitude greater than the $10^{-3.5}$ atm partial pressure of CO_2 found in the atmosphere. Bicarbonate, like chloride, is a negatively charged ion and, hence, moves well underground because it is weakly adsorbed by soil particles.

Iron

The reducing (electron-rich) conditions in a leachate plume can convert relatively insoluble Fe^{3+} to Fe^{2+}, often resulting in high iron concentrations in leachate water samples. Large amounts of iron can also come from materials in the landfill. However, because so much iron exists in natural soils, local conditions unrelated to a leachate plume may cause high iron concentrations. Manganese is also a possible plume indicator because its oxidation-reduction chemistry and its occurrence often mirrors that of iron (Hem, 1970). However, manganese concentrations tend to be much lower than iron concentrations, and the detection limits of instrumentation are approached. Pushing the limits of instrumentation is not desirable when the object is to confirm the presence of a leachate plume, and data of high reliability are required.

Potassium

Solubility controls in natural soils tend to keep dissolved potassium ion concentrations several times lower than dissolved sodium ion concentrations (Hem 1970). However, many materials in municipal landfills, particularly discarded food, are fairly rich in potassium (Ellis, 1980). Although the Na^+/K^+ ratio is variable even in a clean system, the relatively high K^+ concentrations coming out of landfills often lower this ratio in samples taken from the leachate plume. Table II shows this trend for a 3-km-long plume at Babylon Landfill in New York.

Sulfate

Sulfate differs from the indicators already discussed. Although sulfate may leach out of a landfill, substantial biological and chemical degradation often occurs in the most concentrated part of the plume. The result

Table II. Comparison of [Na+]/[K+] Ratio and Total Dissolved Solids Concentration, Babylon Landfill, New York (Geraghty & Miller, Inc. 1981)

Sample	May 1981		August 1981	
	TDS (mg/L)	[Na+]/[K+]	TDS (mg/L)	[Na+]/[K+]
Background Well (outside of plume)	160	7.5	145	6.9
Wells Intercepting Plume (distance from landfill edge)				
3.3 km	241	3.7	222	3.4
1.7 km	335	2.0	310	2.8
1.0 km	576	2.7	730	2.5
0.7 km	847	3.0	671	5.1
0.1 km	1218	1.8	1108	1.7

is that sulfate concentrations in heavily contaminated groundwater are often lower than background concentrations. This is true for the Babylon and Islip Landfills in New York (Kimmel and Braids, 1980) and for five of seven landfills studied in Indianapolis, Indiana (Pettijohn, 1977). In other cases, the sulfate concentration in the leachate plume is higher than in the background, as in two landfills in Indianapolis (Pettijohn, 1977) and in two landfills near Frankfurt, West Germany (Golwer et al., 1976). However, in these cases, the sulfate concentration in the plume is not elevated above the background as much as the total dissolved solids concentration. Data for all these examples appear in Table I. The strongly reducing conditions in leachate appear to provide a good environment for conversion of sulfate to sulfides or other reduced sulfur species.

Other General Indicators

We considered other common constituents for this list of reliable indicators of a leachate plume including phosphorus, calcium, and magnesium. Phosphorus is subject to strong solubility controls, and large amounts of calcium and magnesium occur naturally in many places. Their concentrations in leachate are not particularly informative.

Specialized Indicators

Specific elements or compounds are worth looking for if large amounts are likely to be in the landfill. This type of indicator is specific to a particular site. For example, if laundry water is sprayed on a landfill, boron in the form of borates may appear in the groundwater. Arsenic appears in many pesticides and in a variety of electrical and other manufactured products (Page, 1974). Selenium occurs in photocopier residues and in a variety of photoelectric and electronic devices. Large amounts of these wastes could contribute an easily measurable amount of arsenic or selenium to groundwater. Broken refrigerator and air conditioner coils may contribute Freon to the plume. Zinc and cadmium are chemically quite similar and, therefore, commonly occur together in nature. Both metals have many applications and could show up in landfill leachate. Neither metal, however, can serve as a plume tracer until the presence of common zinc/cadmium minerals is assessed.

GENERAL CHEMICAL INVESTIGATIVE TECHNIQUES

Diagrams

Often, looking at one or a few dissolved chemicals may not help to characterize the leachate plume. Overall patterns in the data may be more important. Several types of diagrams are used to characterize surface and groundwater, and they can help distinguish between contaminated and uncontaminated groundwater (Freeze and Cherry, 1979). Piper diagrams are triangular, with the three corners of the cation diagram representing a water sample with either all Mg^{2+}, Ca^{2+}, or $Na^{+} + K^{+}$ as major positive ions, and the three corners of the anion diagram indicating a water sample with either all Cl^{-}, SO_4^{2-}, or $CO_3^{2-} + HCO_3^{-}$ as negatively charged ions. Because typical water samples have a mixture of major ions, the dot or other symbol representing such a sample is placed within the triangle at distances to the corners that reflect the relative amounts of each major ion. Many dots and, hence, water or leachate samples can be placed within each Piper triangle, with water of common origin or type having a compact group of symbols. The position of symbols representing an unknown water sample in such a diagram relative to those symbols representing samples that are known to be clean or contaminated may help decide whether the unknown sample is affected by leachate.

Whereas the Piper diagram has a constant triangular shape with various water samples being plotted at different places within, the Stiff diagram only represents one water sample, and its shape and size depend on the relative concentrations of major ions in the sample. One advantage of this diagram over the Piper diagram is that both negatively and positively charged ions are represented in one diagram, thus

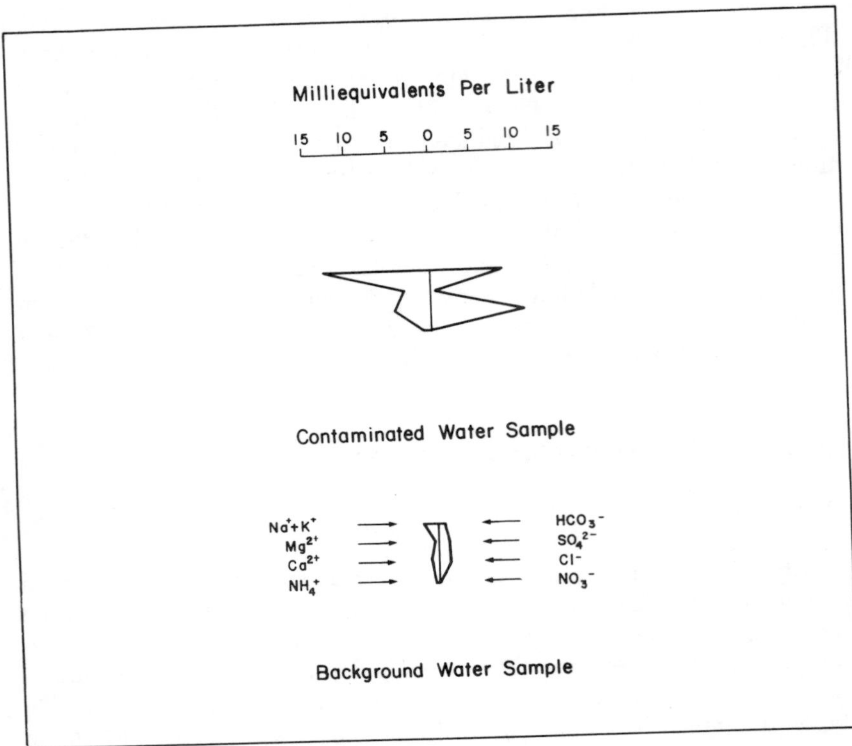

Figure 1. Stiff diagrams for a leachate-contaminated groundwater sample and an uncontaminated (background) groundwater sample taken near the Babylon Landfill, New York. (Modified from Kimmel, G. E., and O. C. Braids. 1980. "Leachate Plumes in Groundwater from Babylon and Islip Landfills, Long Island, New York," Geological Survey Professional Paper 1085, Washington, DC.)

allowing a rough check of the anion/cation balance. The Stiff diagram can take on some interesting shapes if the water sample is from a leachate plume: the concentration of most ions is higher in the plume than in the background, thereby leading to a relatively large diagram. However, the side of the diagram representing sulfate may be indented because the sulfate concentration in leachate is often relatively low compared with that of other ions. Figure 1 shows these shapes for clean and contaminated groundwater near the Babylon Landfill in New York (Kimmel and Braids, 1980).

Degradation By-products

Many chemicals placed in a landfill may be changed through chemical and biological reactions. Chlorinated hydrocarbons can lose some of their chlorines, and chlorinated ethenes can become vinyl chloride and methylene chloride (Wood et al., 1981). Large pesticide molecules can be hydrolyzed, hydroxylated, and in many ways split into smaller entities (Crosby, 1970; Kaufman, 1970; Kearney and Plimmer, 1970). Degradation schemes are possible for many other substances, some leading to hazardous by-products and others to less dangerous materials. An individual survey of landfill materials is necessary.

RELIABILITY OF CHEMICAL DATA

Perhaps the hardest job for an observer of chemical data is to determine whether any of the data are unreliable and should not be used to characterize the leachate plume. It may be tempting to believe that all the data are good, or even equally bad, so that all of it should be considered. However, blatant contradictions in the data set should eliminate consideration of some data.

How does an investigator determine which data, if any, are faulty? Many techniques are possible. Although the following discussion is divided into methods that are fairly inexpensive to use and those that are more expensive, we do not attempt to rank these methods by cost-effectiveness. Site conditions and procedures of sampling and analysis vary so much that in one case a confirmatory test may be extremely helpful, whereas in another case, the same test may not only be useless but may further confuse the data interpretation.

Inexpensive Techniques

Total Dissolved Solids and Specific Conductance

The specific conductance measurement is done in the field, whereas the total dissolved solids (TDS) test is usually done in the laboratory. For a particular site, a good correlation usually exists between these two measurements. It is easy to assess whether a field conductance bridge is working properly if TDS information is available. Although the specific conductance can be done in the laboratory, such a test provides little new insight into the underground chemistry, except possibly to document the conductance change caused by time and handling. Also, there is the risk of considering the field and laboratory tests of specific conductance to be replicates, which they are not.

Hardness—Ca^{2+} and Mg^{2+}

Hardness is the ability of a water sample to precipitate soap. This property of water comes from dissolved divalent cations. Among divalent cations, calcium and magnesium nearly always dominate, so a close correspondence should arise between the hardness test and the sum of Ca^{2+} and Mg^{2+} concentrations. If hardness substantially exceeds this sum and appropriate concentrations of other divalent cations are not present, a contradiction exists. The hardness test and those for Ca^{2+} and Mg^{2+} can confirm each other because they are done by different techniques, using different instrumentation. One faulty instrument will not ruin all these results.

Relationship of pH, CO_3^{2-}, and HCO_3^-

The position of the carbonate (CO_3^{2-}) and bicarbonate (HCO_3^-) equilibrium depends on pH. Equal quantities of each ion should occur at pH 10.3; there is 90% CO_3^{2-} and 10% HCO_3^- at pH 11.3, and 10% CO_3^{2-} and 90% HCO_3^- at pH 9.3 (Garrels and Christ, 1965). The carbonate test is not commonly done because little carbonate occurs, except in unusually high pH samples. The carbonate test should accompany the bicarbonate test for samples with pH values above 9.

Cation/Anion Balance

If data are available for all the major ions, a cation/anion balance can be examined. A major imbalance (~20% or more) in the totals probably indicates a problem in one or more of the major-ion analyses. Offsetting errors, however, could occur, and the cation/anion balance would not reveal them.

Redox Potential/Sulfur and Nitrogen Species

A field measurement of the oxidation-reduction potential often gives a clue to the relative concentrations of the sulfur and nitrogen compounds. As noted previously, a reducing environment favors sulfides over sulfate. The ammonium ion concentration is usually very high in the reducing section of a leachate plume, whereas nitrate (NO_3^-) levels may even be below background levels. The redox potential, however, is not easy to use because the conversions are biologically mediated, so that the sulfur and nitrogen species are not likely to be in equilibrium with prevailing redox conditions. Even so, the redox potential can serve as a rough check.

Expensive Techniques

Analyzing extra samples is an important, although more expensive, way to help determine data reliability. We discuss several strategies if extra analyses can be made.

Replicates to One Laboratory

A laboratory's precision can be evaluated if several replicates of one sample, differently labeled, are analyzed. Low precision (large scatter in the data) is not consistent with high accuracy (closeness to the "true" value). Scattered results for replicates sent to one laboratory indicate a need to examine sampling and laboratory practices.

Replicates to Several Laboratories

If money is available for more than two or three replicates, one or more may be sent to another laboratory. High precision only allows the possibility of high accuracy but does not certify it. Results from more than one laboratory may help to determine whether one of the labs is systematically doing something wrong. Such an assessment depends on careful handling and preservation.

Spiked Samples

High concentrations of contaminants in a leachate sample may interfere with an instrument's performance and may distort the results for several analytes. Adding known amounts of a constituent to extra aliquots of sample and submitting them to the laboratory can help to determine whether matrix effects are serious. If they are, and a revised laboratory procedure cannot be developed, the importance of the affected data should probably be reduced.

CONCLUSIONS

Some of the techniques we use when presented with chemical data for groundwater near municipal landfills have been described. Many of the techniques we discuss, however, are not limited to plumes generated by municipal dumps; industrial landfills, lagoons and other point sources can cause chemical changes underground that are similar to the types described above. Furthermore, the reliability techniques are widely applicable. We have emphasized methods that depend on routine and easily done chemical analyses. Ultratrace analyses can be of interest, but unless quality controls on sampling and analysis are excellent, inadvertent contamination of samples or other analytical problems can confuse the data reader. Information from the relatively routine analyses can answer many of the questions concerning a landfill plume if the investigator collects detailed information about the contents and operation of the landfill and about the hydrogeology of the site. The search for patterns in a data set and the ability to find and discard faulty or contradictory chemical information will help the investigator draw stronger and more reliable conclusions about the nature and extent of a municipal landfill leachate plume.

REFERENCES

Crosby, D. G. 1970. "The Nonbiological Degradation of Pesticides in Soils," in *Pesticides in the Soil: Ecology, Degradation and Movement* (East Lansing, MI: Michigan State University), pp. 86-94.

Ellis, J. 1980. "A Convenient Parameter for Tracing Leachate from Sanitary Landfills," *Water Res.* (GB) 14:1283.

Freeze, R. A., and J. A. Cherry. 1979. *Groundwater* (Englewood Cliffs, NJ: Prentice-Hall, Inc.), Chapter 7.

Garrels, R. M., and C. L. Christ. 1965. *Solutions, Minerals, and Equilibria* (San Francisco: Freeman, Cooper & Company), Chapter 3.

Geraghty & Miller, Inc. 1981. Unpublished data.

Golwer, A., et al. 1976. "Belastung und Verunreinigung des Grundwassers durch feste Abfallstoffe," *Abhandlung des Hessischen Landesamtes für Bodenforschung,* Heft 73, Wiesbaden.

Hem, J. 1970. "Study and Interpretation of the Chemical Characteristics of Natural Water." 2nd ed., Geological Survey Water-Supply Paper 1473, Washington, DC.

Johnson, T. M., and K. Cartwright. 1980. "Monitoring of Leachate Migration in the Unsaturated Zone in the Vicinity of Sanitary Landfills," Illinois Institute of Natural Resources Circular 514, Urbana, IL.

Kaufman, D. D. 1970. "Pesticide Metabolism," in *Pesticides in the Soil: Ecology, Degradation and Movement* (East Lansing, MI: Michigan State University), pp. 73-86.

Kearney, P. C., and J. R. Plimmer. 1970. "Relation of Structure to Pesticide Decomposition," in *Pesticides in the Soil: Ecology, Degradation and Movement* (East Lansing, MI: Michigan State University), pp. 65-72.

Kimmel, G. E., and O. C. Braids. 1980. "Leachate Plumes in Groundwater from Babylon and Islip Landfills, Long Island, New York," Geological Survey Professional Paper 1085, Washington, DC.

Page, A. L. 1974. "Fate and Effects of Trace Elements in Sewage Sludge When Applied to Agricultural Lands—A Literature Review Study," U.S. EPA, Office of Research and Development, Cincinnati, OH.

Pettijohn, R. A. 1977. "Nature and Extent of Groundwater Quality Changes Resulting from Solid Waste Disposal, Marion County, Indiana," Water Resources Investigations 77-40, U.S. Geological Survey, Washington, DC.

Wood, P. R., R. F. Lang, and I. L. Payan. 1981. "Introductory Study of the Biodegradation of the Chlorinated Methane, Ethane, and Ethene Compounds," paper presented at the American Water Works Association Annual Conference and Exposition, St. Louis, MO, June 7-11.

DISCUSSION

***Doug Burgess*, Stone and Webster Engineering:** I have a couple of questions. One, I might have missed it, but why is sulfate that occurs in the plume lower than in the background? And my second question is, if you use a constituent like arsenic or selenium which perhaps is not evenly distributed throughout the landfill, could that fact prejudice the outcome of your survey?

R. Saar: Sulfate can be transformed into organic and inorganic sulfides as a result of reducing conditions in a plume. As far as distribution in the landfill, I did not get into that at all. If you have a pile of salt on one side of the landfill, you find a very heavy concentration of sodium and chloride along one part of a plume and not in another part of a plume. If it is possible to drill hundreds of wells, you will pick up the variation as you are suggesting. However, landfills are very high in entropy and hard to deal with very precisely.

***Harry Freeman*, Office of Appropriate Technology, State of California:** You had a value of 3.3 km for the length of the plume. What have you found is typical, or what is the longest plume you have run into doing this kind of thing?

R. Saar: Several miles, I would say, is a typical long plume. There are so many factors involved with the hydrogeology of the site relative to how fast a plume is diluted. John Cherry spoke about that yesterday. It depends on how high the garbage is piled, the size of the plume, how much water is coming through, and the landfill—whether it is capped and if it is in an area that has a lot of water. I do not think this 3- or 3.5-km plume is very excessive in length. I think you will find that kind of distance in a lot of landfills. There is another landfill just about 20 miles east of that in Islip, Long Island; that plume is not as long. It is a much smaller landfill. Of course, time is another factor. If the landfill is only a few years old, the plume is still growing. But on an average, you can expect ultimately a plume can be miles long.

***Les Dole*, ORNL:** Have you considered looking at ratios, let's say a sulfate/sulfide, as in situ indicators of redox potential? Have you looked at any of the redox couples?

R. Saar: As I indicated before, I have tried to correlate redox potential and either the sulfur species or the nitrogen species, and it does not work out very well. It is just not an equilibrium system in any sense of the word. For example, you may find a fairly low redox potential, but you have a relatively high ratio of sulfate to sulfide. I have not found redox potential to be very useful to predict speciation.

***John Cherry,* University of Waterloo:** Could you comment on the possibilities for use of, say, TOC, COD and things on the organic side as crude inexpensive indicators of plume position?

R. Saar: Indicators like TOC and COD might be useful, as are some of the inorganic analyses. The only problem I have with TOC or COD is that many people are concerned with some of the more carcinogenic chemicals like the chlorinated hydrocarbons, which may comprise only a tiny part of the total organic carbon. A lot of organic carbon can be natural. I think it can be a little deceptive if you depend a lot on a TOC or COD analysis, although it should provide good supplemental data in the way specific conductance measurements do.

CHAPTER 27

MONITORING OF SOLID WASTE DISPOSAL SITES: WHAT FACTORS ARE RELEVANT FOR ELECTRIC UTILITIES?

Chakra J. Santhanam and David E. Kleinschmidt

Arthur D. Little, Inc.
Cambridge, Massachusetts 02140

ABSTRACT

This chapter briefly discusses general methodologies for monitoring solid waste disposal sites for coal-fired electric utilities. An effective monitoring program should result in assessment of environmental effects and soundness of disposal method. It should include site evaluation, site development for monitoring, sampling and analysis, and assessment of generated data.

INTRODUCTION

Technology and Production of Wastes

Management of solid wastes is particularly important for coal-fired utilities. The two major types of waste generated are coal ash and flue gas desulfurization (FGD) waste. Coal-fired utility boilers generate two types of coal ash: fly ash and bottom ash. Fly ash is primarily collected by electrostatic precipitators or fabric filters; bottom ash is usually collected in a water quench tank underneath the boiler. FGD can be accomplished by nonregenerable or throwaway systems, which result in FGD wastes, and regenerable systems, which primarily produce a saleable product (sulfur or sulfuric acid). In 1975 coal ash generation from utility boilers amounted to 52 × 10^6 t; no FGD waste was generated. By 1985 generation of coal ash and FGD waste is expected to increase to 64 × 10^6 and 21 × 10^6 t, respectively, and by 2000 to 85 × 10^6 and 30 × 10^6 t (Santhanam, 1976).

Disposal vs Use Options

Most of the coal ash and all of the FGD wastes now generated are sent to disposal. Considering the expected increase in coal consumption in boilers in the United States, this will probably be the case for many years. Use of FGD wastes is expected to grow, but at a slower rate than FGD waste generation. Over the longer term, an effective way to manage coal and FGD wastes is to use them. There is currently some use of coal ash but no use of FGD wastes in the United States.

All FGD wastes are now disposed of on land. At-sea disposal may be a future alternative if it can be practiced under environmentally acceptable conditions. The principal methods of disposal are (1) ponding; (2) landfilling (including mine disposal, which may be considered a special subcategory of landfilling); and (3) interim ponding followed by landfilling. Table I shows current disposal practices, as obtained in a survey of 176 coal-fired power plants.

Table I. Survey Results of Typical FGD Waste Disposal Methods Utilized at 176 Utility Coal-Fired Power Plants ≥200 MW in the United States[a]

	Number of Plants		
Type of Waste	Pond[b]	Landfill[c]	Interim Pond or Landfill[c]
Fly Ash Only	18	46	6
Bottom Ash Only	29	12	30
Combined Fly and Bottom Ash	69	9	16
FGD Waste Only	5		
Mixed Fly Ash and FGD Waste	7	7	
Mixed Bottom Ash and FGD Waste	1		1
Mixed Fly Ash and FGD Waste (Stabilized)	2	6	
Mixed Fly Ash, Bottom Ash, and FGD Waste	2	1	1

[a]Coal-fired plants (≥80% of their power generated from coal in 1977) that have generating capacities ≥200 MW, with the exception of four plants using FGD systems. Figures represent the number of plants at which each waste type or disposal method is practiced. (Many plants use more than one method.)
[b]Includes direct ponding and interim and final ponding methods.
[c]Includes managed and unmanaged fills and mine disposal.

Environmental Impact

Environmental impacts of FGD waste disposal, and hence any potential threat to human health and the environment, are influenced by three factors: types of waste generated (physical and chemical characteristics), disposal method employed (ponding, landfilling or other) and disposal site characteristics (e.g., soil type, hydrogeology and climate).

Criteria for planning and implementation of monitoring efforts should be ranked to reflect the relative importance of various potential impacts. Specifically, highest priority is usually given to three subject areas that are both characteristically important for utility solid waste disposal and principal regulatory responsibilities: (1) impacts on groundwater quality; (2) impacts on surface water quality from nonpoint sources; and (3) use of potentially mitigative design, management, or control practices.

This chapter briefly describes the broad approaches toward a comprehensive monitoring program at a disposal site (including site evaluation, site development for monitoring, and sampling and analysis) and assessment of the data generated by monitoring.

DATA GENERATION BY MONITORING EFFORTS

The overall objective of a monitoring program should be to develop a methodology for the long-range monitoring of a disposal site. Such an effort should provide both input parameters to predictive models and field verification of environmental effects. The key to developing a cost-effective approach is in tailoring the generic approach to the specific characteristics of the site to be investigated. The geologic and hydrologic settings of disposal sites are extremely diverse. By doing a preliminary evaluation of the site prior to developing a monitoring program, a methodology for handling the complexities and site-specific characteristics of the site can be incorporated into the program before field work is started. Thus the field team has a general understanding of the complexities that may be encountered at the site and an approach for handling them before they occur, which leads to cost-effective monitoring.

A good overall approach incorporates four basic steps:

1. evaluation of the site,

2. site development for monitoring,
3. sampling and analysis, and
4. assessment of data generated by monitoring.

Site Evaluation

The initial phase of site evaluation consists of obtaining and reviewing existing available information pertaining to near-region and site-specific geologic and hydrogeologic conditions. This information is used to determine if any conditions will require special attention in the monitoring program. Geologic information of interest includes site area topographic and physiographic conditions, type of bedrock and surficial materials, stratigraphy and structure of geologic units, and other conditions that could affect site area hydrogeology (e.g., mining and construction). Hydrogeologic information of interest includes stratigraphy, pressure conditions and hydraulic properties of aquifers underlying the site; depth of water table(s) and direction(s) of groundwater flow; location of nearby groundwater recharge or discharge areas; and information pertaining to aspects of the hydrogeologic cycle, including precipitation, evapotranspiration, infiltration, and groundwater recharge.

Geologic and hydrogeologic information pertaining to a particular area can be obtained from a variety of federal, state, and local governmental, public and private organizations. Sources that are often useful include:

- U.S. and state Geological Surveys,
- U.S. and state Soil Conservation Surveys,
- U.S. and state Bureau of Mines,
- National Oceanic and Atmospheric Administration,
- Corps of Engineers,
- U.S. and state Environmental Protection Agencies,
- regional and local technical libraries, and
- local and state water surveys.

After available existing data have been gathered and assimilated for a specific site area, a site inspection is made to collect sufficient information to allow a tentative determination of possible subsurface conditions. The site visit generally involves a comprehensive field inspection of land surface configuration, vegetation patterns, drainage patterns, and soil types. Surficial geologic and hydrogeologic conditions are observed and noted. Local public and private agencies may also be visited at the same time to obtain existing information on near-region and site-specific geologic and hydrogeologic conditions.

The field observations can be combined with the hydrogeologic information obtained elsewhere to develop a tentative estimate of the nature and direction of groundwater and leachate movement. This information is used to develop a monitoring program with the appropriate type, number, and positions of groundwater sampling wells, surface sampling locations, and surface and subsurface soil and waste sampling sites (including liners, if appropriate).

Site Development for Monitoring

After site evaluations have been completed, site development is the next step. Based on the environmental impacts of interest and definition, if any, of the assessment methodology, site development should include a site evaluation plan and a sampling program.

Site Evaluation Plan

The development of a site, including appropriate soil and waste borings, placement of monitoring wells, and other aspects of a complete monitoring program, is best done on the basis of a site evaluation plan that is also tied to the sampling program and the ultimate assessment efforts. A site evaluation plan should be fairly complete and include several key items of information.

Drilling Equipment. Exploratory borings and observation wells may be installed in or through the waste disposal site, in or through manmade disturbances (e.g., mine spoils) around the site, and in natural soil beneath and downgradient of the waste disposal site. Proper planning requires precise specification of the type of equipment that is needed, including significant details on the type of drilling technique used.

Field Evaluation Program. Two major site hydrogeological parameters must be evaluated in any monitoring effort:

1. establishment of the amount of precipitation and runoff that flows through and beneath the disposal site; and
2. attempts at chemical separation of the leachate that is derived from the waste material itself and those from adjacent soil or other material areas.

The field evaluation program usually consists of appropriate soil borings and monitoring wells to permit a precise definition of the hydrogeology of the site, including items to identify physically and chemically the leachate from the disposal site and determine its impact on the environment. Typically, site evaluation plans should specify both the geotechnical and hydrogeologic evaluation plans and objectives.

Geotechnical Evaluation Plan. Geotechnical evaluations are principally focused on borings into the waste at the disposal site and into the surrounding soils to obtain relatively undisturbed samples of materials for chemical and physical testing. Testing will include field tests at the site and laboratory tests to determine various physical and chemical parameters of the waste.

Hydrogeological Evaluation Plan. The hydrogeology effort will consist of additional soil or waste borings, placement of monitoring wells, and the appropriate use of test pits (shallow, 4 × 10 × 12 ft pits) to develop additional information on site stratigraphy. The overall objective of the hydrogeologic evaluation plan is to define the hydrogeology on a seasonal and annual basis to the extent practical.

For illustrative purposes, the kind of site evaluation plan that is typically used is shown in Table II. Detailed procedures on the placement of wells, boreholes, and other site development activities should follow well-defined geotechnical procedures (ADL, 1981a).

The site evaluation plan should also include a comprehensive program of physical and chemical sampling on a periodic basis, which is covered in the next section.

Sampling and Analysis Efforts

Once a disposal site has been developed for monitoring purposes, the next step in generating data for a comprehensive monitoring effort is periodic visits to the site to gather water and, as appropriate, waste samples for detailed chemical and physical analysis.

Chemical Sampling and Analysis

The primary focus of a chemical sampling and analysis program is to obtain data on the extent of migration, if any, of leachate from the disposal site into the surrounding environment. The mechanism of such migration would include both leachate movement into the groundwater and surface runoff from the waste into the surrounding areas. Normally, appropriate monitoring wells upstream of the disposal site, in the dis-

Table II. Typical Exploration and Monitoring Well Installation Program

Exploration Number	Purpose[a]	Estimated Depth (m, ft)	SPT[b] Tests	Shelby[c] Tube Samples	Borehole[c] Permeability Tests	Number of Wells	Well Point Type	Anticipated Subsurfaced[d] Soil Sequence
7-1								
7-2								
7-3								
7-4								
7-5								
7-6								
7-7								
7-8								
7-9								
7-10								
TP7-1[e]								
TP-7-2								

[a]Explorations and well installations are for multidisciplinary purposes. Primary purpose is indicated: G = geotechnical; H = hydrogeological; B = background; C = chemical.

[b]Standard penetration tests (SPT) and associated samples will vary from a few representative samples to continuous sampling in the down-gradient control boring (7-4).

[c]The actual location and depth of Shelby tube samples and field permeability tests will be determined during the field exploration program.

[d]The location of the "Sand Creek" upgradient chemical background monitoring well will be north of the CB&Q rail line and is not indicated on the figure.

[e]Additional machine excavated test pits will be obtained to develop near-surface site stratigraphy and limits of disposed ash and are not indicated on figure.

posal site itself, and a matrix of downstream wells are required for a complete analysis. Collection of appropriate samples to determine the extent of runoff is fairly difficult unless there is a substantial effort to measure runoff effectively from the disposal site. However, the key difficulty is in the interpretation of data from any limited-runoff collection efforts.

Analysis of water samples and solid waste samples would include a range of chemical sampling and analysis procedures and would provide information on the major and minor constituents. Data required normally include:

1. general information such as pH, conductivity, total dissolved solids and total suspended solids;
2. major constituents including (e.g., FGD waste) sulfates, calcium, sodium chloride, and other appropriate ions; and
3. important trace elements of interest in the particular case, typically arsenic, selenium, cadmium, mercury, and thallium.

An important element of a sampling and analysis program is the frequency of sampling required. It is most desirable to have several sampling trips each year. A more cost-effective method would be to time the frequency of sampling at a site to the climatology of the site, which means assessing the climatological data at the site and dividing the year into varying levels of wet and dry seasons. The overall objective of environmental assessment would be to cover the wettest and driest part of the year and as many of the appropriate intermediate seasons as required on a site-specific basis.

Physical Sampling and Analysis

Samples for physical analysis are primarily gathered during site development, but there could be cases where additional sampling of wastes or other materials would become necessary. Physical testing includes moisture content determination, soil classification tests, specific gravity tests, standard Proctor tests, unconfined compression tests, triaxial compression tests, and extended permeability tests. Perhaps the most important properties of interest from physical testing are those related to permeability and stability of the waste disposal site.

Data from the sampling and analysis program provides important baselines for the environmental assessment effort that is discussed in the next section.

ASSESSMENT OF GENERATED DATA

So that data generated by site monitoring can be used as a management tool, they should be tied to an environmental assessment effort to determine the soundness of the disposal method and, if appropriate, to make general projections. The following sequence is commonly practiced:

1. Validated results from the disposal site are compiled in an appropriate format. This includes results from site development, physical and chemical sampling and analysis, and engineering/cost efforts. Appropriate quality assurance and quality control procedures are normally used.
2. Once the validated basis of information for that particular mix of disposal method, waste characteristics, and site conditions (including background information) is obtained, the next step is to develop an understanding of the geologic and geohydrologic framework for that particular combination of site conditions, method of disposal, and waste characteristics. This may include identification of principal flow directions and rates (flow nets) and identification of recharge and discharge areas and rates (water balances).
3. Cause-and-effect relationships for the physical and chemical sampling and analysis results are assessed on the basis of steps 1 and 2 above. The results and assessment could fall into one of two broad categories:

    1. There may be measured effects at the site for this particular mix of waste characteristics, disposal method, and site conditions that are fairly easy to note and evaluate. In such cases, hypothecation requires an empirically based definition of the physical and chemical conditions present at the site that can explain the observed cause-and-effect relationships.
    2. There may be sites where the observed conditions are either ambiguous or have no distinguishing effects within the range of confidence in the reported data. In such cases, one has to rely more heavily on projection techniques for describing the potential cause-and-effect relationships that may be reflected in future observations or that may be applicable to other sites in generic terms.

The latter situation may be substantially more difficult but could be done if a significant amount of background and measured information is available.

4. The last step is to make generic projections of the implications of the effects of this particular mix of waste characteristics, disposal method and site characteristics to numerous circumstances based on a generic mix of waste types, method of disposal and site conditions.

In the development of sound engineering disposal techniques, this method or sequence is normally followed.

The overall sequence of these evaluations can be substantially enhanced by use of interpretation and modeling techniques, as shown in Figure 1. A logical premise for the interpretation technique is an inventory of interpretation and modeling tools available to cover a range of sophistication. In other words, the sophisticated mathematical models of today are a logical extension of the earlier assessment efforts based on cause-and-effect relationships and simple empirical methods to evaluate the combined effect of waste characteristics, method of disposal, and site conditions.

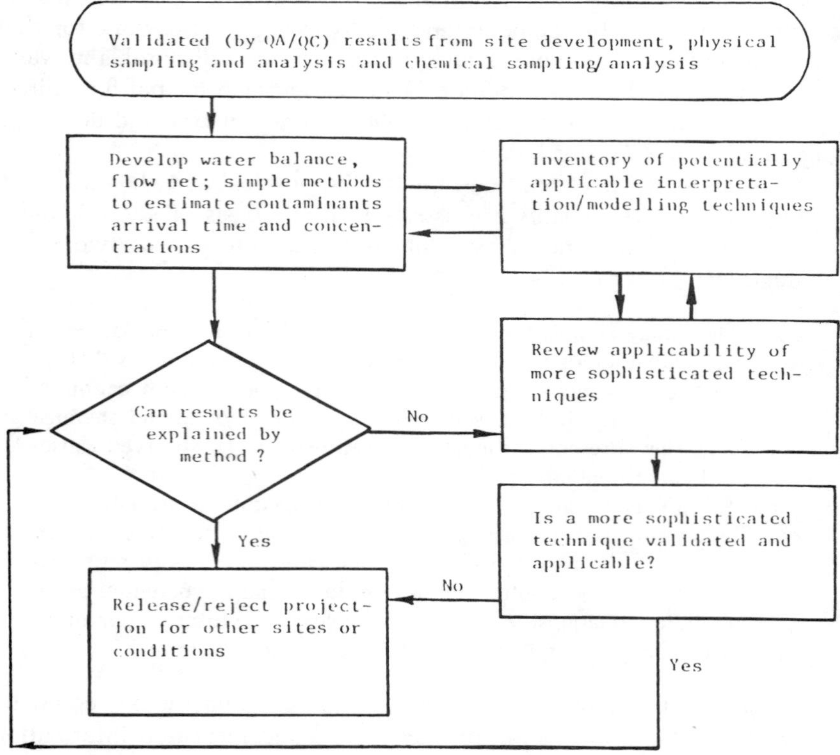

Figure 1. Decision logic for selection and use of interpretation/modeling techniques. (Source: Arthur D. Little, Inc.)

REFERENCES

ADL. 1980. "Candidate Site Selection Report," prepared by Arthur D. Little, Inc. for U.S. EPA, Office of Research and Development, Washington, DC.

ADL. 1981a. "Appendix B: Sampling and Analysis Procedures Manual," prepared by Arthur D. Little, Inc., for U.S. EPA, Office of Research and Development, Washington, DC.

ADL. 1981b. "Appendix A: Hydrogeological and Geotechnical Procedures Manual," prepared by Arthur D. Little, Inc., Bowser-Morner Testing Laboratories and Haley & Aldrich Inc., for U.S. EPA, Office of Research and Development, Washington. DC.

Santhanam, C. J. 1981. "Characterization and Environmental Monitoring of Full-Scale Utility Waste Disposal Sites—Current Status," paper presented at the American Pollution Control Association Specialty Conference on Waste Treatment and Disposal Aspects—Combustion and Air Pollution Control Processes, Charlotte, NC, February 9-11.

DISCUSSION

***R. E. Blanco*, ORNL:** I would like to address the question to the speaker as well as a general question to the audience. We have had several talks on coal waste, ashes and all. I know that coal contains significant amounts of radioactive materials like thorium, uranium, and radium. Has anyone done any monitoring studies on the movement of radium in the plumes?

C. Santhanam: We have analyzed many waste materials for radioactive constituents. In all cases, the levels of radioactivity in the wastes were either at or near the ranges that are considered okay. As far as the base to sites are concerned, you are talking about dilution of those materials to a very much greater level; hence, the difficulty is in being able to discriminate the effect of the contribution of fly ash- and FGD-waste effects from those of the others.

***John Cherry*, University of Waterloo:** A group at the University of North Dakota a few years ago chose several areas that had fly ash and scrubber sludge to monitor. Those particular coals are fairly high in uranium, thorium, and radium. They found no movement of anything including uranium. Not finding radium, thorium, or lead-210 mobile was expected, but we thought maybe uranium would be mobile, so there was no evidence in their study environments of any mobility of radionuclides.

***Dev Sachdev*, Envirosphere Company:** My concern is how the utilities will benefit from the research which you are doing, because the data that you are collecting will be site-specific. The data based on one or ten sites will be applied for all the 176 utilities. What do you recommend? Are you developing some model that will be applicable for different types of utilities, or soil, climate, or clay mineral conditions? What are your recommendations? How will the utilities benefit from this? For example, each utility has different climatic, soil, and groundwater conditions.

C. Santhanam: I'll give you a much broader answer than with reference to one specific program. The U.S. government, of course, is doing a lot in this area; for example, EPA is doing this large study. However, as Dr. Cherry just pointed out, the Department of Energy has been doing a good deal of work in this area. EPRI has been funding strong efforts in analyzing the effects at coal plants. A lot of baseline data are appearing in the literature. There are two methods which you have to take into account. One is that the individual technical staff at every one of the utilities is taking into account all of these efforts in their planning. It is a diffuse method, but it is happening. Item two, the large program that we are running for the EPA will lead to appropriate documents that will be making generic projections. The number of sites in this particular EPA program involves a fairly complete assessment of all the larger coal-fired plants in the country. A matrix of these types, method of disposal, and site characteristics has been included there to permit as many projections as practical. The reports as a result of that program will provide generic data and information to permit the industry to properly enhance our waste management efforts. It is happening at several levels because a lot of work is going on, not only by the EPA but by other agencies of the government, EPRI, and the larger utilities; TVA has a significant program of its own, and so do several others.

***Ishwar Murarka*, EPRI:** Let me add to that answer by saying that there would be some site-specific efforts involved in extrapolating or using the general knowledge, but the advantage of having a large-scale effort was that it will provide the research methods which are needed.

***John Brown*, CEGB:** At the beginning of your approach to monitoring, you listed four factors: evaluation of site, site development for monitoring, assembling and analysis, and assessment of data. It was not clear to me whether you do that before you develop the site to give you a datum, or whether you were merely preparing the ground for what you do after you have used it for a landfill.

C. Santhanam: The effectiveness of your ability to reach conclusions is extremely dependent on your doing a lot of planning ahead of time. In fact, if there is one message I can deliver, it is a complete evaluation of the site, of the type of waste you are handling, and the method of disposal before you put in a single boring or a single monitoring well. I think that is where the most improvement in the assessment of data could come about.

John Brown: But what I am asking is, do you do all four operations before you dump any material at all? Because what is needed, in my view, if you do not have a completely unequivocal idea of what the result of the development will be, is to know what the water quality and movement in the table are about before you start. Now, I am not aware that anybody has yet managed to do this. Usually we end up by filling sites and then putting boreholes in and monitoring effects and using traces and making inferences, but we are now educated enough, in my view, to say, "Shall we look at the site, evaluate it and measure it before we start?" Have you done that?

C. Santhanam: Very often in these types of wastes you do not have the choice. What you have suggested is the right way. In terms of some of the larger plants and the larger waste disposal activities, the trend is exactly what you have suggested, that is, assessing a site to develop broad baseline data in terms of planning a waste disposal facility.

Ishwar Murarka: For the last 30–40 years, the utilities have been disposing of waste. For those existing facilities we could not do the advance planning, but for future new facilities, the licensing requirements will be or are such that advance evaluation will be necessary, at least relative to the intermediate level. Environmental analyses will also be included if necessary. Beginning in 1982, the new facility will follow an advance assessment analysis, and then attention will be directed in the future to whether or not that assessment failed to produce the type of problems encountered.

John Brown: Can I say that at the moment, we would have to do the same thing. We would have to evaluate the site before the event. But as I indicated in my presentation, the way we would approach it is to do laboratory studies of percolate qualities as being the cheapest way of getting that part of the answer. We would then go to the geological library for details of the hydrogeology of the area. What I am saying is that, in some cases, these data may give you enough information to be able to proceed with caution; but if it does not, you are into a kind of monitoring exercise which makes the preliminary examination of the site costly and a gamble. You've actually got to convince somebody it is worth spending $50,000 to actually survey a site to establish what the water movement and the water quality are. And that requires a real act of faith if the preliminary information is not sufficiently encouraging.

Ishwar Murarka: The alternative is to produce liners, and that is even worse.

CHAPTER 28

MONITORING EXPERIENCES AT A FLUE GAS DESULFURIZATION SLUDGE DISPOSAL SITE: WHAT CAN BE ACHIEVED?

David W. Hupe

Michael Baker, Jr., Inc.
4301 Dutch Ridge Road
Beaver, Pennsylvania 15009

Dean M. Golden

Electric Power Research Institute
3412 Hillview Avenue
Palo Alto, California 94303

ABSTRACT

The first full-scale application of the IU Conversion Systems, Inc., system for fixation of utility-produced flue gas desulfurization (FGD) sludge has been monitored at an eastern U.S. power station landfill for 3 years as part of a study sponsored by the Electric Power Research Institute (EPRI). A primary emphasis in this ongoing study is on determining the success of the technique in preventing groundwater contamination.

Monitoring systems designed to adequately detect potential groundwater contamination can be much more complex than general monitoring schemes illustrated in commonly available guideline literature. In this study, a determination of the effects on groundwater is complicated by background influences that made installation of a complex network of monitoring wells necessary. These influences are related to prior disposal activities at the FGD sludge landfill site and several current power station activities in the immediate area. The effects on groundwater quality from the background influences are very similar to the possible effects from the sludge landfill, which makes differentiation of leachate sources a difficult task. None of the effects from the disposal operations noted during the study caused any concern about U.S. Environmental Protection Agency (EPA) Primary Drinking Water Standards.

Careful historical research related to the site, proper attention to the design and positioning of groundwater monitoring well clusters, and monitoring of local point surface discharges, which are known to infiltrate quickly and affect background groundwater quality, have made it possible to make reliable conclusions about the effects of the disposal site without wrongly attributing the background influences to the landfill.

INTRODUCTION

The primary objective of a groundwater monitoring program is to responsibly monitor for contamination that might be produced as a consequence of disposal activities. However, sufficient monitoring to demonstrate, if necessary, that a site is not responsible for contamination caused by another source is just as important.

To determine reliably whether an area's groundwater resources are affected by a disposal site, a monitoring network must be designed that

1. allows one to define the condition of local groundwater as it exists before the start of disposal operations (referred to as background conditions);
2. allows continual monitoring of groundwater that flows into the area and that may have already been affected by some other source of contamination (upgradient conditions);
3. monitors groundwater conditions at the proper locations (direction and depth) with respect to the disposal site that are most likely affected by potential leachate from the disposal site (downgradient conditions).

Figure 1 illustrates a basic monitoring network to accomplish these objectives. Monitoring wells A, B, and C all constitute background wells when monitoring precedes the disposal operations. As disposal operations begin, well A constitutes the up-gradient monitoring point, whereas wells B and C constitute down-gradient monitoring points. The groundwater flows from right to left with respect to the disposal area. Under most actual monitoring conditions, this basic monitoring network is too oversimplified to provide conclusive data, as is illustrated

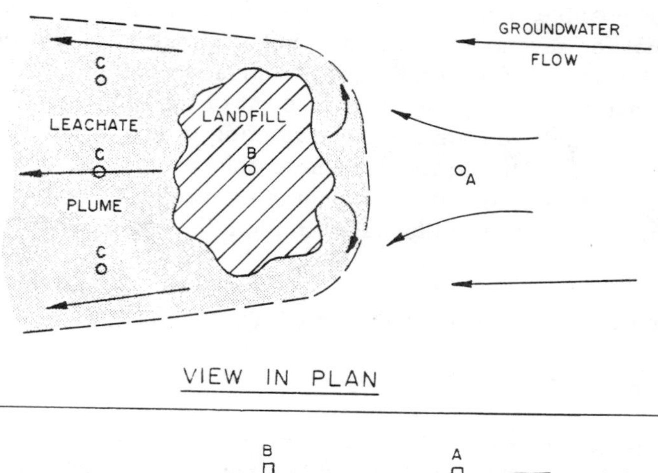

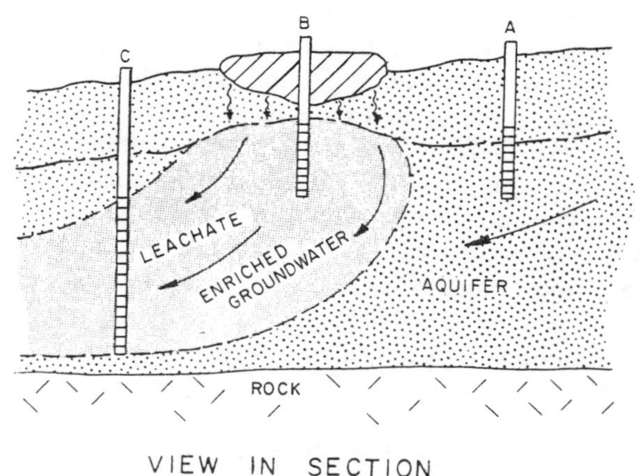

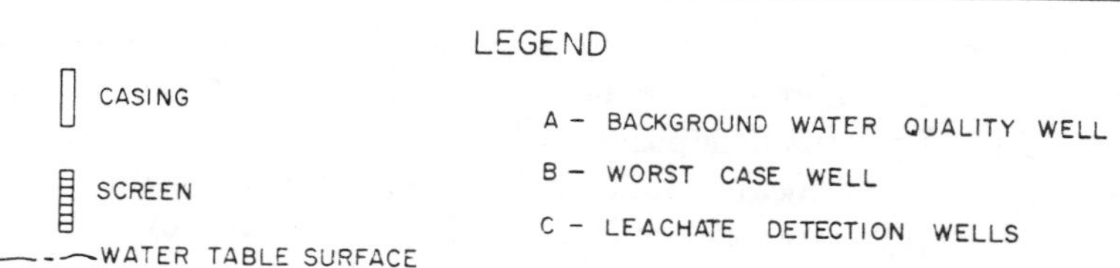

Figure 1. Basic monitoring system for a solid-waste disposal facility.

in the following case study of an ongoing project sponsored by EPRI and conducted by Michael Baker, Jr., Inc.

Situations can easily arise where conclusively determining whether an area's groundwater resources have been affected by a disposal site is not possible. The results of groundwater monitoring can be inconclusive for several reasons. Inadequate attention to the design of a monitoring network, insufficient investigation of relevant background information, and improper management of the sampling activities and personnel involved are some contributing problems within one's control. However, uncontrollable circumstances can also lead to uncertainty in the interpretation of monitoring results. These types of potential problems are also illustrated in the following case study.

GROUNDWATER MONITORING CASE STUDY

Purpose of Study

The EPRI-sponsored study is being conducted at the Conesville Power Station of Columbus and Southern Ohio Electric Company, located in east-central Ohio. FGD sludge, which has been fixed or stabilized according to the system developed by IU Conversion Systems (IUCS), Inc. (Poz-O-Tec® system), has been landfilled at the Conesville facility since January 1977 (Hupe and Shoemaker, 1979). When this study was started, the operation was the first full-scale application of the IUCS system for treating FGD sludge and, as such, presented an excellent opportunity to evaluate the pros and cons of the system for the utility industry. Evaluating the success of protecting groundwater from contamination has been a primary emphasis in this study.

Development of Monitoring Well Network

Influence of Site Geology and Offsite Conditions

The Conesville station and, more particularly, the FGD sludge disposal area are located on the floodplain of the Muskingum River (Figure 2), where thick deposits of glacial outwash, consisting primarily of highly permeable sand and gravel, were deposited in the underlying deeply entrenched ancestral valley of the Muskingum River. The sand and gravel beneath the site is about 30 m thick, capped by a much less permeable, relatively thin alluvial silt layer. Other than the surface silt, no definable low-permeability layers interrupt the thick sand and gravel aquifer.

At the start of the study, groundwater was assumed to flow generally from the hill areas east of the plant west to southwest toward the Muskingum River. Although groundwater discharge to the Muskingum

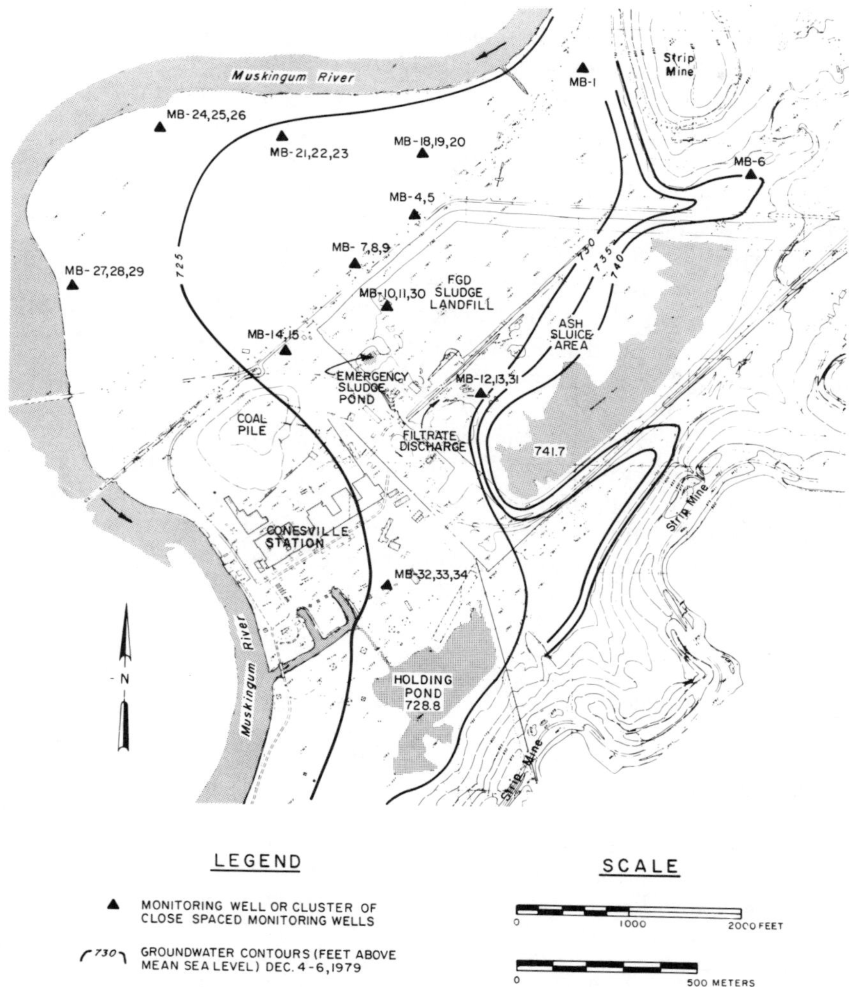

Figure 2. Site map—Conesville Power Station.

River was considered to be the usual case, intermittent recharge from the river was expected in response to quick rises in river stage. Additional recharge to groundwater was expected to occur in swampy areas fed continually by local streams to the northeast and southeast of the disposal site.

There has been extensive strip mining on the hillsides east of the plant. Background groundwater quality effects were anticipated from direct groundwater flow from these hills and from infiltration of affected runoff in the swampy areas discussed above. The river was expected to exert a large influence on local groundwater quality.

Influence of Plant Development and Operation

The present FGD sludge disposal site occupies part of the area that has served as an ash sluicing site for the plant since the 1950s. During a past expansion of the ash pond in the early 1970s, much or all of the near-surface low-permeability silt was excavated from the impoundment area and used for construction of the existing perimeter dikes. The present ash pond is located immediately adjacent to the sludge disposal area. The two disposal areas were separated by the construction of an ash dike.

Because of the long-term use of the site for ash disposal, removal of the silt that served as a natural liner, and the large influx of water to the present ash pond, the ash pond was expected to have a large background influence on the direction of flow and quality of groundwater near the FGD sludge landfill. A groundwater "mound" was assumed to have formed around the existing ash pond, affecting the natural groundwater conditions.

Monitoring-Well Layout and Design Philosophy

Well Locations. Monitoring well locations (see Figure 2) were chosen on the basis of:

1. probable directions of groundwater flow, as defined by preliminary water table contours developed from water levels noted on available drill logs coupled with inferences drawn from visual inspections of the area;
2. the assumed mounding of groundwater beneath the ash pond and expected significant background quality influences;
3. input from adjacent property owners and the potential for well damage considering existing land usage; and
4. the expected background quality influences from recharge related to the Muskingum River and the stream draining the strip-mined areas east of the plant.

Monitoring wells MB-1 and MB-6; the cluster of wells MB-12, MB-13, and MB-31; and cluster MB-32, MB-33, and MB-34 were positioned to act as background monitoring points. Neither well MB-1 nor MB-6 was expected to be affected by the ash pond or the FGD sludge landfill.

Well MB-1 was expected to reflect the influence of the Muskingum River on local, clean groundwater. Well MB-6 was expected to indicate the potential influence of adjacent strip mining on clean groundwater. The cluster of wells MB-12, MB-13, and MB-31 were located to allow direct monitoring of the influence of the ash pond on groundwater. Wells MB-32, MB-33, and MB-34 were installed to primarily monitor effects from the holding pond.

The cluster of wells MB-10, MB-11, and MB-30 was established to monitor expected worst-case leachate conditions produced by the landfill. The remaining clusters of wells (seven clusters) were positioned downgradient of the landfill to monitor for potential leachate.

Clustering of Wells. Because the low-permeability surface silt overlying the thick glacial outwash aquifer had been removed from the disposal areas, and no other shallow impermeable layer existed below the site, downward leachate movement was assumed to be essentially unhindered. Available literature indicates that leachate plumes may sink as distinct bodies toward the bottom of an aquifer. However, thick (30 to 60 m) aquifers tend to have more pronounced shallow and deep flow systems (Fenn et al., 1977).

An individual shallow well at the various downgradient monitoring locations was considered to be inadequate on the basis of these site conditions, because a leachate plume might not be penetrated by the well if the leachate were to sink in the aquifer as speculated. A deeper well or a well penetrating the full thickness of the aquifer would probably intercept the leachate plume. However, a water sample taken from such wells could potentially be diluted because of intermixing of leachate from a thin plume with unaffected groundwater from above or below the plume so that the leachate might go undetected.

The system finally used consisted of three monitoring wells in a cluster, each with relatively short (3 m) screened sections and different finished depths in the aquifer. Monitoring for a shallow flow system was emphasized. The shallowest well in each cluster typically penetrated the water table by 1.5 m. The screened section of the second well was placed 3 m below the bottom of the shallow well. These two wells were considered to be an appropriate means for monitoring successive zones in the shallow groundwater flow system. The final monitoring well in a cluster was drilled and screened, when possible, to near the base of the aquifer as a check against the shallow flow system premise.

Construction. The wells were drilled to the desired depth by using continuous-flight hollow-stem augers. Flush-joint polyvinyl chloride

(PVC) pipe (51 mm i.d.) was used as a well casing. The bottom of the casing was slotted with a hacksaw to form a screen. As the casing was assembled, the completed string was lowered to the bottom of the hole through the augers. The augers were gradually withdrawn while fine gravel was carefully fed to the void between the inside of the augers and the outside of the PVC casing to form a gravel pack around the slotted part of the casing. After the augers had been raised to a point just above the slotted casing, a bentonite seal was placed above the gravel pack; bentonite pellets were used to prevent the drilled hole from acting as a conduit for downward leachate movement. The remainder of the hole was backfilled with the native auger cuttings, and a second bentonite seal was placed at the surface to prevent surface water inflow. Complete costs for constructing the monitoring wells were $41/m or about $2200 per cluster.

Monitoring Difficulties

Background Influences

Monitoring of the Conesville landfill has been complicated by several past and present background influences on groundwater quality, which will be discussed later. The method of constructing the disposal site has also complicated monitoring. The background influences are similar in chemical quality to that of leachate potentially produced by the landfill. Also, the process for fixing the FGD sludge and overall management of the landfill results in a significantly lower potential for the production of leachate relative to the background influences. As a result, if leachate is produced, detecting or differentiating it from the background influences is difficult because of their similarities in quality and because the quantity of landfill leachate is so negligible (Hupe, 1981).

Ash Pond. The past and existing ash ponds constitute the primary background influence. This influence was expected from the beginning of the project, and monitoring to characterize this influence was provided, as explained previously, by installing the monitoring well cluster of wells MB-12, MB-13, and MB-31 (see Figure 2). The influence has caused changes in groundwater quality that exceed the limits of proposed EPA Secondary Drinking Water Standards, but not those of the mandatory Primary Drinking Water Standards. Because the natural surface silt "liner" that occurs near the plant was removed during expansion of the ash pond in the early 1970s, the infiltration rate for leachate from the ash pond was increased significantly. Leachate from the ash pond has been estimated to be produced at a rate of 0.1–0.2 m^3/s. This rate is many hundreds of times greater than the potential production rate of leachate from the FGD sludge landfill (Hupe and Shoemaker, 1979). Monitoring results to date indicate that the influence of the ash pond on groundwater quality extends beneath the FGD sludge landfill and, in fact, reaches the Muskingum River. Differentiation of leachate produced by the landfill, if any, from that originating as a result of long-term ash disposal is difficult because of masking by the much greater quantity of ash pond leachate. Differentiation of leachate sources is further complicated because dewatered FGD sludge and fly ash are blended during the fixation process. Therefore, the leachates from either source are similar.

Filtrate Discharge. As part of the sludge fixation process at Conesville, excess water is vacuum-filtered from the sludge. The filtrate is usually returned at utility stations to the scrubbing system for reuse. However, at Conesville, this filtrate, which is highly mineralized, is now discharged to the ash pond. The utility plans to return this filtrate to the scrubbing system in the future.

This discharge, which was discovered after monitoring was started, constitutes a second background influence on local groundwater quality with respect to the EPA Secondary Drinking Water Standards. By coincidence, this filtrate is discharged near the monitoring well cluster within the ash pond area. The influence of this discharge has been reflected at these wells due to infiltration through the settled ash. However, because of the common nature of the filtrate and FGD sludge landfill leachate, the significant mineralization of the filtrate, and the small quantity of potential landfill leachate, any leachate produced by the landfill can be easily masked by the background influence of the filtrate.

Emergency Sludge Pond. After monitoring was begun, discussions with power plant personnel revealed that the pond used during emergency conditions to temporarily store unfixed FGD sludge had been unlined during the period from startup of operations (January 1977) to September 1978. While the emergency pond was unlined, leachate from unfixed sludge in the pond and supernatant could easily percolate into the underlying permeable outwash and measurably influence local groundwater quality. The cluster of wells MB-14 and MB-15, situated downgradient from this pond, has been affected by this leachate. However, a well cluster was not planned at a location immediately upgradient from the emergency pond, which would facilitate differentiation of leachate from the emergency pond from that from the FGD sludge landfill. During these investigations, improvement in water quality downgradient from the emergency pond was noted and attributed to the lining of this pond.

The utility has experienced difficulties in reclaiming the unfixed sludge from the emergency pond on many occasions. The pond liner has been disrupted because of the method necessary to clean out the pond. The pond sometimes had to be used again before the liner was repaired. Therefore, periodic background influences on groundwater quality can again be expected with intermittent operation of the emergency pond.

The leachate from the emergency pond is similar to potential leachate from the fixed sludge landfill. Because estimated quantities of leachate from the emergency pond are also much greater when the liner is disrupted than that of the landfill, any leachate from the landfill is easily masked in the area of the emergency pond as it is by filtrate discharged to the ash pond.

Well Maintenance

Differentiation of sources of leachate, as described above, would be impossible had the well clusters in the fixed sludge disposal area and the ash sluice area not been installed. The fact that the well clusters are still intact can only be attributed to the cooperative attitude and attention of the power plant personnel. Although additional clusters would have been beneficial for the differentiation of leachate sources, well clusters in such disposal areas are difficult to maintain. Well casing was repeatedly added to the disposal area well clusters as the waste levels rose. Without a significant amount of manual labor, adding material (fixed sludge or ash) around small-diameter casings with heavy equipment without kinking or breaking the casing is difficult. Such activity resulted in the recent loss of the shallowest well in the fixed sludge landfill area.

As the water level of the ash pond rose, an island was formed by inundation of the slightly higher area of ash containing the ash pond monitoring well cluster that was not accessible during one sampling run. A dike extending to the wells had to be constructed before the following sampling run to permit safe access. Power station personnel have requested that both the FGD sludge area and ash pond area monitoring well clusters be relocated in the near future because of planned construction activities that will destroy these wells.

Wells placed in the neighboring fields were susceptible to destruction by farm equipment. To prevent this destruction, most of the perimeter wells were drilled intentionally along property lines, which are watched more carefully by the farmers. The utility also paid the farmer whose field is situated west of the disposal areas a set price per well per year to permit drilling of the wells and continued access to the well cluster sites.

Steps to Resolve Difficulties

The interpretation difficulties generally have been overcome through a combination of investigative methods. Careful investigation of available literature, subsurface data, waste stream water quality data, old site maps and wastewater flow charts, and discussions with plant and IUCS personnel proved invaluable in the planning stages of monitoring and throughout the study. Such an investigative step is necessary because of the vast amount of insight that can be gained.

Continual monitoring of water table levels and periodic contouring of these data helped to define probable groundwater flow directions. The hand-interpreted groundwater contours were confirmed by Battelle Pacific Northwest Laboratories staff, who used a predictive numerical model for groundwater flow calibrated for the site (Bond, 1980). Later tracing of the thermal effects of discharges to the ash pond by remote monitoring of groundwater temperature in the wells also helped to substantiate conclusions drawn about the expected directions of groundwater flow. This information was extremely important when we were faced with the problem of determining sources of groundwater contamination that were detected at only certain monitoring locations.

Determining the quality of surface point discharges that were found to affect background groundwater quality with respect to the landfill was another invaluable step in resolving interpretation problems. Samples of the filtrate discharged to the ash pond, and surface runoff from the FGD sludge landfill (which seeped into the permeable sand and gravel underlying the landfill until the area was completely lined) were tested repeatedly in the laboratory for an indication of worst-case, unattenuated conditions. Water quality data for the ash pond supernatant supplied by power station personnel also gave additional background information.

Samples of the fixed sludge were core-drilled from the landfill and used to generate artificial leachate samples. These data were useful in determining potential leachate quality for direct comparison with monitoring well water quality data.

Finally, the water quality data for each well were graphed according to individual chemical parameters. The graphical representations made analysis of trends in water quality much easier than comparing raw data from laboratory result summaries.

Unresolved Problems

The background influences on groundwater quality from leachate related to the emergency sludge pond and the filtrate now discharged to the ash pond are fairly localized in extent. Their influences should be limited to groundwater beneath approximately the southern one-third of the FGD sludge landfill and the area immediately downgradient to the west and southwest (around the cluster of wells MB-14 and MB-15) because of the direction of groundwater flow. However, the background influence of the ash pond on groundwater quality has affected all the downgradient monitoring wells near the landfill. Detection of leachate, if any, produced by the landfill will not be possible by using the monitoring well network unless the leachate causes the chemical constituents to rise to significantly higher levels than the background levels already attributable to the ash pond.

SUMMARY AND RECOMMENDATIONS

Several important points can be drawn from the experiences gained during this study:

1. Without sufficient background and upgradient monitoring, false conclusions can easily be reached on whether a disposal site is causing groundwater contamination.
2. Careful site historical research and attention to day-to-day operations are extremely important because they can significantly affect the design of the monitoring program and later interpretations.
3. Regardless of the level of attention to the monitoring program, use of the monitoring wells to unquestionably determine whether an area's groundwater resources are affected by a solid waste disposal facility may be impossible because of background influences.
4. Pressure-vacuum lysimeters placed beneath a disposal site in the unsaturated zone before the start of disposal activities may be the only way to adequately monitor under conditions where there is significant background interference such as has been experienced at Conesville.

A few general recommendations about groundwater monitoring programs are:

1. Groundwater monitoring schemes and programs must be designed, conducted, and evaluated by qualified professionals because of their site-specific nature.

2. Shortcuts on costs to establish groundwater monitoring networks and failure to allocate qualified personnel or obtain necessary equipment should be avoided. Corrective measures such as liners can be far more costly if they cannot be proven unnecessary.

3. A steady staff of trained personnel should be maintained for monitoring activities. Changes in routine, procedures, and record-keeping can be detrimental to the program.

REFERENCES

Bond, F. W. 1980. "Modeling the Fixed FGD Sludge Landfill, Conesville, Ohio–Phase I," EPRI CS-1355, Electric Power Research Institute, Palo Alto, CA.

Fenn et al. 1977. "Procedures Manual for Ground Water Monitoring at Solid Waste Disposal Facilities," EPA/530/SW-611, U.S. EPA, Cincinnati, OH.

Hupe, D. W. 1981. "Monitoring the Fixed FGD Sludge Landfill, Conesville, Ohio–Phase II," EPRI CS-1984, Electric Power Research Institute, Palo Alto, CA.

Hupe, D. W., and S. H. Shoemaker. 1979. "Monitoring the Fixed FGD Sludge Landfill, Conesville, Ohio–Phase I," EPRI FP-1172, Electric Power Research Institute, Palo Alto, CA.

DISCUSSION

***John Brown*, CEGB:** One of the objectives in fixing waste, as I understand it, is not only to make it easy to handle but to reduce its surface area so that the actual rate at which materials can leach from it is considerably reduced, by comparison, say, with a fly ash which has fairly fine surface and easy access to water. I ask whether anybody took a sample of the FGD sludge and tested its permeability and leaching characteristics in a lab column? It would seem to me if the ash pond is going to have a big influence, that one might have been able to predict what the incremental contamination from the FGD was. Do you have some laboratory results?

D. Hupe: We knew about the permeability, or the supposedly low permeabilities, of the FGD sludge when we started from information that is commonly available. We have been taking samples of the sludge throughout the program and doing permeability and leachate tests on it to determine approximate quantities and quality of leachate. Regrettably, we could not do this before we started.

***David Wilson*, Harwell:** Could you briefly describe what the IUCS process actually does and, from your work, could you tell us how effective it is?

D. Hupe: The IUCS process takes bleed from the scrubber system after it has been thickened to approximately 30% solids, vacuum filters the excess water from the sludge, and blends it with fly ash and lime, which produces a material that, within a short period of time, hardens sufficiently so that it can be handled by normal earth-moving equipment. The permeability was reported by IUCS to be on the order of 10^{-7} cm/s. We, in our permeability testing, did come up with a few higher permeabilities, on the order of 10^{-5}, but some of the more recent testing has resulted in very low permeability. Overall, the system seems to be very good. We have not been able to detect leachate from permeation through the landfill. So, in effect, it is doing what it is supposed to be doing.

Michael Dahlberg, **DOE, Pittsburgh:** Can you give me more description on the liner that was used?

D. Hupe: There is no liner. The liner is, in fact, the fixed FGD sludge.

Michael Dahlberg: I would like to know how much leaching there was into the Muskingum River, and what effects it might have on the river?

D. Hupe: Well, there was no leaching to our knowledge from the landfill. That has been part of the problem—trying to detect it at all because of the background influences.

Dev Sachdev, **Envirosphere:** In Figure 2 you showed that the groundwater elevation rose above the existing elevations. During those days, was this FGD sludge being dumped, and what was the elevation in that area? Did you take any samples for that?

D. Hupe: I think what you are referring to is the groundwater contours directly below the ash pond, and to our knowledge the groundwater surface does not extend up into the FGD landfill. It drops off that quickly from the ash pond before it gets beneath the FGD landfill. It does not wet the bottom of the landfill.

Aarne Vesilind, **Duke University:** One thing that has impressed me about the last two papers and also others I have heard on FGD is that people do not seem to make any effort at distinguishing what precisely these sludges are. Relative to the question from our friend from overseas, who asks why do we not go into the lab and find out something about these sludges—this is not precisely my field, and I am asking the question: Are these sludges so much different from plant to plant that they behave differently, or are they chemically similar so that we do not have to go and characterize them at every facility?

D. Hupe: I do not think that you can assume at all that the sludges are alike from one plant to another. It all relates back to the coal and additives that are used during the scrubbing process. They are going to be different.

Aarne Vesilind: Then the obvious follow-up question is why don't we do these kinds of tests, and why do you and the previous speaker and others that I have heard just simply go speaking as if an FGD sludge is an FGD sludge and there are no differences?

D. Hupe: Well, we did characterize the fixed FGD sludge by conducting leachate tests on the material which was cored from the landfill.

C. J. Santhanam, **A. D. Little, Inc.:** A lot of work has been done in terms of characterizing both coal ash and FGD sludges. In fact, literally, there are thousands of pages of data. A lot of background information is also available in terms of column tests and the like. To give you a perspective on that, the earlier effort in terms of assessing the environmental impacts of these groups of materials involved column sludges, lab work, and projections. The attempt in terms of the assessment effort that I described relative to what is going on at Conesville is a further building block, an extra stage in the development because the environmental effects are complex. No two FGD sludges are exactly the same; trace elements are often significantly different. More important than that, the chemical mobility of various materials, particularly trace elements, also depends on the environmental site conditions that you encounter. For example, a trace element might be highly mobile at one pH and at a pH that differs by 1.5 it may not be. The complication arises at several levels, but there is a huge amount of background data available that has been gathered in the past seven years.

CHAPTER 29

FIELD DATA COLLECTION AT ABANDONED WASTE DISPOSAL SITES: PROBLEMS IN DATA GATHERING

Benjamin J. Mason

Ethura
P.O. Box 1280
McLean, VA 22101

ABSTRACT

The role of the environmental researcher in field investigations at abandoned hazardous waste sites is presented. Problems of legal, safety, and quality assurance aspects are discussed from the perspective of one of the major environmental studies undertaken in recent years—the Love Canal field studies conducted by the U.S. Environmental Protection Agency (EPA). Insight gained through 2 years of work on several abandoned waste sites located in the eastern United States is provided.

INTRODUCTION

Abandoned hazardous waste site investigations present a challenge to the environmental scientists that few researchers encounter in the course of their careers. The investigations are usually based on litigation or at least potential litigation. Activities normally taken for granted can thus become monumental problems if not adequately addressed and documented. Familiar sampling plans may be severely altered because of political or public relations pressures. Attorneys normally approach a problem from a purely judicial basis, with little empathy for the scientist's concerns for "good" science.

For about 2 years the author has been actively involved with the investigations at several abandoned hazardous waste landfill sites. This chapter draws on that experience in developing the concepts and approaches outlined.

SETTING

An abandoned hazardous waste disposal site can occur in many settings. Some, such as the farm sites in southern Missouri where dioxin-contaminated wastes were deposited in pits located on several farms, are difficult to locate and may go unnoticed for years. Others, such as the Kin-Buc landfill in Edison, New Jersey, or Love Canal in Niagara Falls, New York, are in urban areas where problems are often noticed shortly after they surface. General awareness of the problems create public pressures to "do something fast," thus creating a situation that can become explosive if not handled properly by the authorities. Tasks that are simple to perform in the comfortable old clothes most field people wear become exceedingly difficult in the protective clothing required for sampling contaminated materials. Simple record-keeping becomes a chore because of rigid procedures dictated by the rules of evidence of the legal system. Time, costs, and risks all escalate to the point that few scientists want to become involved in the entire investigation process.

SAMPLING

Purpose

Sampling at the abandoned disposal site is designed to provide data for three purposes (JRB, 1980):

1. identifying the problem and determining the appropriate remedial response;
2. developing the background information necessary to support litigation at the abandoned site; and
3. determining the extent of exposure to the populace and the environment and recommending actions to eliminate or reduce that exposure.

Each purpose has its own set of requirements and goals. Because these requirements and goals often conflict with one another, they lead to difficulties in developing a clear, definitive purpose for a particular study. An example of this conflict can be seen in the case where the attorney and the scientist have different needs.

The attorney only needs to show that (1) a toxic chemical is present; (2) it has migrated offsite or has a potential to migrate offsite; and (3) substantial danger to health is imminent. This can be accomplished with a few grab samples taken at appropriate locations. The scientist, on the other hand, would like to have enough samples to develop a measure of reliability along with some measure of a baseline or control. Strong leadership is required to ensure that both purposes are satisfied.

Some of the purposes intermesh, allowing one set of samples to serve multiple purposes, which saves time and money when properly obtained. The engineer can use the results to develop the remedial plan; the attorney uses the results for showing migration; and the environmental scientist can use the data as part of an integrated exposure assessment. An integrated team approach is required to meet these objectives.

Sampling Plan

The clearly defined objective leads to development of a study plan to meet those goals. Where peer-reviewed protocols have been developed before the study, this process is greatly simplified. [One outcome of the major EPA study at Love Canal during the summer of 1980 was the development of a set of protocols that can be used as part of future studies (GCA, 1980)]. Determination of the exact plan to be used requires that the team assemble background data on the site. These data should include all available prior sampling data, maps, aerial photographs, and reports. It is beneficial, if not essential, for some of the team members to visit the site before developing the plans. These data should be reviewed to the point that hypotheses of likely routes of migration can be developed. Statistical designs should be an integral part of all sampling to be done, thus allowing conclusions to be drawn from the data.

In cases where few or no data are available, preliminary site investigation will be necessary. These preliminary investigations often provide a "quick and dirty" look at the site. My experience has been that these "quick and dirty" approaches invite serious trouble. Simple statistical designs should be used in these preliminary studies. By using proper designs, the preliminary data can be evaluated, a reliability can be placed on the numbers, and the data can often be used as a part of the main study design. The use of these simple designs provides protection to the scientist in those cases where the data are prematurely leaked or are used to make administrative decisions before the final study is completed. In general, this approach is safer and often more economical.

Phased study designs may also be the best approach for another reason. The status of our knowledge about the behavior of the mixture of waste materials found in the abandoned sites is so rudimentary that phasing is necessary to make the most economical use of the resources available. Techniques such as Kriging (Barnes, 1980) can be used to develop a reliable data base for evaluating pollution migration. This technique, developed originally by the mining industry in France, allows the scientist to collect data in phases while working toward a predetermined level of reliability.

Some types of data that are useful for developing the plan are shown in Table I. Careful evaluation of these data, combined with a site visit, allow the investigating team not only to identify data gaps but also to

- select the parameters of the investigation,
- identify routes of migration,
- locate the boundaries of the disposal area,
- plan for safety equipment, etc.,
- determine the location and number of samples needed,
- obtain the desired equipment,
- arrange for access to sampling areas,
- plan contacts with local officials and the general population, and
- arrange for analyses.

Sample Collection

The planning process does not differ substantially from that required for any field study, but implementation of the plan brings the scientist into the realm where major problems begin to surface.

Record-Keeping

A scientist working in analytical or developmental chemistry is acutely aware of the legal requirements for accurate, permanent logbooks. This same requirement moves into field studies where the environmental scientist acquires data that may be used in a court of law. Care must be taken to record observations that may be helpful in interpreting data, verifying the location at which samples were taken, and contributing to decision-making.

One of the biggest problems with this approach is that, if one is part of a governmental team, these field notes are often open to the public

Table I. Types of Background Data Useful for Developing the Sampling Plan

Type of Data	Source
Waste Generation Data	Files of owners and operators, production records, cargo manifests, employers, public agencies
Public and Private Files	Health departments, environmental control agencies, hospitals, zoning and planning departments
Published Data	
Site Specific	Journal articles, newspapers, EPA and state reports
General	Geologic data, hydrologic data, soils data, ecological data, public health data, demographic data, maps
Aerial Imagery	National archives, U.S. Soil Conservation Service, NASA

through the Freedom of Information Act clauses, if they are not part of the legal proceedings. You will hear the term *discovery* used by attorneys. This simply means that the legal process gives the parties in a litigation the right to acquire copies of all records, correspondence, data, and reports. The scientist must be aware of this and should not record caustic comments in places likely to be revealed later. Some scientists are known for their records, and others for their lack of records. When one becomes involved with site evaluations, one *must* record one's actions. Photographs of sample sites are vital records in case of litigation.

This recording process goes beyond just keeping a logbook. Every sample taken must be subjected to tagging, logging, and a chain-of-custody trail from the moment it is collected until it is finally analyzed in the laboratory. The sample must be physically in your possession or locked in storage with controlled access at all times. When the sample must leave your possession, a written receipt is required. For the geologist, core samples may have to be archived for extended periods before they can be discarded. Analyses, such as those done by gas chromatography/mass spectrometry, may require quick extraction and storage of extracts under a controlled-access storage program.

Results obtained from samples not subjected to this process are often thrown out completely or challenged in the courts. In one such process, the results of an entire study, costing thousands of dollars, was thrown out because the samples had been turned over to a courier, placed in the back of a pickup truck in unlocked ice chests, and transported along a delivery route covering several other facilities. The driver left the vehicle unattended in a parking lot at several stops and overnight at a motel. The custody of those samples was known, but the security was not maintained; therefore, a study that was vital to litigation was lost. The process is a nuisance to work with, but it is essential to the evidence collection process.

Safety

A sample collected from a wildland setting far removed from any pollutant source can be taken in a relaxed, open manner. You may wear gloves more for comfort than for safety. When the abandoned waste site investigator begins operations, this whole picture changes. If detailed analyses are not available, the investigator must approach his work with considerable caution. Respirators, protective clothing, and medical backup are required. Where samples are likely to be taken in confined spaces or close to highly toxic fumes, the researcher must have self-contained breathing apparatus (SCBA) available for use; in fact, he may be required to don the SCBA before even approaching the sampling site. The safety of the sampling team, as well as concern for public safety, should be foremost in your mind at all times.

Experience gained during work on the Love Canal indicates the wisdom of pre- and poststudy physical examinations. Several problems were detected before the study that could have created serious liability questions if discovered at a later date. Tests have been designed to help verify exposure to organic chemicals. These tests can best be administered by physicians trained in occupational medicine. (Contact the American Occupational Medical Association, 150 N. Wacker Drive, Chicago, IL 60606, 312/782-2166 for information.)

Cross-Contamination and Decontamination

Analyses for chemicals associated with abandoned waste sites must be able to detect extremely low levels of the chemicals. Cross-contamination can become a major problem during investigations where low levels are sought. Rigid quality assurance procedures can greatly reduce the problems. Wherever possible, disposable sampling equipment can be used to assist in eliminating much of the cross-contamination encountered in these sampling operations.

During the Love Canal studies, a procedure was developed for ensuring that the quality of the sampling was maintained (GCA, 1980). All glassware was cleaned, rinsed in spectrographic-grade solvents, and baked in an oven equipped with a filtered air source. Precleaned Teflon®* liners were used in all jar or bottle lids. All plastic or rubber

*Registered trademark of E. I. du Pont de Nemours Company, Inc., Wilmington, DE.

seals were avoided. Blanks of all reagents, solvents, and sample media were processed through the entire sample-shipment-analytical chain. As many as 20% of the samples should be quality assurance samples to provide the reliability needed in litigation.

Decontamination in the field is often difficult under the best conditions, but a hazardous waste site is especially troublesome. A decontamination station or pad complete with waste collection facilities may be necessary at very large sites. High-pressure hoses are helpful for working with augers and drilling equipment. A useful piece of equipment for decontaminating small pieces of equipment is a stainless steel fruit-tree sprayer. These sprayers, which can be cleaned and maintained easily, are readily available. They are pressurized, which allows removal of materials such as soil.

The decontamination sequence used for soil sampling equipment at Love Canal is shown below. This list gives an idea of the detail required if field decontamination is necessary. Acetone is used to dry the equipment before the final cleaning with methylene chloride:

- *Remove Soil:* city water wash with scrub brush to remove soil;
- *Remove oily chemical residue:* rinse with used acetone; wipe with Kemwipes; rinse with acetone;
- *Final decontamination:* distilled water rinse; rinse with spectrographic-grade acetone (save for use in rinses described above); rinse with spectrographic-grade methylene chloride;
- *Drying:* air dry; and
- *Protection:* wrap openings with precleaned aluminum foil and bag in protective coverings.

Sample Transportation

Samples collected for organic chemical analyses must be preserved by packing in ice. At Love Canal ice chests equipped with hasps and locks were used. The samples were kept in the custody of the sampling crew until they were turned over to the sample bank for processing and shipping. Custody of the samples is the primary concern of the investigators and attorneys working on the abandoned waste site study.

PUBLIC RELATIONS

The role of a field investigator in abandoned waste site sampling can be an effective avenue for reducing public concern. This can only occur if the team leader is provided with information and is given correct instructions on the type of responses to give to the public. An uncontrolled, talkative person can do much to create fear and panic in the general public, whereas a team leader with all the facts can help reduce public concern. Large operations such as the EPA Love Canal study can best be handled by a public information officer. In those large programs where public concern has already been aroused, it is best to rely on a good public relations effort and ask the team members to refer all questions to the appropriate officer. This approach ensures that a single source of information is provided to the public, thus reducing cross currents.

A second aspect of public relations is that acquired data will likely be used in litigation. Care must be exercised in discussing technical aspects of the investigation. The legal theories being developed, opinions about the severity of a problem, and statements that appear to be conclusions are all likely to be revealed in court proceedings and should be avoided when dealing with the public.

CONCLUSIONS

Field data collection at abandoned waste sites results in many problems, but excitement and a sense of accomplishment are often associated with such studies. Good technical people are needed to assume an active role in all aspects of such studies. The better the scientist, the more likely the results will stand the tests imposed by the judicial system. Much of the difficulty can be avoided if common sense is used, and the concepts mentioned in this presentation are considered.

REFERENCES

Barnes, M. G. 1980. "The Use of Kriging for Estimating the Spatial Distribution of Radionuclides and Other Spatial Phenomena," Tran-Stat #13, PNL-SA-9051 Battelle Memorial Institute, Richland, WA.

GCA. 1980. "Quality Assurance Plan: Love Canal Study. Appendix A. Sampling Procedures," GCA Corporation, Technology Division, Bedford, MA

JRB. 1980. "Training Manual for Hazardous Waste Site Investigations," JRB Associates, Inc. McLean, VA.

DISCUSSION

***David Wilson,* Harwell:** I would like to make a comment on the protective clothing that you were talking about. Obviously, protective clothing is necessary, but for what you showed and what I have seen elsewhere, I am a little worried about inconsistency in procedures. What you showed was very good protection for the body; you showed people wearing suits with heavy gloves and boots, and a respirator with the forehead and neck exposed. My experience of working in these sorts of conditions is that it is almost inevitable that you are going to brush your gloves across your head. I think if you are going to have protective clothing that works, you need to have formal dressing and undressing prcedures similar to those in the nuclear industry, which are rigidly enforced by the personnel.

B. Mason: I agree wholeheartedly on that. You will notice that the crew that was in the sewer had the hoods on. We started out with a lot more rigid procedures than what we ended up with. After we had been there awhile, it became apparent we were not running into the levels of contamination that everybody had anticipated, and we began to relax some of the rigid requirements that we had in the early days of the work. But the storm sewers, because of the presence of dioxin contamination that was there, required that procedure be held through the whole sampling. One other thing I might mention along the safety line is that the crew must have a physical examination before they start, and they must have a physical when they end. We had 65 people, mostly university students, working on the crews. We caught six incidents where, if we had not found out the health problem and it came up after the sampling program, we would have been in a tremendous liability position. Several of these were liver function disorders and, of course, with the organics, that is one of the characteristics. I would strongly recommend complete physicals for the working crew even though it is expensive. The cost far outweighs the potential problems that you could run into further down the line.

Aarne Vesilind, **Duke University:** You mentioned the potential master's theses and doctoral dissertations from these data. When will these data be available? Will they be available to the public and how can I get hold of them?

B. Mason: You will have to talk to some of the people in EPA. I have heard repeatedly that it will be out this month, and the date seems to be focusing toward the end of October. Once that is released, it is my understanding that all of the data will be available either through the Freedom of Information Act or it will be made available from things like the Agency reports. I would strongly recommend making contact with the EPA people.

CHAPTER 30

APPLICATION OF PATTERN RECOGNITION TO THE EVALUATION OF CONTAMINATION FROM OIL SHALE RETORTING

Robert R. Meglen

Center for Environmental Sciences
University of Colorado at Denver
Denver, Colorado 80202

Gerald A. Erickson

Infometrix, Inc.
Seattle, Washington

ABSTRACT

From a trace element point of view (and others, as well), the single most important characteristic of the oil shale process is the large amount of material that must be processed. To obtain a million barrels of oil per day, 480 million ton/y of shale must be processed. If the operation uses an aboveground retort, the retorted shale will be disposed of on the surface; if an in situ operation is used, it will be underground. Thus, rather prodigious amounts of trace elements will be handled, and potential contamination of surface and subsurface waters may occur as a result of leaching. A large program to monitor water quality near a test underground retort has produced large quantities of data. An interpretive analysis of these data, using multidimensional pattern recognition techniques, has been undertaken.

This chapter describes how pattern recognition techniques may be used to assist in selecting parameters to be measured, measurement techniques, sampling frequency, location of sampling sites, and related activities.

INTRODUCTION

The purpose of a water quality monitoring program is to detect specific effects of the monitored activity. An anthropogenic excursion must be detected before decisions can be made about changing operating procedures or implementing remedial actions. The schematic diagram shown in Figure 1 depicts the ideal information flow in a monitoring program. Effective design of a monitoring plan depends on a clear statement of the monitoring objectives. If the stated objective is too broad, the selection of parameters to be measured, frequency of measurement, and location of sampling sites cannot be optimized.

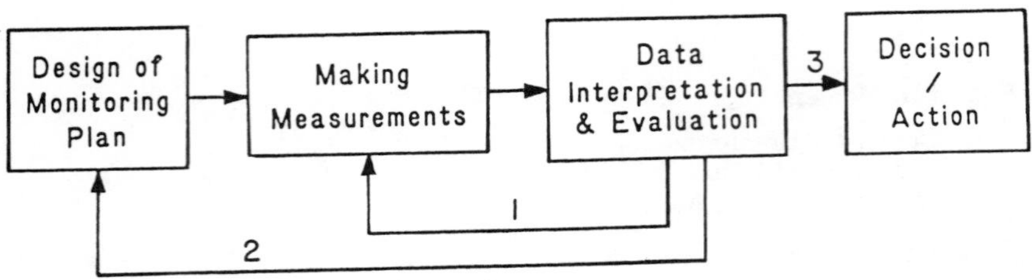

Figure 1. Ideal information flow in a monitoring program.

Recent advances in measurement techniques and instrumental analytical chemistry have improved our ability to obtain sensitive and selective measurements of water quality parameters. Automation and multicomponent instrumental techniques now permit rapid determination of many parameters on a large number of samples in a fraction of the time previously required for much smaller efforts. When a clear idea of what the target parameters should be is lacking, planners choose frequent measurement of many parameters in the hope that "the obvious" will emerge. Thus, the newly acquired data generation affluence produces massive quantities of numbers to be processed and evaluated. Little attention has been given to better interpretive techniques. In the absence of improved data evaluation protocols, more data tend to cloud the issues, rather than clarify them.

Decision-makers require an interpretive summary of accumulated data to make effective policy judgments. When massive data bases have been generated, interpretation and evaluation are complicated. Large tables of data are not adequate to depict the multidimensional character of the data. Subtleties of small changes are obscured by the presence of large, invariant features. Complex multiparameter relationships are multidimensional and not easily perceived by the human mind, which is used to thinking in two or three dimensions.

The work described here attempts to improve the data evaluation step in the monitoring program by applying artificial intelligence techniques to the massive data base. The first objective is to enhance the information feedback loops labeled 1 and 2 in Figure 1. Examination of the existing data may reveal that sampling frequencies or sampling sites should be changed. The data may also contain information that would permit alteration of the number of parameters required to detect anthropogenic excursions. Examination of the data may indicate that some analytical techniques are not sufficiently sensitive to provide useful indicators and, therefore, must be improved or replaced with new measurement techniques.

The second objective of this work is to examine techniques that could assist in preparing interpretive data summaries (No. 3 in Figure 1) required by decision makers. A rational approach to the selection of parameters for summarizing findings for presentation would enhance the communication between the data generators and data consumers.

PROBLEM DEFINITION

The data base under study was generated by a research and development organization in response to requests by federal and state agencies for water quality data. The broadly stated objective was to determine whether "significant contamination" of natural waters was occurring as a result of operation of an experimental, modified in situ (MIS) oil shale retort. The monitoring plan devised by committees consisted of frequent measurement of 74 water parameters before, during, and after the retort was burned. The specified parameters were measured for samples obtained from streams, alluvial wells, bedrock wells distant from the retort, satellite wells near the retort, and water obtained from within the retort chamber itself. These parameters follow:

1. Cations: Li, B, Mn, Sr, Al, Na, K, Ca, Mg, Ba, Zn, V, Mo, Si, Se, Hg, Cd, Fe, Th, Cr, Co, Ni, Cu, Ag, Zr, Ti, Sb, Sn, Ga, Pb, Be, Yt, Sc, As, and NH_3;
2. Anions: SO_4, HCO_3, Cl, F, NO_3, NO_2, CO_3, SO_3, S, CN, S_2O_3, Br, and P_4;
3. Water quality parameters: specific conductivity, pH, total alkalinity, Kjeldahl-nitrogen, phenols, total dissolved solids, dissolved O_2, turbidity, TOC, COD, DOC, BOD, odor, fecal coliforms, total

coliforms, surfactants, hardness, volatile solids, pesticides, and temperature;

4. Alpha and beta radiation, and ^{226}Ra.

Table I illustrates the magnitude of the data presentation task. Conventional data display techniques used to illustrate the time variation of 74 parameters for all categories and sites would require more than 2000 plots. Additional insight could be gained by examining parameter-vs-parameter correlations. However, distinguishing behaviors between categories of sample sites or individual sample sites would require many additional plots. Although one could argue that many of these plots would be "uninteresting," the selection of plots to be examined introduces a bias that could preclude the discovery of significant unanticipated relationships. The approach described in this chapter attempts to improve the water quality data evaluation and its display. Applying artificial intelligence techniques assists in the identification of important features for human attention and graphical display.

Table I. Number of Plots Required to Depict Results

Number of Plots to Examine for 74 Measured Variables vs Time	
10 streams	740
8 alluvial wells	592
4 bedrock wells (distant)	296
8 satellite wells (near)	592
1 retort well	74
Total time plots	2,294
Two-at-a-Time Element Correlations	
All Sites and Categories	2,628
By Category	13,140
By Site	81,468
Total	99,530

METHODOLOGY

The multidimensional pattern recognition techniques used in this work have been described previously. A series of unsupervised and supervised learning algorithms (ARTHUR) was developed (Duewer et al., 1975a,b; Harper, 1977) and used to analyze the data. Table II provides an outline of the procedures followed in this work. Although a detailed description of the full data treatment is not given here, it will be provided in a future publication. A brief qualitative description of the techniques and a few examples of the information generated are provided.

Table II. Outline of Data Analysis Tasks

I. Exploratory Data Analysis
   A. Data input
      1. Screen out bad data points
      2. Convert sampling data to time scale—include as a feature
      3. Determine extent of missing and below-detection-limit data
      4. Use filling models (mean fill or Gaussian fill) to replace missing data
      5. Autoscale data
      6. Create clean subsets with minimal missing data for testing
   B. Exploratory data analysis
      1. Perform cluster analysis and examine output for logically identifiable groupings among features (parameters), samples, etc.
      2. Apply factor analysis, interpret factors
      3. Evaluate clusters and factors for consistency with known relationships
      4. Plot data for selected features and for display of important relationships
   C. Evaluation
      1. Identify natural sampling groups (e.g., sites, wells, and categories) from cluster analysis
      2. Identify factors in light of chemical interpretation
      3. Derive criteria for testing classification scheme on complete data set and determine stratification within data

II. Applied Pattern Recognition Classification Model Development
   A. Run basic correlation
   B. Apply supervised learning methods and evaluate classification
   C. Reexamine stratification and classification criteria, and repeat B

Before application of the algorithms, representative test subsets of the data were constructed from samples containing few missing data. One test subset was prepared by replacing missing data with the mean value for that category. Another test subset was created by replacing missing values with a randomly selected value from a hypothetical Gaussian distribution of values about the mean for that category. Other test subsets of data were composed of samples chosen to test various hypotheses about the representativeness of the categories and the validity of the data. Each measured parameter was scaled to produce a mean value of zero and unit standard deviation. This autoscaling procedure gives equal statistical weight to all measured parameters. (This is a desirable attribute when one wishes to identify significant relationships among variables that differ in absolute magnitude.)

Several multivariate methods were used. Factor analysis uses all the measured variables (features) to examine interrelationships in the data. It accomplishes dimension reduction by minimizing minor variations in the data so that major variations can be summarized. Thus, the maximum information from the original variables is included in a few derived variables or factors. Qualitatively different areas, where little generalization can be made between one area and another, are referred to as separate factors. An example of a factor may include contributions of several measured variables such as COD, phenols, TOC, oil, and grease. Another factor may include Mg, Na, Ca, K, or bicarbonate. We would then attempt to ascribe some physical or chemical significance to these factors, perhaps calling one an "organic" factor and the other a "soluble inorganic." Thus, several measured variables are grouped according to their behavior in the data. In this example, the similarly acting variables appear obvious. Such intuitive factor-analytic methods, which have been the traditional method of handling complex systems, do not require complex calculations. This intuitive approach uses the ability of humans to recognize patterns. The disadvantage of intuitive pattern recognition is that human bias or preconceptions are introduced when deciding which variables should be grouped and considered a single factor. The branch of artificial intelligence known as pattern recognition attempts to eliminate this bias. By determining similarities of behavior within the data set through calculation of correlation coefficients and other measures of similarity, the unbiased factors emerge. The measure of the degree of generalization found between measured variables and each factor is called the factor loading. Specifically, one computational technique used in this work, principal component analysis, enables us to identify these derived variables (factors), which are useful in discriminating among preburn, postburn, sampling sites, and other components. By plotting the first two or more principal components against one another, we can classify (discriminate between) behaviors among sites, wells, aquifers, and other elements. After determining which features are important, one can select which of the more than 2000 possible feature vs time plots are valuable to examine.

Another method used to characterize similarities of behavior (among sites and among variables) is cluster analysis. This technique is conceptually halfway between the intuitive approach and the full-factor analytic method. It serves as an additional aid to identifying variables that are related enough to be placed under the same label. Cluster identification, principal component analysis, factor analysis, and determination of factor loadings further help in selecting important variables for graphical display. Selected plots that illustrate the methodology and its utility in depicting classification and separability are shown in Figures 2 through 5.

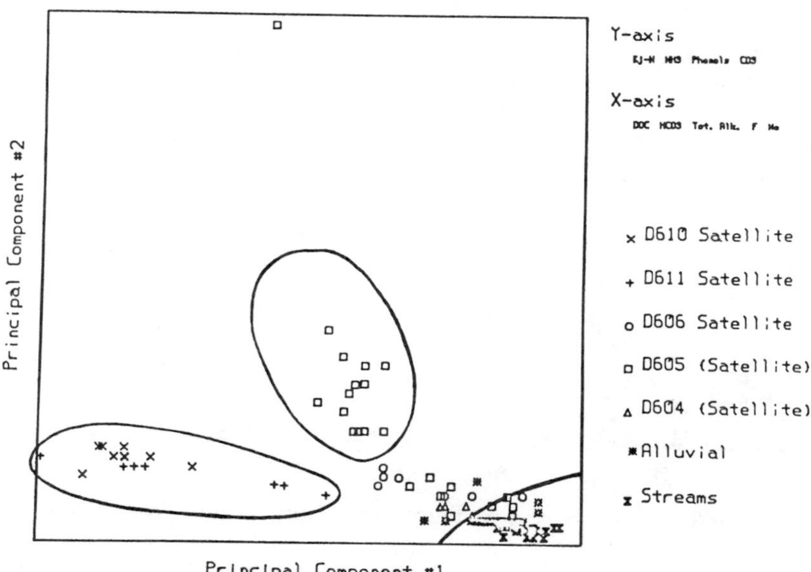

Figure 2. Principal component plot showing separability of some categories.

RESULTS

Figure 2 shows a plot of the first two eigenvectors from a principal component analysis of the data. Such plots permit a reduction in dimensionality by representing in two dimensions most of the variance contained within the total feature space. The principal components (PC) consist of linear combinations of all the measured variables. The major contributions to the linear combination in PC 2 are Kjeldahl nitrogen, ammonia, phenols, and carbonate. Very small contributions to the total variance are caused by other measured variables. Similarily, PC 1 consists mainly of dissolved organic carbon, bicarbonate, total alkalinity, fluoride, and sodium. This plot illustrates one set of principal components that permits distinguishing among various sampling categories. Most alluvial well and stream samples are indistinguishable in this rep-

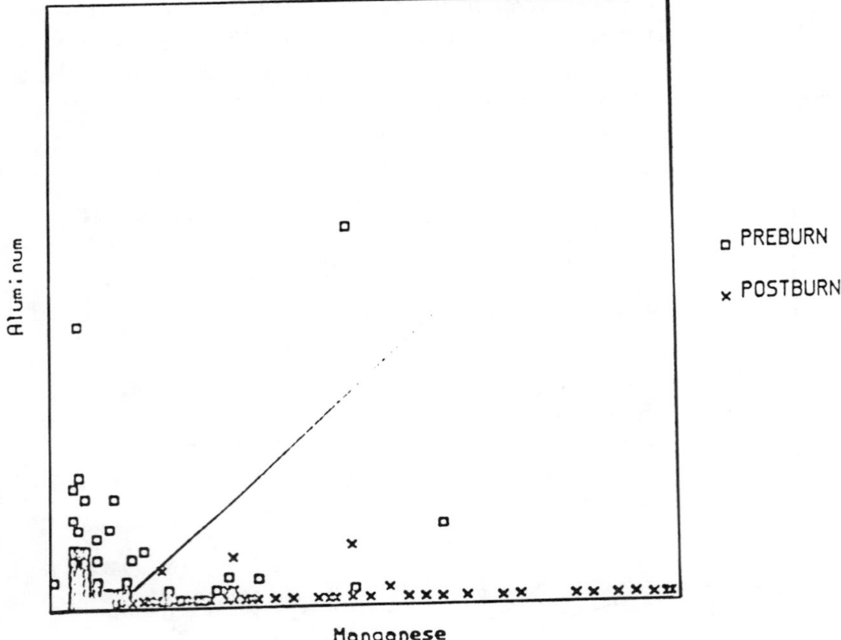

Figure 3. Preburn and postburn separability: feature-feature plot.

resentation and lie in a dense cluster at the lower right of this plot. Samples from satellite wells D610 and D611 lie in a region separate from other sample categories. Most of the samples from satellite well D605 lie in a different region of this plot. The one point located at the upper center suggests that this sample may be an analytical outlier. Similarly, plots for other test data sets and additional principal components help in identifying key features that distinguish between samples by site or category.

Figures 3 through 5 illustrate feature-vs-feature plots identified as being important in separating various samplings. Figure 3 shows that plotting aluminum vs manganese is effective in distinguishing between samples obtained during the pre- and postburn periods. (The implication is made that burning of the retort may have affected some measured variables.) These elements could, therefore, be important in identifying the effect on water of a retort burn. Preburn samples showed large variations in aluminum and small variations in manganese, and postburn samples showed small variation in aluminum and large variation in manganese.

Figure 4 shows that a plot of magnesium vs specific conductivity can

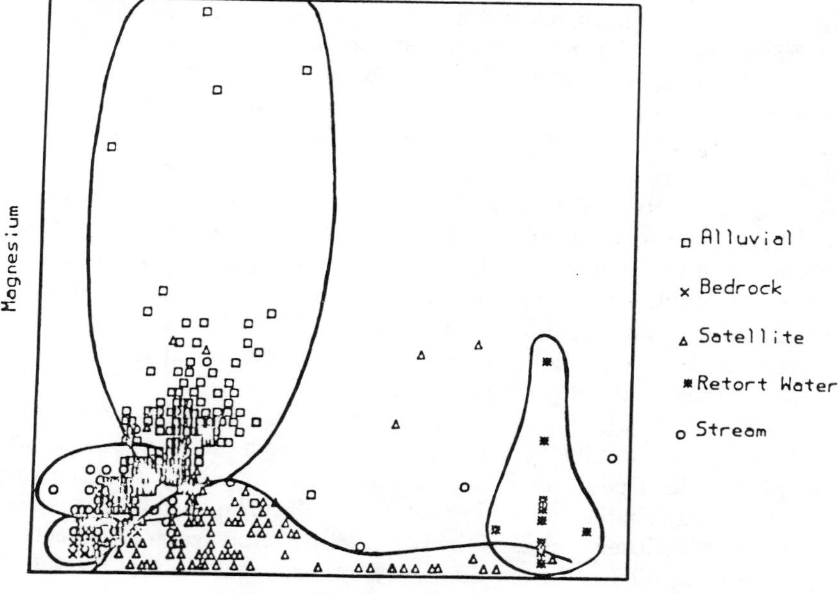

Figure 4. Separability of sample categories: feature-feature plot.

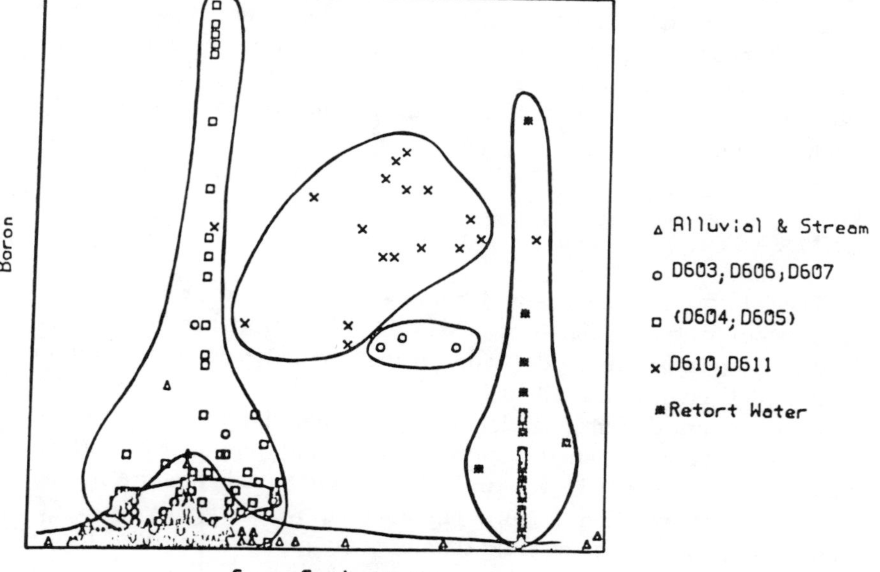

Figure 5. Separability of sample categories: feature-feature plot.

be used to distinguish retort water from other sample categories. Although some overlap exists between categories, clear patterns of behavior distinguish alluvial wells, satellite wells, and retort water. The tendency of some satellite well samples to "move" toward the retort water suggests an underground communication between the waters from these two sampling sites. Subsequent investigation showed that two satellite wells (D604 and D605) were located in a mined area that underwent a sill pillar collapse shortly after retorting began. The exposure of this area and its wells to the retort flame and its condensation products could account for the anomalous behavior of these samples. Identification of anomalies from these plots illustrates the utility of the method and the essentiality of the human interaction in the interpretation of the identified patterns.

Figure 5 shows that a plot of boron vs specific conductivity successfully separates satellite wells D610 and D611 from other satellite wells. Satellite wells D604 and D605 (involved in sill pillar collapse) are again distinguishable from other satellite wells. The variance of boron in these wells closely parallels the boron in the retort water as expected.

The results obtained from this study indicate that most of the information permitting separation of samples by categories and by site is contained within the data obtained for 12 of the 74 measured parameters. They are Li, B, Mn, Sr, Al, SO_4, specific conductivity, pH, total alkalinity, Kjeldahl nitrogen, and phenols. Therefore, this group of parameters should continue to be measured in future monitoring programs in the oil shale region. Plots of these parameters vs time are valuable in showing the effect of retorting on the natural waters of the region.

SUMMARY

The interpretive analysis of accumulated data on the surface water and groundwater quality near a MIS oil shale retort has been undertaken. Results of this study, when using applied pattern recognition techniques have been used to (1) improve and simplify the presentation of the multidimensional data for dissemination and use; (2) examine the need for continued monitoring of the large number of parameters now measured; (3) determine how many parameters are needed to distinguish whether a burn has affected water quality; and (4) determine whether the existing data contain any unanticipated results that have not been detected by conventional data handling techniques.

During the final stage of this work, we are quantifying the accuracy of the classification models derived from the exploratory data analysis. We will be able to specify the amount of variance information contained within each parameter and the degree of certainty with which conclusions can be drawn from a reduced set of measurements. The application of pattern recognition techniques to these data will provide a more robust basis for designing future monitoring programs. This work will help to design a more cost-effective plan by including the measurement of parameters that clearly provide useful water quality information.

REFERENCES

Duewer, D. L., J. R. Koskinen, and B. R. Kowalski. 1975a. "Documentation for ARTHUR, Version 1-875," Chemometrics Society Report No. 2.

Duewer, D. L., B. R. Kowalski, and T. F. Schatzki. 1975b. "Source Identification of Oil Spills by Pattern Recognition Analysis of Natural Elemental Composition," *Anal. Chem.* 47:1563.

Harper, A. M., D. L. Duewer, B. R. Kowalski, and J. L. Fasching. 1977. "ARTHUR and Experimental Data Analysis: The Heuristic Use of a Polyalgorithm," in *Chemometrics: Theory and Application,* ACS Symp. Ser. No. 52, B. R. Kowalski, Ed. (Washington, DC: American Chemical Society).

DISCUSSION

***Steve Stow,* ORNL:** I think it is obvious as to why you chose many of those 74 parameters. There were elements such as zirconium, scandium, and tin which are known to be especially immobile in these conditions. I just wonder if you could comment on the rationale for selection of these elements and the exclusion of perhaps some more mobile elements like rubidium? Second, what analytical technique was used for these very, very low abundance elements?

R. Meglen: I agree with you 100%. I did not select the 74 elements on this table. These were selected by the company that was required to produce the information to get permits from the state. These were all mandated in that way. Ralph Franklin says he would like to elaborate.

***Ralph Franklin,* DOE:** It was not really the company, but this is the kind of thing that frequently grows out of permitting operations where you have, in this case, a state agency that has to write the permit and represent the public to assure them that the groundwaters are not being contaminated. Frequently, these kinds of suggestions come out of public meetings where somebody says I think you should measure antimony and silver, or I think you ought to measure lead and chromium. Consequently, you end up with an extremely long list to satisfy everybody that happens to be there. Usually it is a rather cavalier attitude with regard to costs. And, of course, what happened here is that the state was getting rooms full of data that they did not even have time to look at. So this is an attempt to see if we could apply some of our scientific talent to lead them out of that wilderness.

***Robert Saar*, Geraghty & Miller:** I have two questions. First, on your principal component plot, you talked about linear combinations of various species. Are those just sums of concentrations or are they coefficients on those? Second, how did you come about those coefficients if there are any?

R. Meglen: One of the key points about the principal component analysis and this kind of plot that differs from the plots you are using is that it is not just the concentrations that are perceived as important in recognizing these patterns. First, if one looks just at columns of data, one tends to look at the concentrations that are largest. The principal component analysis uses normalized data; that is, if one is looking at fluctuations of relatively small concentrations (parts per billion), they have the same weight as do the larger concentrations. This is one of the reasons why I came up with some of the elements that you did not have in your list as being particularly important. These are represented equally on that plot. The coefficients for those particular components in the linear combinations come out of distance measurements in so-called multidimensional space. They come from multiple correlations among all of the elements simultaneously in that 74-parameter space. So there are statistically determined coefficients on each of those, the magnitude of which is proportional to the contribution that it gives to preserving the separability among the various species.

By the way, that was one of the most useful pieces of information that came out of the principal component plot; namely, it enables us to say with what certainty we would be able to distinguish between two different wells, pre- and postburn. When I come up with the reduced list of elements that may not have to be determined, we can then go back to these coefficients and the principal components plot and say that if you want to have 95% certainty of detecting the difference between one alluvial well and another alluvial well, you will have to determine these 22 elements. If you are satisfied with 90% certainty, you only have to have these 11 elements and so on.

Robert Saar: My second question is a more general pattern recognition one. More and more of our clients are being required by regulations to cap landfills or some other remedial action. If we were observing one well, we expect to see at some time in the future a change, hopefully an improvement, in the quality of the water. The problem we have is when we look at a graph, at what point can we say that the improvement of the remedial action is taking effect? Do you have any method or insight on where we can say yes, improvement is starting to take affect?

R. Meglen: This is one of the things that we are looking at as well. In a sense, you are asking for a reverse flag, that is, something that would indicate when contamination is beginning to subside. By looking at the major elements as well as trace elements in this multidimensional space there is the possibility of recognizing the onset of a major change. For example, when you couple information from calcium together with the information from lithium and perhaps boron and two or three other trace element concentrations, it becomes a more sensitive indicator of change than is any single parameter. That is one of the things we are looking at very carefully, and right now, it looks very promising.

***Aarne Vesilind*, Duke University:** One of the big advantages of computerized manipulation of data is that you have a chance to find if many of your data are skewed; for example, if you measure calcium and magnesium, you find that your hardness is less than the two of those combined. I have some real questions as to the accuracy of one of those three measurements. Did you have a chance to run through such internal checking of your data, such as total dissolved solids and specific conductance? And if you did, what fraction of your data did you throw out?

R. Meglen: The very first part of this procedure is to identify missing data and data that were below the detection limit. The below-the-detection-limit information has to be put into the program in some way, and it is important how you put that in. In that process, the so-called preprocessing of the data, one looks for outliers. We used a reduced set of data, that is, throwing out data that were inconsistent. We produced eight sets of data from the original set, selecting out those samples that seemed reasonable. From these data, we paired down the sets and looked at the plots. We then later threw back all of the data, and that is where we can definitely identify the outlier.

***Rufus Chaney*, USDA:** I had the overwhelming feeling that although I certainly admired the computer approach, somehow it lost track of the issue of risk analysis or detection of significant deterioration. There is a lot of difference between hunting for changes and the powerful work you showed us. But there is a lot of difference between that and the kinds of things that the health officials want to know. Does your program focus on any of the water standards? Did you avoid, for example, some of the wells that are so far away that it will be 10 years before contamination could get to them, or maybe 100 years, and possibly should not even be considered in the overall risk analysis?

R. Meglen: First, I am not doing a risk analysis, but more importantly, the reason for doing this work is to provide the information to people like Larry Gratt and others who are doing risk analysis. I can't supply them with 75,000 determinations and expect them to make heads or tails of them. What I would like to do is provide them with a list of the elements that we have found to be important in detecting a change. Then those are the elements they will look at with respect to the water quality criteria and risk analysis. Thus, they can use these data in their risk analysis scheme and evaluate impacts. With regard to the wells that are very distant, I think you would be surprised in looking at some of these plots. Some of the effects that have occurred are some distance away. There has been some sort of communication over distances of 1.75 miles away in some of the parameters, something we're very interested in. There appears to be some movement other than directly to the bed rock—fracturing and so on—that is appearing only 2 years after the retorting was terminated.

SECTION 6

ENVIRONMENTAL CONTROL SYSTEMS: TECHNIQUES AND TECHNOLOGY

CHAPTER 31

OVERVIEW

P. Aarne Vesilind

Department of Civil and Environmental Engineering
Duke University
Durham, North Carolina 27706

Closing the loop for the recovery and reuse of societal wastes is still a dream. The movement is there—ponderous, uncertain, capricious—but there is, nevertheless, a net movement toward the elimination of wastes, which will mean minimizing the adverse environmental impact. In its own way, each chapter in this section contributed to this process.

The European practices in solid waste disposal and resource recovery are reviewed by Hansen, who surveyed northern European countries. Incineration is an increasingly attractive option, with heat recovery approaching economic parity with processing. Of equal interest is the gradual but steady elimination of landfilling as an alternative. Additionally, source separation is considered to be consistent with incineration and heat recovery practices. This could be a bellwether in our own planning for solid waste management.

A significant contribution to the technology of materials processing was made by Hasselriis. The prediction of process recoveries and materials purities from various configurations of processing unit operations is a major development. This knowledge leads to a possible optimization of the processing plants and thus minimal operating costs. Optimization is performed by using a matrix with separation efficiencies of the individual unit operations, and the efficiency depends on the process configuration and the materials properties.

For wastes that cannot be readily processed and recycled, and in particular wastes with hazardous properties, high-temperature incineration provides efficiencies in the 99.99% range, as is shown by Freeman. The role of such incineration in the wise management of solid waste will undoubtedly continue to grow.

For wastes that perhaps should have been incinerated, but instead were dumped surreptitiously into landfills, remedial action may be necessary. The procedures and philosophy of such action are reviewed by Mutch and Siok. Remedial and mitigative measures at abandoned and active landfills involve a wide spectrum of alternatives that (1) seek to reduce or eliminate leachate generation, (2) contain and collect the leachate, or (3) provide for waste excavation and disposal.

For wastes that are to be disposed of on land, concern about hazardous properties is of major public and professional importance. Brown describes a test that uses bioassay techniques to determine their mutagenic activities. The test can become a valuable tool in the design of future landfill and land-farming operations.

Finally, Purcell looks at the question of priorities in solid waste management and concludes that many areas are of prime importance. The very act of setting priorities implies a certain urgency, and all the chapters in this section reflect this fact. We no longer have the luxury of time—or vast space—to either postpone or export the solid waste problem. The immediate environmental considerations and urgencies discussed in this chapter attest to the immediacy of a crisis in waste management.

CHAPTER 32

INCINERATION IN EUROPEAN SOLID WASTE MANAGEMENT

Jens Aage Hansen
Environmental Engineering
Aalborg University, Denmark

ABSTRACT

Through a limited letter survey, the processing of municipal solid waste (MSW) was assessed for Switzerland, West Germany, Sweden, Norway, and Denmark. Incineration was the dominating technology, whereas landfilling without processing is decreasing. Whether this is a general European tendency cannot be judged from the data gathered.

Incineration was evaluated in terms of resource recovery, environment, and economy. Although this study was based mainly on Danish experiences, some more-general observations are presented. Incineration with heat recovery and simultaneous materials recovery by source separation seems compatible. A few projects on central separation proposed in Norway and Denmark suggest incineration of residues amounting to about 60% of solid waste sorted. Environmental impacts are poorly known, but good methodology and equipment are available; efficient training of personnel and skilled operation may significantly improve environmental control. At European energy prices, incinerators could be operated at zero cost where district heating is feasible—at least in temperate climates.

Materials recovery from MSW may require a more thorough examination than presented thus far. A simple economical comparative evaluation would not suffice, but would only show that, on a marginal price basis, paper recovery is inferior to burning at an existing incinerator.

INTRODUCTION

Incineration plays a significant role in solid waste processing. This seemingly firm position is not the milestone of a harmonious technological development, but rather a result of painful attempts to solve problems that were often overdue at the time the installation was started.

At the turn of the century, Hamburg and Copenhagen introduced refuse incineration in an attempt to prevent the risk of cholera. However, these first installations were rather simple and not carefully maintained or renewed when worn out. Not until the middle of this century did solid waste incineration in Europe gain real momentum; interest is now increasing because of the scarcity of landfilling areas. Volume reduction became the key issue of the 1950s and 1960s. At that time emphasis was still on disposal at a low cost, but during the 1970s, concern about environmental protection and resource recovery increased. This situation provides a background for reevaluating existing installations and discussing incineration as a possible integral element of new solid waste management schemes.

PROCESSING IN EUROPE

Letter Survey

A questionnaire was prepared and sent to a few knowledgeable persons in some selected European countries. Basically, these persons were known to the author through cooperation within the International Solid Wastes Association.

The questionnaire asked for information about processes such as incineration, pyrolysis, and composting and operations such as source or central separation. Information was also requested on ultimate disposal or use. However, difficulties arose because the term municipal solid waste (MSW) is used for different waste categories, depending on collection routines and national traditions. Although household waste may be well and often similarly defined, MSW may include office and light industry wastes in proportions that vary from country to country. Thus, the information collected may be valid when comparing relative importance of different processes for MSW treatment, whereas absolute amounts of waste or total extent of landfilling should not be judged from the numbers presented in Table I or the curves shown in Figure 1.

Therefore, in the present study MSW basically means household wastes in addition to the extra quantities of similar wastes that are collected by the same municipal practice.

Of the nine countries surveyed, seven responded. Only five of these could be included in this comparative evaluation.

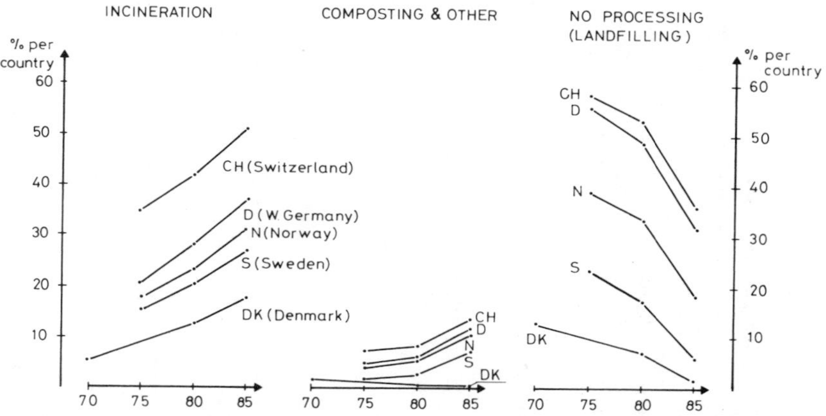

Figure 1. Processing of municipally collected solid wastes in five European countries. Ordinates are percentages per country [e.g., CH would, in 1985, incinerate about $(51 - 37) \times 5 \simeq 71\%$ of their total MSW].

Source Separation and Collection

Glass separation and recycling present a problem because the return bottle system works differently, and sometimes very efficiently, in various countries. For example, a well-functioning return bottle system for beverages would leave little glass for household waste source separation. Similarly, a combined system of return bottles and centrally located containers for one-way bottles would require a specially designed questionnaire to obtain comparable information. Consequently, glass recycling after source separation is not included in Table I.

Paper separation and recycling on a per-capita basis seems to be increasing in the five countries studied. France and Italy provided inconclusive information on this item, partly because the collection is not always municipally organized. In Switzerland about 90% of the recycled paper originates from households, whereas in Sweden the household contribution to total recycled paper is expected to increase from 25 to 30% within the next 5 years. The total per-capita amount

Table I. MSW Management in Five European Countries

Country and Population in 1980	Year	Paper Recycled (kg/y per person)	Total MSW Collected (kg/y per person)	Bulk Incineration[a] (%)	RDF (%)	Pyrolysis (%)	Com-posting (%)	No Process[b] (landfilling) (%)
Denmark, 5.1×10^6	1970		290	27(–)	0	0	10	63
	1975	34						
	1980	43	395	63(>90)	0	0	1	36
	1985	85	435	89(~100)	0	0	2	9
Sweden, 8.3×10^6	1975	54	295	31(78)	0	0	0	69
	1980	70	300	38(83)	0	0	6	58
	1985	78	300	46(90)	0	0	24	30
Norway, 4.1×10^6	1975	25	310	12(33)	0	0	10	78
	1980	30	320	14(44)	0	0	7	79
	1985	50	345	21(83)	6	0	11	62
West Germany, 61.4×10^6	1975	42	490	15(90)	0	0	2	83
	1980	51	445	23(98)	0	0	2	75
	1985	55	440	29(100)	0	3	2	66
Switzerland, 6.3×10^6	1975	60	285	72(69)	0	0	14	14
	1980	71	350	68(67)	0	0	11	20
	1985	71	380	71(76)	0	0	8	21

[a]Numbers in parentheses indicate installations with energy (heat) recovery.
[b]Landfilling is found by subtraction (100% less any processed MSW).

of paper recycled does not differ significantly between these two industrialized and organized countries. The question, therefore, remains whether a nonexploited paper resource could be found in the households in Sweden and in the offices, shops, and industry in Switzerland.

The total amount of MSW collected, as indicated in Table I could be used for comparison among years within each country and for estimating the absolute amount of MSW processed. Comparison of MSW per capita among countries is not advisable because of the lack of consistency in defining MSW.

Processes Applied

Figure 1 seems simple and clear. Incineration is increasing, whereas "no processing" (landfilling) is decreasing. At the same time other technologies (pyrolysis and composting) are being considered, but have not yet been significantly established, nor are they expected to become so, even in 1985. This may or may not be a trend applicable to other European countries.

In France (not included in either the table or figure) incineration is certainly a well-established processing technology, and heat recovery is consistently applied. About 35% of all MSW was incinerated in 1980, but no trend was reported. Composting is supposed to cease in Paris and suburbs, but a national trend is not reported.

In Italy incineration seems insignificant (i.e., about 10% of the household waste produced is burned). No new incinerators are being constructed, and small, older ones are being closed. The situation, according to the questionnaire, seems rather static and landfilling seems to be the predominating method of disposal.

Composting has been established at a certain level in Norway and seems to stay at the level (10–11% of MSW). In Sweden this processing technique is increasing and is backed economically by government subsidies. In Germany and Denmark composting is expected to remain at a low 2% of MSW production.

Pyrolysis is considered on an experimental scale and only in West Germany (about 3% of MSW in 1985). None of the other countries have reported preparations in this respect; Denmark closed an experimental facility about 3 years ago.

This background information indicates that incineration should be examined more closely. Resource recovery, environment, and economy are evaluated below.

RESOURCE RECOVERY

Incineration

A typical incinerator in the five countries studied is a bulk-fired furnace with boiler and heat recovery. Recovered heat is used in a district heating system. A typical Danish incineration system is illustrated in Figure 2, which shows flows of waste, heat, and process residues. Input is 100,000 t/year of municipal solid waste. Outputs are (1) 10^{14} cal actually sold for district heating; (2) 10^{14} cal actually lost (e.g., by cooling, transmission, and inefficient conversion of the theoretically available calorific value of solid waste; (3) 16,000 t of coarse slag to be used for construction of roads and parking lots; (4) 6000 t of fine-grade slag and ash used for concrete slab production; and (5) 4000 t of scrap iron for which landfilling is the most realistic alternative.

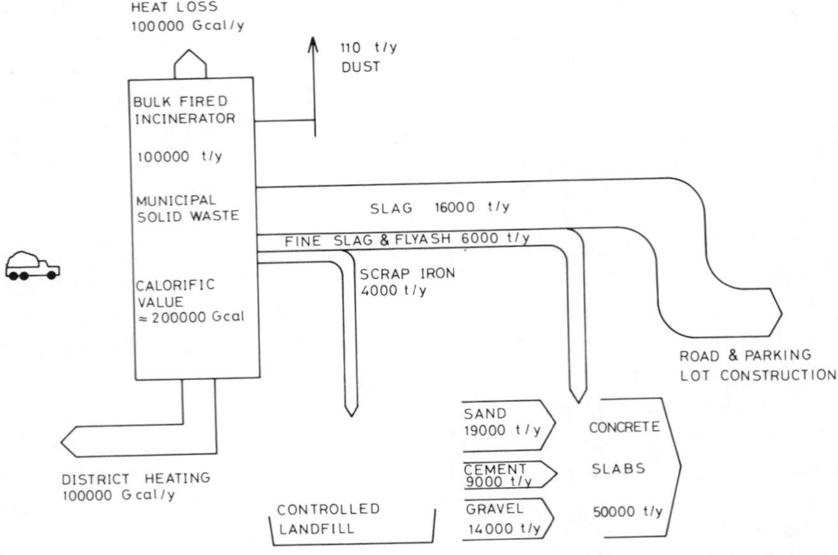

Figure 2. Flows of material and energy at a typical, modern Danish incinerator connected to a district heating system.

Is the situation depicted in Figure 2 realistic? The letter survey does not allow evaluation on a European scale. On the Danish scale the depiction does apply, but management is difficult. Reasons for such difficulties are obvious. The building and construction sector is at a low level of activity, resulting in a poor slab and slag market, and delivery of heat generally requires careful planning of plant location and pipe extensions and coordination with competing heat producers (e.g., existing or new electricity power plants).

In Denmark tail-end-separated scrap iron has little or no market value. The existing steel refining industry sees a limited potential of iron from household wastes, and the potential risk of contamination by trace metals is significant on the scale of steel refinery production.

Source Separation

Increased recycling of materials such as paper, cardboard, glass, and metal is an obvious challenge when discussing resource recovery. Advocates of increased materials recovery would often see incineration as a detrimental alternative; first, because the consumer may prefer to use only one waste bin, and second, because the incinerator manager may want all paper and cardboard for fuel.

Through full-scale experiments with household separation in the past decade, citizens can likely be motivated to join such schemes (Birkerød, 1975; Genvinding Fyn, 1981). Paper, cardboard and glass are the most obvious items, and experience shows that these can be successfully separated. Of the potentially available amount (about 30–40% paper and 6–8% glass in bulk household refuse), only a fraction will be separated at the source because of varying degrees of participation (e.g., some 70% of households join effectively) and poor constitution of waste material (e.g., wet wrapping paper goes to the waste bin). However, materials recovery through source separation does seem feasible, and the extent to which this practice is compatible with the "established" bulk collection and incineration technology becomes questionable.

Some very recent Danish observations to this effect (Genvinding Fyn, 1981; Økoconsult, 1981) state the following:

1. Source separation of paper and cardboard would decrease delivered district heat by 1–2% annually.
2. Source separation of all potentially available paper would be threefold the above estimate, but such efficiency is deemed unrealistic.
3. Source separation of glass as well as paper and cardboard may involve unchanged calorific value of the solid waste collected.
4. Source separation would tend to increase processing cost per ton because of over-capacity and relatively higher fixed costs. However, the change seems insignificant and, at best, some extra processing capacity is achieved without extra investment.

These results indicate that source separation matches incineration with little or no difficulty.

Central Separation

Central separation is a technology still in an experimental stage. The Italian Sorain-Cecchini technology has been in operation for several years in Italy and is now being proposed for full-scale operation in Norway and Denmark. Other dry- and wet-separation systems have been proposed, but none are yet in full-scale operation.

Table I and Figure 1 suggest that, among the five countries listed, only Norway intends to build a central separation plant during the next 5 years. This is shown indirectly under "refuse-derived fuel" (RDF), which will be one of the separation lines in the Cecchini plant suggested for Oslo. In this context, RDF might imply incineration of 60–70 wt %, while 30–40% will be sorted for materials recovery, comprising in this particular case, (1) cattle feed (fodder pills), (2) plastics, (3) paper pulp, (4) iron, and (5) cullets.

For the recovered products severe difficulties are still inherent. Basically, energy recovery and cost of operation may be assessed exactly, whereas the marketability of, and hence the revenue from, the recovered products remains questionable. For example, the economy of the Cecchini technology is linked partly to the prices of fodder pills and paper. The fodder, however, does not necessarily meet the legislative requirements (e.g., due to heavy metal contaminants), and the paper pulp may require deinking and other refining that would heavily influence the net revenue.

Whatever the prospect of central separation technology for materials recovery, incineration seems not only compatible with, but may be even indispensable for, a sound economy of a central separation management scheme.

ENVIRONMENTAL PROTECTION

Although incineration of MSW is not a new technology, this study emphasized its environmental evaluation. Reasons for this are that thermal processing in general (e.g., for energy production), and thus the stress on the environment, has increased; an awareness of environmental protection has also increased during the past two decades. Focal topics for the environmental evaluation will be stack emission and ultimate use or disposal of residues from incineration.

Particle Emission

Particle emission is often controlled by electrostatic precipitators. Only such equipment, or more recent developments such as fabric bag filters, would realistically be compatible with emission criteria set by environmental protection agencies. Danish emission control practice suggests 150 mg/N m^3 (at 7% CO_2) as a criterion for particle emission from major incinerators. In West Germany, 100 mg/N m^3 (at 11% O_2) applies, whereas in Italy only 50 mg/N m^3 (at 7% CO_2) is allowed.

Figure 3 presents recent Danish data on incinerator stack emission of particles. Some interesting information is available from the plots and curves:

1. The distribution seems log-normal. Thus, the median rather than the mean (average) would characterize the environmental load from an incinerator.
2. The statistical distribution of emission concentrations, when compared with a single fixed-emission criterion, clearly demonstrates the inadequacy of a single criterion. To be operational, an emission standard must specify period and frequency of sampling as well as an acceptance criterion (e.g., if not more than two out of ten single

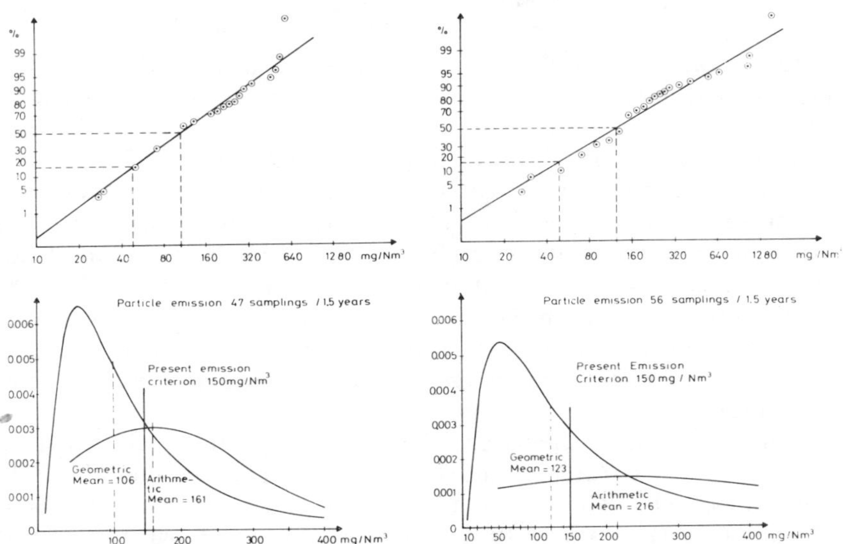

Figure 3. Particle distributions in stack emissions from modern Danish solid waste incinerators furnished with electrostatic precipitators.

observations exceed a certain concentration, the emission quality may be accepted). Unfortunately, such specified standards are not yet widely used, if at all established (this holds true also in other environmental areas such as water pollution control).

3. The data in Figure 3 suggest that electrostatic precipitators are capable of providing satisfactorily low particle concentrations (about the median value). However, high emission concentrations occur, and reasons for this could be found in solid waste composition, lack of homogenization, or poor control of feeding operation, furnace, and plant performance.

There seem to be many challenges to achieving better control of particle emission from incinerators. Improvements on existing equipment and operation methodologies, rather than looking for new technology, seem to be obvious routes toward such goals.

Heavy-Metal Emission

Emission standards for heavy metals are not normally available. The concern, however, is rather obvious in the scientific as well as the lay society.

Heavy metals in the environment deserve attention. Cadmium, lead, and mercury offer the "classical" examples of negligence and detrimental effects to humans or other species exposed. As an example, significantly increased human intake of cadmium is generally agreed to be unacceptable, because the interval between actual (30–60 μg/d per person) and tolerable (70 μg/d per person) intake is small. A major cadmium intake route is food, and there is strong evidence that increased cadmium in soil results in proportionally increased plant uptake and, hence, intake by food. Consequently, the inflow of cadmium to soil should be strictly controlled, which basically means control of atmospheric and fertilizer input. In Denmark, 1 ha of agricultural land receives 2 and 3 g/year of cadmium by these routes. Totally (3×10^6 ha of agricultural land in Denmark), this means 15 t/y of Cd, and it implies an annual increase of soil cadmium (now 660 g/ha) at a rate of 0.8% (i.e., a doubling time of only 130 years). Such an increase is not acceptable.

How much does incineration influence atmospheric precipitation of cadmium to soil? Tables I and II suggest that incineration of MSW emits about 50 kg Cd/100,000 t MSW (i.e., 1 t Cd/y from MSW incineration in Denmark in 1985), which could be compared with the total agricultural land precipitation of 6 t Cd/y. Although long-range

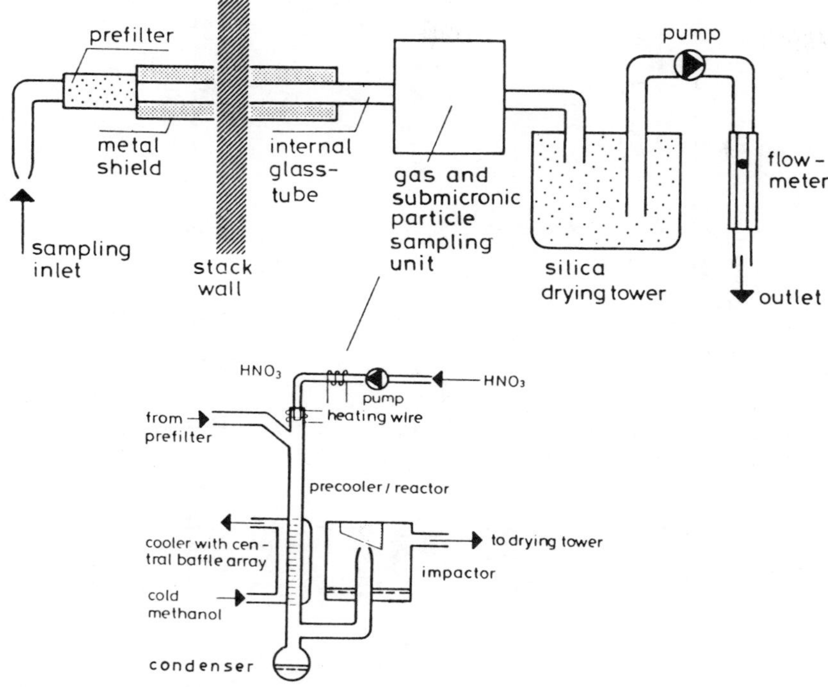

Figure 4. Stack sampling setup and details of the gas and submicronic particle sampling unit GSPSU developed by Søby.

transportation must be taken into account, the incinerator contribution seems significant and worthy of some consideration. The apparatus shown in Figure 4 and developed by F. Søby, together with the provisional results reproduced in Table II, presents a provisional contribution in this respect.

Figure 4 gives the principles of the stack sampling setup used for both particle and heavy metal sampling. The flue gas is sucked through a quartz-fiber filter, supplied by Dansk Kedelforening (Danish Boiler Association), and through a glass tube to the GSPSU operating at 0.6 N m^3/h. In the GSPSU, the gas stream is mixed with HNO_3 (100 mL/h) vaporized in an electrically heated glass tube into which preheated HNO_3 is injected by a peristaltic pump. The particles and any vaporized metals are allowed to react at high temperature before entering the cooler. By cooling the gas stream with cold methanol in the heat exchanger, a state of supersaturation is reached, causing the HNO_3 vapors to condense, preferably on particles.

Table II. Sampling of Particles and Heavy Metals at the Aarhus Incinerator

Date and Time	Particle Collection by Quartz Filter (mg/N m³)		Heavy Metals Collected in Prefilter and GSPSU (μg/N m³)					
			Zinc		Cadmium		Copper	
	DKF Isokinetic[a]	Søby Anisokinetic[b]	Prefilter	Total	Prefilter	Total	Prefilter	Total
April 3, 1981								
10-11 a.m.	253	250	31.0	32.6	20	23	160	270
11 a.m.-12 p.m.	35	40 (11 a.m.-1 p.m.)	2.2	2.4	27	30	80	115
12-1 p.m.	41		2.5	2.8	34	35	110	120
1-2 p.m.	215	40	1.8	2.1	28	30	260	275
June 17, 1981								
5-6 p.m.			3.7	4.1	62	64	260	305
7-8 p.m.			2.0	2.5	15	15	300	300
June 18, 1981								
9-11 a.m.		480	1.6	2.0	14	14	170	200
11 a.m.-1. p.m.		280	3.4	3.7	50	50	420	430
1-3 p.m.		180	3.0	3.1	36	36	150	190
6-7 p.m.		90	5.6	6.1	70	78	70	210
June 19, 1981								
10-11 a.m.		60	4.2	5.0	58	58	130	135
11 a.m.-1 p.m.			3.1	3.3	60	66	180	240
1-2 p.m.		60	4.0	4.2	72	72	220	235

[a]DKF isokinetic = Dansk Kedelforening (Danish Boiler Association) standard isokinetic sampling equipment. $\phi = 10$ mm; 3 m^3/h.
[b]Søby anisokinetic = gas and submicron particle sampling unit developed by Søby (in preparation). Anisokinetic factor $\simeq 1.8$, compared with DKF isokinetic suction; $\phi = 6$ mm; 0.6 m^3/h.

Some of the enlarged particles are deposited inside the cooler, and the rest are trapped in the impactor operating at a cut diameter of 0.2 μm. The large number of droplets formed in the cooler generates a very large area of deposition for metal vapors.

Since the metals studied are located in the particle surfaces, an acid digestion is performed during operation, and samples are thus prepared for chemical analysis by either atomic absorption spectrophotometry or anodic stripping voltammetry.

The apparatus is an all-glass construction, and the acid medium used reduces sample collection to emptying the condenser and impactor flask, and washing the sampling line preceding the GSPSU with diluted HNO_3.

Penetration in this equipment was assessed with a ^{65}Zn-tagged polydisperse uncharged CuO aerosol with a maximum particle diameter of 0.05 μm and ^{203}Hg vapor. In both cases about 95% was retained in the GSPSU. A full description of the GSPSU and its calibration are given by Søby (1982).

Table II summarizes some provisional results obtained using the described GSPSU equipment. Results are still too scarce to allow conclusions, but a few preliminary observations may be adequate. As already mentioned, the total Cd emission seems to be 50 kg/100,000 t MSW, assuming an average emission of 50 μg/N m^3 (cf. Table II). To reduce this total, one feasible procedure, and possibly the most obvious and efficient one, would be a ban on cadmium wherever unnecessarily used.

The supposedly improved recovery of metals through the refined GSPSU control equipment is significant for copper, less so for zinc, and insignificant for cadmium. Because of the anisokineticity indicated in Table II and the limited number of samplings so far, it is premature to generalize as to recommendable future sampling equipment and routines. However, with the methods and equipment now available, heavy-metal stack emissions can be more truly assessed. Results by other investigators (Coleman et al., 1979; Greenberg et al., 1978), would provide background information for such efforts.

Slag and Ash Utilization

The use of slag and ashes, as indicated in Figure 2, is also of environmental concern. In Denmark, certain criteria have been established for slag used as a filler substitute for gravel. Although the physical characteristics of sorted slag and ashes have proved satisfactory in a structural sense, some questions are still open as to environmental impact. Again, heavy metals are of prime interest. A typical interim permission issued by the Danish agency for environmental protection reads as follows (excerpts only):

- pH $\geqq$9.5,
- total alkalinity $\geqq$ 1.8 eq/kg dry matter,
- mercury <0.0002 g/kg dry matter,
- cadmium <0.01 g/kg dry matter,
- lead <2 g/kg dry matter.

Limitations are also imposed on the thickness of layers applied and distance to drinking water wells.

The scientific basis for such regulations is still poor, and work is in progress to improve this situation. However, although landfilling under especially strict and separate control used to be the dominant practice, utilization is now the usually accepted Danish provision for incinerator slag.

Miscellaneous and Missing

Substances other than particles and heavy metals are of environmental concern. Hydrogen chloride, sulfur dioxide, and nitrogen oxides should be mentioned (e.g., as contributors to increased acidity of precipitation). Also, some hydrocarbons may be of interest when evaluating air quality. To what extent incineration of municipal solid waste contributes to the emission of these substances is yet to be monitored in most of the European countries. At best, certain emission studies are in progress whereby at least absolute and relative contributions may eventually be assessed.

The heavy metal concern is one important reason for reluctance to use compost on cropland which, in turn, removes an important economic incentive for processing by composting, at least in Denmark.

ECONOMY

The economic evaluation of solid waste management will be limited to a Danish situation. Only processing by incineration will be included.

Collection

Collection costs often dominate the total solid waste management budget (e.g., 80% has been suggested by several American authors applying operations analysis to solid waste collection problems).

For comparative purposes collection costs were investigated in the Aarhus area, for which processing is assessed below. As a crude estimate of collection costs, 600 Danish kroner/t (i.e., $85/t) can be used. This collection is from a combined city/suburban area, where both dustbins (city area) and paper bags (suburban area) are used, and where the whole operation is conducted by one corporate management.

Processing

The incinerator at the city of Aarhus is used for the cost assessment. A breakdown of the costs, investments as well as operation and maintenance, is given in Table III. Also, the processing elements of this installation are shown in the table, and their relative and absolute significance may be judged.

Particular features are the sludge and pathological facilities. Sludge is delivered, dewatered, and dried before being incinerated with other waste; this operation is estimated to give zero net yield of heat. The pathological utility (hospital waste) is a separate unit, which is considered to contribute insignificantly to the total energy and cost budget.

In Table III, the annual cost of operation is calculated on the basis of the expected lifetime (capital recovery period) for each major item of investment. Thus, the annual expense listed in the table is calculated by:

$$\text{Annual expense} = I\frac{i}{1-(1+i)^{-n}}$$

where I = investment at time zero, $n = 0$
i = interest rate applied for discount calculations (e.g., 0.15)
n = life expectancy of investment (i.e., number of years)

The suggested discount rates require extensive evaluation before general acceptance. In this case, 8% may reflect a "real interest" rate excluding inflation, whereas 15% may resemble a market interest rate

Table III. Base Cost Breakdown for a Danish 100,000-t/y Bulk-Fired MSW Incinerator

	Cost ($10^6)	Σ ($10^6)	Recovery Period (years)	Annual Expense at Interest Rate ($10^6)	
				8%	15%
Investments					
Buildings, Roads, Plantation	3.4				
Design	1.7	5.1	25	0.48	0.79
Furnaces, Boilers, Cooling, Auxiliary Equipment	2.6				
Pathological Facility	0.1				
Sludge Facility	0.4				
Electrostatic Precipitators	0.6				
Mechanical Design	1.6				
Ash and Slag Sorting	0.7				
Monitoring Equipment	0.1				
Miscellaneous	0.9	7.0	15	0.82	1.20
Ash and Slag Depot		0.2	15	0.02	0.03
District Heating Connection, Including 4-km Pipe and Heat Distributor		2.7	15	0.32	0.46
Operation and Maintenance (O&M)					
Salaries and Wages	0.7				
Electricity and Fuel	0.2				
Maintenance	0.6				
Taxes	0.07				
Monitoring	0.02				
Miscellaneous	0.1	1.69	1	1.69	1.69
Total Annual Cost Including Capital and O&M				3.33	4.17
Cost per ton as is, $/ton				33	42

[a]Prices refer to ultimo 1980 and the Bruun & Soerensen incinerator installed north of the city of Aarhus. One U.S. $ rated at 7 kroner.

available to a corporate loan taker. On this basis, prices of $33 and $42/t, respectively, are derived for processing by incineration, if prices are fixed at a 1980 level and income from sale of heat is not considered.

Accounting for income from heat rendered to an existing district heating system necessitates a fee formula. In the case of Aarhus, the formula reads as follows:

$$W = A\frac{F}{H} + P$$

where W = income (kroner/y)
A = amount of heat delivered at a heat distribution station (gcal/y)
F = average fuel oil price (kroner/t)
H = net calorific value of oil multiplied by a heat conversion factor equal to 0.89 (gcal/t)
P = cost of electricity for pumping water through the district heating system (kroner/y), in this case obviously operated by the incinerator management

The above agreement between the managements of the incinerator and the district heating gives an estimated W/A = 279 kroner/gcal (i.e., $40/gcal) for 1982. This, in turn implies an average fuel oil price of $340/ton. In 1980 the same price was set at $200/ton.

An assessment of the net economic result of incineration at a 100,000-t/y scale is attempted in Table IV. To account for the energy price in-

Table IV. Revenues Accrued from District Heating by a Bulk-Fired 100,000 -t/y[a] Incinerator in Aarhus, Denmark

Year	Energy Price Increase over Other Prices (%)	Sale of Hot Water ($/t)	Sale of Slag and Ash ($/t)	Sale of Tail-End Scrap Iron ($/t)	Revenue[b] ($/ton) at Interest Rate as Indicated	
					8%	15%
1982		40[c]	0	0	7	−2
1983	5	42	0	0	9	0
1984	5	44	0	0	11	2
1985	5	46	0	0	13	4
1986	5	48	0	0	15	6
1987	5	51	0	0	18	9
1988	2	52	0	0	19	10
1992	2	56	0	0	23	14

[a]The nominal capacity of the incinerator is 15.2 t/h, assuming a heating value of 2500 kcal/kg. Realistically, this means 100,000 t/y, allowing for maintenance. In 1980, 93,000 t was actually burned, while heat sold amounted to 95,000 gcal (i.e., ~ 1 gcal/t MSW).

[b]Revenues mean income less all expenses for operation, maintenance, and capital recovery. Thus, solid waste delivery fees are not included on the income side.

[c]Based on the actually set price for fiscal year 1982 for hot water delivered at 279 kroner/gcal (~$40/gcal).

crease over that of other commodities, a cash inflow is suggested as shown (i.e., 5%/y extra increase on energy in 5 years, followed by 2%/y subsequent years). This incremental scale, which was suggested by Shelef (1981); seems as arbitrary as any other. Some extra increase rate should be applied to the energy (heat value) component.

At any rate of discount for capital recovery (Table IV), the incinerator management will likely be operating at a net plus revenue basis within a short time. This is the case without any revenue from ash, slag, and scrap iron even if ash and slag depositing are paid for as before. Further, this net revenue does not include any fee collected upon delivery of solid waste at the incinerator.

Summary

Solid waste management will be at a price that is dominated by cost of collection. At best, processing by incineration may pay for itself when heat is recovered efficiently. Net positive revenue from successful incineration may even pay for some of the costs involved in landfilling of certain residues or contribute to the costly collection, and thus minimize total cost, of MSW management.

CONCLUSIONS

A minisurvey by correspondence showed that incineration is the dominating MSW processing methodology and is increasing in several European countries. Inherently, landfilling is decreasing. Incineration combined with heat recovery seems to provide strong economic incentive.

From the five-country assessment, separation and materials recovery is not reported as a significant activity. This statement includes source as well as central separation methodologies. Increased paper recycling, however, seems manifest per se; in the absence of central sorting, this must be through source separation. A recent Danish assessment suggests that source separation is an entirely feasible activity in combination with processing by incineration, including heat recovery. In the tentatively proposed cases (e.g., in Norway and Denmark) for central separation, incineration seems to actually account for processing about 60% of the collected MSW.

Because incineration is dominantly planned or already in use, the environmental impact from stack emissions and residue use or disposal should be more carefully assessed. Such an environmental assessment would require improvements of data, monitoring equipment, statistical procedures for monitoring and quality control, improved education and training of personnel, and more efficient plant operation schemes.

The minisurvey included too few countries to provide any conclusive information on general trends. It may be interesting to extend the survey after more careful planning to reveal significant developments in MSW management. The International Solid Wastes Association may consider this a worthwhile task.

Separation and materials recovery would seem an important case for thorough assessment. If overshadowed by an economically successful and comprehensive incineration technology, some important material recovery schemes may well be detrimentally postponed or even eliminated. Only after an extremely thorough evaluation and strong evidence of nonutility should such decisions be made.

ACKNOWLEDGMENTS

Colleagues in other countries were extremely helpful in providing assistance and data from their home countries. These people are L. Bonomo, Italy; R. Braun, Switzerland; J. Defeche, France; T. Halmø, Norway; Jan Hult, Sweden; and W. Schenkel, West Germany. Also, Peter O. Jonsson, Denmark, assisted. Danish data were provided by Hans Falster and Aage Hansen, Dansk Kedelforening. Frederik Søby performed the measurements in Aarhus using his own personally conceived and built equipment. The staff at the Aarhus incinerator and of the office of the city engineer provided data and took time to assist in the assessment of their own case. Jesper Duwe Nielsen compiled data, drafted diagrams, performed the base work for statistical evaluation, and collected information on use of ash and slag. Birte Torstveit typed and cheered. The author hereby expresses his gratitude to all these persons and hopes that the many contributions are justly used.

REFERENCES

Birkerød. 1975. "Reuse Experiments in 1974–75," Municipality of Birkerød, Denmark.

Coleman, R., et al. 1979. "Sources of Atmospheric Cadmium," EPA 450/5-79-006, U.S. EPA, Office of Air Quality Planning and Standards, Research Triangle Park, NC.

Genvinding Fyn. 1981. "Source Separation," Enviroplan, Copenhagen, Denmark.

Greenburg, R. R., et al. 1978. "Composition and Size Distributions of Particles Released in Refuse Incineration," *Environ. Sci. Technol.* 12(5):566-573.

Økoconsult. 1981. "Economy of Refuse Incineration with Heat Recovery and Increased Recovery of Paper, Cardboard, and Glass," prepared for the Ministry of the Environment, Copenhagen, Denmark.

Shelef, G. 1981. "Recovery and Utilization of Thermal Energy from Incineration," in *Proceedings of the International Municipal Conference* (Milan, Italy: International Solid Wastes Association, Italian Section).

Søby, F. 1982. "Opsamling og Submikrone aerosoler (Collection and Analysis of Submicronic Aerosols, in Danish). PhD Dissertation, Chemical Department, Technical University of Denmark, Lyngby.

DISCUSSION

***Doug Burgess*, Stone and Webster Engineering Corp.:** One problem that we have in the United States if we ask householders to voluntarily segregate their trash is getting them to do so. One of the slides that you showed, the one that had some data on the amount of glass that you recovered, implied to me that you might get five out of seven householders to voluntarily participate in a segregation program. Is that correct?

J. Hansen: It is correct, and there is other evidence from several studies that indicates that the participation of the household is paramount.

***Arthur Purcell*, Technical Information Projects, Inc.:** You mentioned the 130 years doubling time as being not good. Is there a number that is considered acceptable for cadmium?

J. Hansen: No, there is not. Our problem is that the models suggested so far are not very satisfactory as was indicated also by Rufus Chaney. Without firm guidance in terms of total cadmium intake, the Food and Drug Administration people will tend to say any increase should be avoided. For example, they would ask to totally ban cadmium applications to the soil. We have positive evidence that if you increase soil cadmium by a factor of 2, then in the plants you grow, you will also increase the cadmium level in the plants. Therefore, they would be cautious, and they would try to, wherever possible, lower the cadmium input to soil.

Arthur Purcell: Is 660 g/ha natural background of what has accumulated to date?

J. Hansen: It's a very universal figure. I think, in my part of the world and for all depository material in low-land situations, 0.2 ppm is the figure that you have.

***David Wilson*, Harwell:** I would like to make a few comments on the paper. What we have heard is an excellent exposition of the position in Denmark, and I would like to make the point that Denmark is, from the information you will see in my book and elsewhere, rather atypical rather than typical of Europe. Denmark, Switzerland, and Japan are the only countries in the world where incineration accounts for more than 50% of the solid waste disposal. In most European countries the figure is about 20–30%, and in my country, it was 10% a few years ago; I guess it is going down pretty fast. The reason for that is high capital costs, which for new plants are generally greater than the new source recovery plants. We have also had considerable problems with very high maintenance costs and poor market possibilities.

Going to another subject, I am concerned about the misinformation on cadmium. Cadmium pigments used to be fairly prevalent. They are now confined to, if you like, market uses where the properties of cadmium pigments are necessary, and that includes plastics that have a high processing temperature. This means that more than 90% of cadmium pigments are now used in high-density polyethylene, and in styrene copolymers—in things like electric drill casings and so on. I would suggest that no yellow ring binders now being sold would contain any cadmium at all. We have done some work on cadmium leaching from pigment plastics in landfill sites, which is one of the two sources sited for this disposal environment. Our results suggest that there will be no new significant complications, so the mere presence of cadmium is not a reason for looking for material. There are two questions that I would like to ask. One, do the consumers pay you $40/kcal steam that you deliver to them in the summer months? Second, there is considerable concern expressed on the leaching of soluble metals from incinerator slag. Have you considered this in your application?

J. Hansen: Incineration data in Switzerland, Denmark, and Japan were based on 4 years of bookkeeping for this particular plan, so I rely on these numbers. Even though the capital investment is pretty enormous, it does pay for the city of Aarhus to have borrowed this money and used it this way. There were two questions at the end; the first one was?

David Wilson: Do they pay $40/kcal heat all the year round?

J. Hansen: Yes, they do pay that because there is a very strict contract between the district heating management and the incinerating management that the heat value be calculated on the basis of the market price of oil. And this means that they do pay $40/kcal delivered.

David Wilson: What do they do with it in the summer months?

J. Hansen: In the summer months? That was taken care of in my saying that only half the heat theoretically available when feeding refuse into an incinerator is delivered as heat; the other half is lost, partly due to the fact that during the summer you cannot utilize all of it. I should say that this is a management question. You can supply a small area in the winter and a big area in the summer. The system is adaptable. In the city of Aarhus, they can utilize half of the theoretical heat values of the refuse. It is just a fact, and other people may not be that successful or lucky to have a situation like that already established. Leaching tests were performed throughout 3 years to establish guidelines for utilization of incinerator ash. It is based on this evidence that the regulatory agency now gives permission to use it. I should say that these tests were not satisfactorily carried out because there is some confusion as to whether incinerator fly ash was used in the testing. Presently, this permission is given only for the use of the slag. I think we need another year of leaching studies.

***R. E. Blanco,* ORNL:** On your leaching tests, have you run any tests after the fly ash or the slags were incorporated into concrete? Does the concrete immobilize the cadmium? We have always been interested in perhaps adding complexing agents and inorganic complexing agents for holding hazardous materials; that's why I am asking the question.

J. Hansen: As indicated, the tests were run on a mixture of fly ash and slag, and actually the permission is only for slag. Prematurely, I would guess that we could pretty easily prove that this is not a major problem, considering the adsorptive capacity of the soil underlying these slags. But we do not have the hard evidence of what is really happening, or how many hundred years we will leach, and what is the interaction of the soil. This is yet to be proved.

CHAPTER 33

OPTIMIZATION OF SOLID WASTE SEPARATING SYSTEMS

Floyd Hasselriis
Forest Hills Gardens
New York 11375

ABSTRACT

Municipal solid waste (MSW) can be processed to yield products such as refuse-derived fuel (RDF), metals, and glass and to remove contaminants in the products. The economic viability of a recovery plant depends on the yield and value of the products and the cost of disposing of unusable or contaminated materials. The value of the products depends on the amount of contaminants, particularly glass.

The loss of valuable materials and the quantity of products depend on the arrangement and effectiveness of the system components and operations. Optimization of solid waste separation systems can be simulated mathematically by analyzing the system as a matrix of operations, using the separation efficiencies of the individual operations, which in turn depend on performance of the equipment and MSW properties such as particle size distribution and aerodynamic properties.

With sufficient data, equipment can be designed and arranged to achieve optimum yield of products with acceptable contaminant levels. The optimum economic balance can also be achieved by relating capital and operating costs to the product revenue resulting from different degrees of process refinement.

INTRODUCTION

Processes for recovering energy, metals, and other resources from MSW have been developed intensely during the last decade. The effectiveness of operations such as shredding, air classifying, magnetic separation, screening, and trommeling for recovering the combustible fraction and removing dirt, glass, and metal contaminants has been demonstrated. Techniques for processing the noncombustible fraction to recover metals, and glass have also been demonstrated.

After the oil embargo was declared, the emphasis shifted from recycling paper, metals and glass toward recovering the energy component of refuse and toward removing glass as a contaminant rather than as a product. Recovery of ferrous metal was successful from the beginning, but the value of the product was marginal because of erratic market conditions. Aluminum recovery received much attention but, because of its low concentration in MSW, has not been worthwhile.

The first full-scale demonstrations revealed numerous problems and produced extensive data. These demonstrations proved that MSW could be processed into RDF, which could be burned in utility boilers designed for coal firing, in semisuspension and on grates, in the form of fluff and densified pellets. Powdered RDF could also be cofired with oil in boilers designed for coal. Atmospheric emissions were controllable to acceptable levels, and sulfur emissions were less than from coal-fired boilers.

Although many demonstrations were successful, many more failed financially and technically. In most cases the RDF product was unsatisfactory for burning in existing boilers because of the high ash and moisture content of the RDF. Also, the boilers selected were built in a highly competitive era to match closely the requirements of specific coals. At the time of conversion to cofiring RDF, these boilers were already in trouble with available coals. Although improvements in processing could have been made to reduce ash and moisture, the expense of retrofitting these plants was, in most cases, not undertaken by the public agencies and private firms that sponsored the plants.

On the other hand, many plants are operating today, using second- or third-generation processes. The city of Ames, Iowa built a small facility to produce and burn RDF, generating power in the municipal power plant; major problems were identified and corrected (Adams, 1979). The city of Madison, the pioneer in shredding for landfill, patiently developed a plant, evolved from privately developed technology, that produces a satisfactory RDF for burning in utility and industrial boilers (Barlow et al., 1980).

The Black Clawson Hydrasposal Process for producing power from MSW, now operating in its second generation in Dade County, Florida, benefitted from the experience gained from operation of the first full-scale plant in Hempstead, New York (Plato, 1981). P&W spent millions of dollars in upgrading the Hempstead plant in preparation for putting it back into service. The Monroe County and Hooker Chemical systems are now in operation.

The processing of refuse into energy is entering a new developmental era in which optimized processes will be built to serve community needs.

PROCESS OPTIMIZATION

Process optimization is a logical method by which mechanical operations are evaluated and assembled into systems that perform in the best way from an overall point of view so that the total cost to society, including intangibles, is the lowest in the long run. The process analyzes the consequences of all potential decisions and weighs the sum of all benefits against the sum of all costs to get a practical system that works well enough to suit everyone, is reasonably reliable, and has the confidence of the community and financial investors.

Table I lists the major considerations in finding the optimum method of processing waste into energy, most of which are beyond the scope of this chapter. We will concentrate on finding the optimum process for producing a satisfactory RDF with maximum yield and minimum residue to be landfilled.

Evolution of Solid Waste Processing

Given enough time and the willingness to continue despite failure, humans evolve better products and systems. After a few trials have been made and the experience has been studied, we tend to do things more nearly right the next time. A review of the performance of RDF processing plants shows that a learning curve has been followed and that a degree of optimization has occurred during the last 10 years, resulting in RDF of increasingly high quality.

A high-quality product is not enough; it must be produced without excessive cost, without wasting valuable materials, and without high residue-disposal costs. The product is good enough if it can be sold to a steady customer at a price that is beneficial to producer and customer, whether the product is fuel, steam, or power.

When preparing a fuel from MSW, skimming off a clean "light fraction" that is fairly dry and low in ash content is relatively easy. However, the rest of the MSW residue would have to be disposed of in a

Table I. Major Considerations Determining Optimum Process

Size of Facility	Economy of scale vs transportation; political factors; energy user size and reliability; power generation and electric rates
Type of Boiler	Remote or dedicated; cofiring or refuse only; mass-burning stoker; traveling grate; semisuspension firing; suspension firing
Processing	None; minimum; maximum
Shredding	None; one stage; two stages; grinding
Refuse Fuel	Raw MSW; coarse; fine; RDF; wet; powder
Ash (%)	35; 25; 20; 15; 10
Moisture (%)	35; 25; 15; 10; 5
Metals Recovery	None; from ash; mixed; separated; cleaned
Glass Removal	None; one stage; two stages; three stages
Process Residue	None; minimum to maximum
Transportation	
MSW	Direct delivery; transfer stations
Residues	To small, medium, or large landfill
Boiler Type	Stoker; moving grate; fixed grate; slagging
Furnace	
Size	Very large; large; medium; small
Temperature	Low (810°C); medium (930°C); high (1370°C)
Pressure	Low (1380 kPa); medium; high (13,800 kPa)
Superheat	None; low; medium; high (1000°F)
Coal Type	Lignite; bituminous; anthracite
Combustion Time	1 h; 20 min; 7 s; 1 s
Refuse/Fossil (%)	100; 10–100

landfill at a cost that may subtract substantially from the fuel revenue.

Decreasing the residues and increasing fuel yield result in increased contamination of the fuel and passage of the disposal problem to the fuel consumer.

Between these extremes lies the optimum or, in fact, many optima, depending on the specific circumstances. Where landfill cost is still cheap, as little processing as possible, followed by dumping the residues, would be best. Where landfill cost is high, products should be refined so that a maximum amount is recovered and sold as fuel or recycled as metal, glass, or paper. The cost of money and operating and maintenance costs of the plants are also major factors that will slant the decision toward more or less processing.

Selection of equipment and sequence of operations requires detailed knowledge of how the process configuration influences the final product and the costs of production. To obtain this knowledge, extensive testing of the operations is needed. Such knowledge is difficult to obtain without building complete plants for testing with real MSW so that the performance of the entire process can be documented.

Even when data are available from testing these operations, they are specific to the material used to run the tests. The MSW stream varies from minute to minute, through the week and seasonally. Also, household, commercial, and industrial wastes differ and fluctuate with time.

With all these problems, the industry has taken time to evolve and mature. After 10 years of development, however, both the information and the methods that can be used to optimize these systems now appear to be available.

Evolution from Pilot Plant to Processing Plant

Although a full-size operation would provide data, a pilot plant can provide more flexibility for testing different arrangements and obtaining a broad range of data. Combustion Equipment Associates (CEA) built and tested several generations of RDF plants in East Bridgewater (Hasselriis et al., 1981). The first generation produced a 10% ash fuel at the expense of losing half of the fuel. A second-generation plant, using a flail mill, trommel, air classifier, and secondary trommel, succeeded in achieving a higher recovery of combustibles but produced a product with an ash ranging from 15 to 25%. Since the target for Bridgeport was <15% ash at all times, a third-generation process had to be developed.

Many different configurations were tested at a pilot plant in Minneapolis. The ash content of the MSW during processing was analyzed. The shredder smashed glass into sandy grit, which could not be removed from the RDF in subsequent operations. When the trommel was placed first, the glass, along with part of the combustibles that would not need to be shredded, could be removed to bypass the shredder. The shredded stream was mixed with the glass-rich stream and fed to the air classifier, where the glass would fall out as heavies. Because the fine mix of glass and sand was found to fly in the classifier, a secondary trommel was provided to remove it. Using the air classifier as a dryer helped in releasing the sand and glass in the secondary trommel.

This process was installed in Bridgeport (Hasselriis et al., 1981) and produced a fuel product with an ash level of 10–15%. Part of the combustibles dropped by the secondary trommel could be recovered by additional processing. Tests indicated that a 10% ash fuel could be produced at high yields with this process.

THE PROCESS AS A MATRIX OF OPERATIONS

The process of separating valuables and contaminants from MSW can be described mathematically as a matrix of operations, each of which performs a splitting function on the various components of the MSW.

Table II and Figure 1 shows a composition of the feed to be processed. The operations are listed in Table III, showing the destinations of streams leaving each operation. The table of split fractions describes how the components of the feed are diverted at each point of the matrix. Some streams are recirculated for reprocessing.

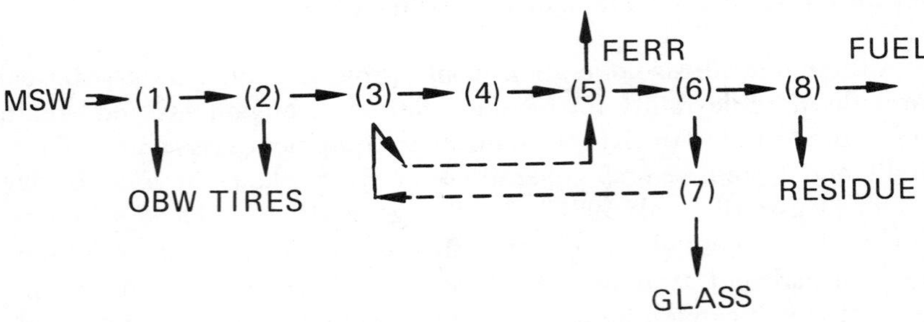

Figure 1. System schematic.

After removing oversize bulky waste (OBW) and tires, the feed enters the trommel, where most of the glass is dropped; the glass is again dropped by the air classifier. As processing continues, each machine performs similar splitting operations on each type and size of material in the stream.

Table II. Feed Specifications

	Entry Point	Feed Rate
Paper	1	62.7
Plastics	1	1.4
Food/Yard	1	6.0
Ferrous	1	7.6
Nonferrous	1	0.8
Glass	1	9.0
Inerts	1	3.2
Moisture	1	16.1

Table III. Operation Configuration and Split Fraction Table

		Destination		Fractions Going to First Destination					
No.	Operation Name	First	Second	Combined	Moist	Inert	Ferrous	Nonferrous	Glass
1	OBW	2	0	0.97	1.00	1.00	0.98	0.99	1.00
2	Tires	3	0	0.98	1.00	1.00	1.00	1.00	1.00
3	Primary trommel	5	4	0.35	0.35	0.50	0.80	0.90	0.80
4	Shredder	5	0	1.00	1.00	1.00	1.00	1.00	1.00
5	Magnet	6	0	0.95	0.99	1.00	0.08	0.98	1.00
6	ADS	8	7	0.90	1.00	0.85	0.85	0.90	0.13
7	Recovery	0	3	0.15	0.15	0.20	0.15	0.10	0.95
8	Secondary trommel	0	0	0.06	0.06	0.65	1.00	0.00	0.65

Processing equipment can be tested with real MSW to determine what happens to various components of the feed, such as paper, plastics, food, metals, glass and miscellaneous inerts.

The combustible, ash, moisture, and inert content of the various species in MSW can be tabulated and input to a computer, which can then convert the input MSW composition, in terms of paper, plastics, cardboard, ferrous, glass, and nonferrous items into combustible, inherent ash and inerts. The computer program can accept information about the feed and the processing operations and then process it to produce a complete flow stream analysis (Table IV) and composition analyses.

With this program set up, various feed material compositions may be input, and variations in end-products can be obtained. Splitting functions can be changed easily to determine what splits are needed to obtain a desired result. Changes can also be made in accordance with data on the performance of the equipment under the loading conditions generated by the program.

The splitting functions are not independent of the feed material or the operations. Algorithms can be installed in the computer program to calculate the splits, or they may be changed manually.

Although a computer program can perform these operations swiftly, the garbage-in product-out relationship is no more reliable than the input. Hence, valid data should be developed from testing real MSW with full-scale equipment operated at full loading. If this is not available, properly developed pilot-plant data can be used as a first approximation.

Table IV. Summary of Flows in Each Branch (ton/h)

Stream	Combined	Moist	Inert	Ferrous	Nonferrous	Glass	Total	Product
Entering								
0–1	53.4	26.2	9.8	7.6	0.8	9.0	106.8[a]	
Internal								
1–2	51.8	26.2	9.8	7.4	0.8	9.0	105.0	
2–3	50.7	26.2	9.8	7.4	0.8	9.0	104.0	
3–5	19.3	9.2	5.0	6.0	0.8	7.5	47.8	
3–4	35.9	17.0	5.0	1.5	0.1	1.9	61.4	
4–5	35.9	17.0	5.0	1.5	0.1	1.9	61.4	
5–6	52.4	25.9	10.0	0.6	0.8	1.2	84.1	
6–8	47.2	25.9	8.5	0.6	0.8	1.2	84.1	
6–7	5.2	0.0	1.5	0.1	0.1	8.2	15.1	
7–3	4.5	0.0	0.1	0.1	0.1	0.4	5.2	
Exiting								
1–0	1.6	0.0	0.0	0.2	0.0	0.0	1.8	OBW
2–0	1.0	0.0	0.0	0.0	0.0	0.0	1.0	Tires
5–0	2.8	0.3	0.0	6.9	0.0	0.0	9.9	Ferrous
7–0	0.8	0.0	1.3	0.0	0.0	7.8	9.9	Glass
8–0	2.8	1.6	5.5	0.5	0.0	0.8	11.2	Residue
8–0	44.4	24.4	3.0	0.0	0.8	0.4	72.9	Fuel

[a]For two processing lines.

OPTIMIZING RDF PRODUCTION

The historical optimization of RDF production processes is depicted in Figure 2, which shows the ash content of the RDF produced by five plants plotted against the "yield," or percentage of MSW delivered as RDF, and flowsheets for these plants are shown in Table V.

The St. Louis demonstration had the simplest process, resulting in high RDF yield (87%) and high fuel ash (28% on a dry basis) (Fiscus et al., 1977).

The city of Ames, Iowa process at first delivered 84% RDF with an ash content of 20%. The MSW at Ames is relatively high in paper and low in moisture. The revised Ames process installed two screens, one to remove grit from the infeed to the second shredder and another to recover the combustibles from the first-screen underflow, resulting in a reported reduction in ash content to as low as 10%, at a yield of 72%. This process modification illustrates the benefits of two-stage separation to refine the fuel and recover combustibles.

The points appear to fall between two curves. The lower curve shows the improvement in ash content that can be accomplished by removing more of the high-ash materials, without recovering the com-

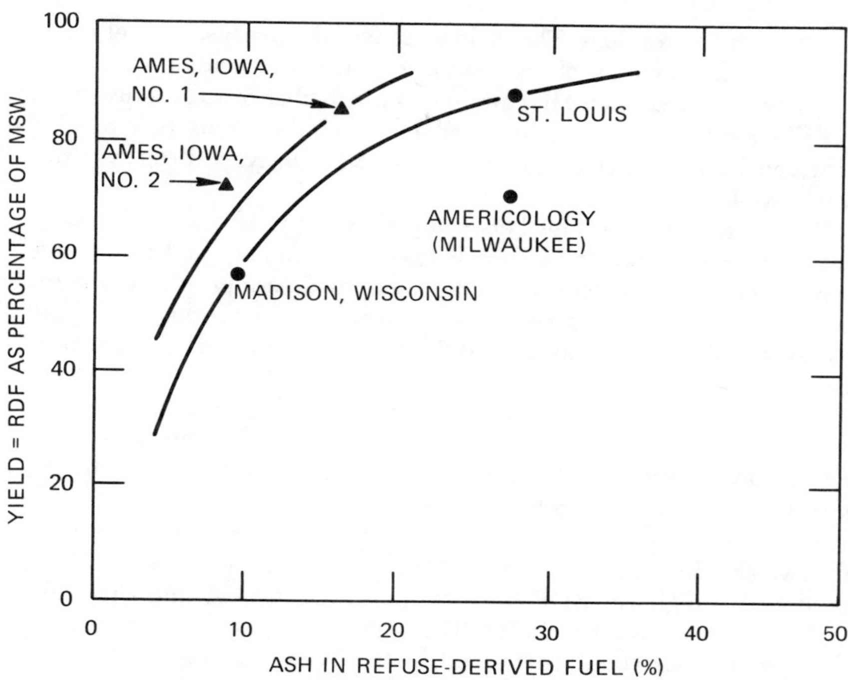

Figure 2. Percent ash in RDF vs yield of RDF as a percentage of MSW for various processing plants. From Adams et al. (1979), Barlow et al. (1980), Fiscus et al. (1977), and Mallan (1981).

Table V. Flow Diagrams of Fuel Preparation Systems[a]

	RDF/ MSW (wt %)	Ash (% dry)
St. Louis Demonstration[b]		
shred → mag → air classify → RDF; (air classify) → residue	87	28
Americology, Milwaukee[c]		
shred → mag → air classify → shred → screen → RDF; (air classify) → residue	69	27
Ames, Iowa, No. 1[d]		
shred → screen → shred → RDF; (screen) → screen → RDF; (screen) → residue	72	10
Madison, Wisconsin[e]		
bag breaker → trommel → shred → air classify → RDF; (trommel) → low-grade RDF; (trommel) → residue	58	12

[a]Magnetic separation omitted in some cases for simplicity.
[b]Fiscus et al., 1977.
[c]Mallan, 1981.
[d]Adams et al., 1979.
[e]Barlow et al., 1980.

bustibles in the residues. The yield suffers at the expense of fuel quality, and a large amount of residue must be taken to landfill.

The upper curve shows the product of plants that process refuse with initially lower ash and moisture content and the yields that can be achieved by using two-stage processing of the heavies to recover the combustibles.

The general relationships between processing yield and fuel quality determine the economic balance between fuel quality and processing and operating costs, which can be used to determine the optimum process. However, it is necessary to understand the reasons why these plants performed the way they did and what improvements are possible.

The Ames process was greatly improved by installation of disc screens, partly because of the relatively low moisture content of Ames refuse. On the other hand, the same type of screen installed in Milwaukee, Wisconsin proved to be ineffective.

The Americology flowsheet, with the air classifier located between the two shredders and screens located downstream from the shredders, produced a high-ash RDF because glass was ground into the high-moisture refuse (Mallan, 1981). The high ash content severely limited the amount that could be burned in the Wisconsin Electric boilers, which were sensitive even to the available coal, resulting in discontinuing use of the RDF. Modifying an existing plant is difficult and expensive. Efforts have been made to use this fuel in a stoker-fired boiler.

The city of Madison process, which uses a slow-speed flail mill to break cartons and bags, followed by a trommel to remove glass, dirt, and other inerts, produces a fuel that has an ash content of 10–15%. The price of obtaining this clean fuel is the 35% residue removed by the trommel. This residue could be burned in a stoker-fired boiler.

Removing Inerts Before Shredding

The systems described above have remarkably different relationships between fuel ash and yield, or recovery, of combustibles from the MSW for several reasons: (1) the systems with high-speed primary shredders smash the glass into small particles, which are hard to remove from moist material; (2) when the MSW is relatively dry, screening can be effective in removing this grit; and (3) those systems that have two-stage recovery screens can yield more fuel.

Figure 3 shows the particle size distribution of raw and shredded MSW (Ruf 1974). Before shredding, the glass is concentrated in the

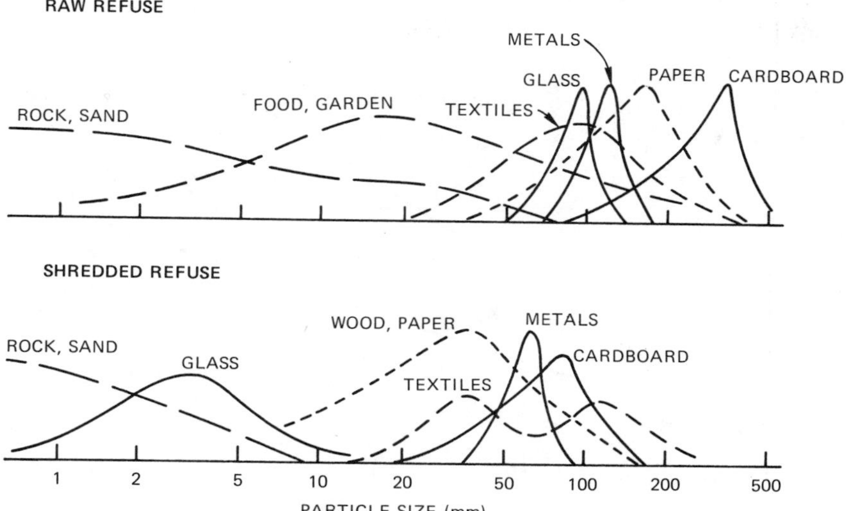

Figure 3. Particle size of various species in municipal waste before and after shredding. From Ruf (1974).

100-mm size, along with metal beverage containers. After shredding, the glass is spread over a wide size range, approaching that of the sand.

Note that the textiles have been reduced in size but that a large fraction remains unshredded. To eliminate problems with rags, the optimum grate and hammer configuration for the shredder must be determined.

MSW Moisture Content vs Fuel Ash

Figure 4 shows a correlation of more than 50 data points, showing a relationship between RDF moisture and the ash content of the fuel, from tests run by Americology in Chemung County, New York. Other data obtained by CEA show the same trend. This relationship is consistent with the difficulties that have been encountered in removing inerts, especially dirt and grit, from high-moisture refuse.

When high-moisture refuse is shredded without removing the glass first, the glass is broken into fine grit, which embeds itself in the moist paper and is hard to remove. The best remedy for this problem is to screen out the glass before shredding, as mentioned above, and to process the underflow to recover the combustibles.

The dirt in MSW, particularly from yard waste, also tends to stick

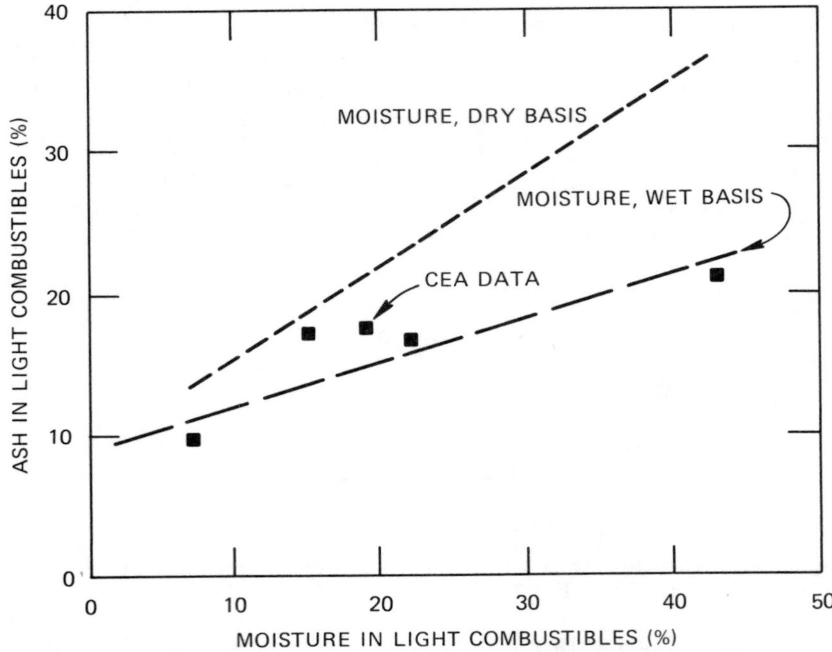

Figure 4. Correlation of data relating moisture content of shredded light combustible fraction with ash content (source: unpublished data from Chemung County, New York, courtesy of S. P. Lawler).

to moist paper and should be removed as much as possible before shredding. Drying the shredded MSW has been found to help in removing dirt from the fuel, as in the Bridgeport process illustrated above.

The new generation of the Hydrasposal process (Plato, 1981) uses primary trommels to remove glass and grit before wet processing (pulping) of the refuse, thus minimizing the problems associated with erosion and high ash in the fuel product. This process is very effective in cleaning the metals and glass products, minimizing the amount of residues requiring landfill, and reclaiming the maximum amount of combustible as fuel.

CONCLUSIONS

Experience gained during the last decade in building, operating, and testing plants to process refuse into fuel products provides an extensive data base from which to design the optimum process for a given site.

Much has been learned about the sources of ash in the fuel product and the ways available for removing these contaminants. In addition, there is a better understanding of the methods that can be used to process the residues of the fuel-refining operations so that a maximum amount of combustibles can be directed to the fuel product, while cleaning the ferrous product and minimizing landfill. This information can be used in computer programs to simulate process performance so that the optimum design can be developed.

REFERENCES

Adams, S. K., et al. 1979. "Flow Stream Characterization, the City of Ames, Iowa Solid Waste Recovery System," ERI/Iowa State University, Ames, IA.

Barlow, K. M., G. L. Boley and M. L. Smith. 1980. "Design, Evaluation and Operating Experience of the City of Madison Energy Recovery Project," paper presented at the ASME National Waste Processing Conference, Washington, DC, May.

Fiscus, D. E., et al. 1977. "St. Louis Demonstration Final Report," EPA-600/2-77-155, U.S. EPA, MERL/ORD, Cincinnati, OH.

Hasselriis, F., H. Master, C. Konheim, and H. Betzig. 1981. "Eco-Fuel™ II: The Third Generation," Proceedings of the International Conference on Prepared Fuels and Resource Recovery Technology, Nashville, TN.

Mallan, G. M. 1981. "The Americology Resource Recovery System," Proceedings of the International Conference on Prepared Fuels, Nashville, TN.

Plato, C. 1981. "The Advantages of the Hydrasposal Process as Applied to the Hempstead Resources Recovery Plant and the Dade County Solid Waste Recovery Plant," Proceedings of the International Conference on Prepared Fuels, Nashville, TN.

Ruf, J. A. 1974. "Particle Size Spectrum of Raw and Shredded Municipal Solid Waste," PhD Thesis, University of Florida.

DISCUSSION

***David Wilson*, Harwell:** What is the best way to separate solid wastes?

F. Hasselriis: I've actually seen them all myself and have been involved with them. I think there is a very, very good argument that refuse should be processed to some extent before burning. For example, the primary trommel, which lets out glass and metals, permits better processing and is a very valuable machine. The question of what to do next is pretty wide open. A lot of paper can be directly recovered once you have done that. It'll never be a large fraction, but it may be valuable depending on the market. A lot of the plastics can also be recovered that way and could possibly be worthwhile, but what's economic can only be determined by the people who run the plant. It is difficult to plan because we can't predict markets, so I think something in the middle is probably the best way.

David Wilson: We started a few years behind you, and we introduced the trommel perhaps a few years before you. Could I just ask one question on that? How do you prevent textiles from fouling up the trommel if you haven't done any size reduction beforehand? This is a considerable problem that we have found, particularly with ladies' stockings.

F. Hasselriis: From very, very real experience we struggled with them for a good solid year up in Massachusetts, and we finally found the right size hole that didn't give any trouble. If the hole was too big, the rags wound up on them, and if we put the holes far enough around so the rag couldn't tie a knot on the other side, the problem went away. It's amazing how you can discover something as simple as that and solve the whole problem. It's just a question of the correct size.

***Jens Hansen,* Aalborg University, Denmark:** In terms of paper quality, I think that you may have a higher quality if you separate at the source. Also, I think its very pertinent, at least in a number of European countries, to have households involved in this separation process, instead of throwing the whole thing into the same garbage can. There are both public opinion and paper quality at stake here. What is your comment?

F. Hasselriis: I certainly agree with that. You now have messed everything up. And if people have an idea that they would like to recover paper and to keep its purity, then let those people keep their paper. And if those people who will be neat and will keep it, that's fine. However, you'll always have the other ones, and you'll still have to get rid of their refuse. I think it's just not possible to recover clean paper out of one of these processes.

***Ray Blanco,* ORNL:** I was very interested in your comments on what the major problems were. I was quite surprised that you did not mention corrosion, particularly in the off-gas system, as being a major problem.

F. Hasselriis: Corrosion in boilers has been a real problem in mass burning boilers. The reason for the corrosion was the oscillation between oxidative and reductive conditions in the combustion. In other words, it was difficult to provide the correct amount of air in the combustion of material. When incomplete combustion occurred, corrosion resulted. What's necessary is to be sure that the correct proportions of air and fuel are placed throughout the boiler. This is very difficult unless you shred and process the refuse. If you get poor combustion, you will have corrosion. In Europe, they have reduced the corrosion by making the boilers gigantically large. In fact, the boilers have gotten so large that perhaps the temperatures can't be kept high enough to burn the chloride-containing materials.

Ray Blanco: I thought chloride was the major problem itself?

F. Hasselriis: Well, in that type of furnace it was, but the moment you are burning the material properly there is no corrosion problem that I know of. I think it has simply disappeared. Incidentally, the refuse contains the materials that will counteract most of the unfavorable chemical reactions. The refuse contains enough alkaline materials to absorb a great deal of the sulfur, for example, absorb and bring it down as ash. The same process will also take care of the chloride to a large extent, if you can mix it properly.

Ray Blanco: Apparently, none of these types of incinerators use liquid scrubbers for the off-gas system because there the chloride is quite a problem.

F. Hasselriis: Well, that type of incinerator might be a hazardous waste incinerator but not a municipal incinerator. They did use wet scrubbers in the early days thinking that would solve the problems of municipal incinerators. You are correct—they had terrible corrosion problems. They also found that type of scrubber is hopeless for collecting submicron particles. Therefore, they had to abandon wet scrubbers. Now, they have electrostatic assisted gravel bed and dry scrubbers, but these are not perfected yet.

CHAPTER 34

INCINERATION: WHAT ROLE?

Harry M. Freeman
California Office of Appropriate Technology
Sacramento, California

ABSTRACT

Hazardous waste generation rates appear to be increasing, and disposal of these wastes in landfills is becoming increasingly unpopular. Other more environmentally acceptable treatment and disposal options are needed, one of which is incineration. The advantages of incineration are that it detoxifies organic wastes; it greatly reduces the volume of wastes that must be ultimately placed in landfills; and it eliminates any long-term liability for the waste. The disadvantages are that it is expensive relative to current disposal costs; it produces emissions possibly containing trace levels of hazardous constituents; and it provides for only limited recycling of the materials contained in the waste streams. This chapter explores the state of the art of using incineration to destroy hazardous wastes, and evaluates the role of incineration in the nation's approach to hazardous waste management.

INTRODUCTION

"Incineration is a relatively well developed and well-understood technology. Properly executed, it can accomplish safe destruction of primarily organic hazardous waste, permanently reducing large volumes of waste materials to non-toxic gaseous emissions and small amounts of ash and other residues. Incineration can often provide an optimum, permanent solution to hazardous waste management with minimal long-term ecological burden." Thus said the U.S. Environmental Protection Agency (EPA) in its May 1980 report, "Hazardous Waste and Consolidated Regulations." However, incineration is only one of many approaches to managing hazardous chemical wastes. As with any technology, there are advantages and disadvantages. This chapter discusses incineration and some of the results of a study of alternatives to the land disposal of hazardous wastes that was recently completed by the California Office of Appropriate Technology.

INCINERATION AND PYROLYSIS TECHNOLOGIES

Incineration reduces the volume or toxicity of organic wastes by exposing them to high temperatures. When organic chemical wastes are subjected to temperatures of 800–3000°F, they break down into simpler and less-toxic forms. If the wastes are heated in the presence of oxygen, combustion occurs. Incineration systems are designed to accept specific types of materials and to vary according to feed mechanisms, operating temperatures, equipment design, and other parameters. The main products from complete incineration include water, carbon dioxide, ash, and certain acids and oxides depending on the waste in question.

If the wastes are exposed to high temperatures in an oxygen-starved environment, the process is called pyrolysis. The products of the process consist of a range of less-complex organic compounds, which may be recovered or incinerated, and a char or ash.

Hazardous waste incineration and pyrolysis systems include:

1. Single-chamber liquid system: brick-lined combustion chamber in which liquids are burned in suspension. In addition to being the primary parts of an incineration system, these units are also used as afterburners for rotary kilns.
2. Rotary kilns: versatile, large refractory-lined cylinders capable of burning virtually any liquid or solid waste. The unit is rotated to improve turbulence in the combustion zone.
3. Fluidized-bed incinerators: stationary vessels in which solid and liquid waste is injected into an extremely agitated and heated bed of inert granular material. The process promotes rapid heat exchange and can be designed to scrub off-gases.
4. Multiple hearths: large refractory-lined shells in which waste sludge is burned on several hearths. Multiple hearths are usually used to burn sludge from municipal waste treatment plants.
5. Molten salt combustion: waste is injected beneath a bed of molten sodium carbonate for combustion. The molten bed, in addition to transferring heat to the wastes, also scrubs halogenated acids from the off-gas.

6. At-sea incineration: wastes are burned in a conventional single-chamber incinerator mounted on a ship. Potential impacts from air emissions are reduced because combustion takes place away from land.
7. Cement kilns: organic wastes are burned at 2600–3000°F during the production of cement. Offgases are neutralized by alkaline-cement production processes and ash is contained in the cement.
8. Coincineration: a term used for burning organic wastes as supplemental fuel in fossil-fuel boilers.
9. Pyrolysis: waste is converted into combustible gas through the applications of heat in an oxygen-starved environment.
10. Plasma arc torch: solid and liquid organic wastes are pyrolyzed into combustible gases by exposure to a gas that has been energized to a plasma state by an electrical discharge.
11. High-temperature fluid wall: wastes are pyrolyzed through the application of radiant heat at 4000°F. During the process, an inert gas is injected to coat the wall of the reactor and prevent destruction from the high temperatures.

A summary of incineration technologies is shown in Table I.

Following are some of the advantages of incineration:

1. It is typically a very versatile process capable of accepting a wide variety of wastes.
2. It can be designed to recover the energy value of wastes through the production of steam, electricity, or combustible gases.
3. Most incineration processes represent the evolution of years of experience in burning waste and fuels. The processes are not exotic.
4. It is capable of handling extremely large volumes of wastes. A capacity of 4 ton/h of waste (8000 Btu/lb) is not uncommon.
5. Large land area is not required.

Disadvantages include:

1. Equipment is expensive. In most instances, disposal via incineration is several times more expensive than disposal via landfills.
2. The ash that results from an incineration process is usually treated as a hazardous waste relative to its disposal.
3. Extensive air pollution control equipment is usually required to control gaseous and particulate products of combustion.
4. Air pollution permits are at times difficult to obtain.

High-temperature incineration can, in a matter of seconds or minutes, destroy or significantly reduce what might otherwise take months

Table I. Incineration Technologies

Type	Process Principle	Application	Combustion Temperature (°F)	Residence Time
At-Sea Incineration	Waste is burned in a liquid-injection incinerator or rotary kiln on board a ship in the ocean.	Any solid or liquid organic waste	1200–3000	Seconds to hours as necessary
Cement Kilns	Waste is cofired in a kiln with constituents of cement during the manufacture of cement.	Liquid organic wastes; especially attractive for chlorinated wastes	2600–3000	Up to 10 s for gases; up to hours for liquids
Coincineration	Hazardous wastes are mixed with refuse/sludge to improve combustion characteristics. Not unique technology.	Wastes that have low energy values	Depends on type of incineration	Depends on type of boiler
Fluidized-Bed Incineration	Waste is injected into an agitated bed of heated inert particles. Heat is efficiently transferred to the wastes during combustion.	Organic liquids, gases, and granular or well processed solids	1400–1600	Seconds for gases and liquids, minutes for solids
High-Temperature Fluid Wall	Waste is broken down into its elements through pyrolysis with radiant heat in a porous carbon reactor.	Liquids and granulated solid wastes	4000	Milliseconds

Molten Salt	Wastes are injected into a bath of molten sodium carbonate. Resulting waste gases react with the salt, lessening need for scrubbing.	Liquids and solids with low ash content	1400–1850	Several seconds for gases, longer for liquids and solids
Multiple Hearth	Wastes descend through several grates to be burned in increasingly hotter combustion zones.	Sludges and granulated solid wastes	1400–1800	Up to several hours
Plasma Arc Torch	Wastes are pyrolyzed using gas stream activated by electrical discharge.	Liquids and solids	90,000	Fraction of a second
Rotary Kilns	Waste is burned in a rotating, refractory lined cylinder.	Any combustible solid, liquid or gas	1500–3000	Seconds for gases to hours for liquids and solids
Single Chamber/ Liquid Injection	Wastes are atomized with high-pressure air or steam and burned in suspension.	Liquids and slurries that can be pumped	1300–3000	0.1–1 s
Starved-Air Combustion/ Pyrolysis	Wastes are heated with insufficient air for combustion to occur. Equipment similar to incinerators.	Potential use for purely organic wastes (only C, H and O)	300–1200	Seconds for gases to hours for solids

or years to degrade in a landfill. It can eliminate certain health and environmental hazards associated with landfill disposal, reducing the need for new landfill capacity, and can eliminate the possibility of problems resurfacing in the future. However, there are several major concerns that must be addressed:

1. Air pollution: although extremely high destruction efficiencies for wastes have been documented, there are always some contaminants emitted to the air.
2. Energy requirements: many mixed wastes are only marginally combustible, requiring a supplemental fuel. This results in an additional use of possibly scarce fuel.
3. Worker safety: exposure of workers to hazardous materials could lead to occupational illnesses. However, the risks are not greater than those in similar heavy industries.
4. Esthetics: even if no significant pollution occurs from incineration, there is usually a rather large steam plume, which some find esthetically unacceptable.

Air Pollution

EPA and others have repeatedly determined that total emissions from incineration can be very small. In several tests conducted throughout the country, EPA found that waste destruction of 99.99% was not uncommon and destruction up to 99.999% and greater could be achieved for selected wastes. Even in such highly successful burns, however, some residual amounts of the chemical remain.

There has been extensive research regarding emissions of conventional pollutants from incineration, but only limited data are available regarding the emission of trace-level pollutants and environmental and health effects. Although there have been no reported incidents in which properly controlled incineration has represented an immediate threat to environmental quality, additional monitoring data are needed.

Energy Requirements

Many organic wastes have high Btu values, but it takes additional fuel to burn some organic wastes. This could represent an increase in energy consumption, although the increase would be small relative to general energy use. In large operating treatment centers, the hard-to-burn wastes are usually blended with highly combustible wastes to improve overall combustibilty. This approach reduces the need for external fuel and results in a unit that uses fuel only for startups. In some instances heat can be recovered from a process, offsetting some of the energy needed for operation.

Worker Safety

The exposure of workers to the chemicals being handled at incineration facilities is a serious issue that must be addressed, but it is not unlike the exposure of workers in other aspects of the waste management industry.

Esthetic Concerns

Steam plumes represent pollution to many people. No health effects are involved, but the issue would have to be addressed by a treatment facility, either through education programs or through the installation of equipment to remove the plume.

ROLE OF INCINERATION

Where does incineration fit into a state's hazardous waste management program? A hierarchy of treatment options adopted by the Office of Appropriate Technology in developing recommendations for hazardous waste management within California follows:

1. *Waste reduction* involves changes in industrial processes so that fewer hazardous by-products are produced.
2. *Waste recycling and resource recovery* reduce dependence on land disposal and conserve raw materials.
3. *Physical, chemical, and biological treatment* processes render material completely innocuous, reduce toxicity, or substantially reduce the volume of material requiring disposal.
4. *High-temperature incineration* destroys the many organic materials that cannot be effectively recycled or treated.
5. *Solidification/stabilization of residuals before landfill* uses chemical fixation or encapsulation techniques to "solidify" wastes and make them less mobile.

HAZARDOUS WASTE INCINERATION IN CALIFORNIA

Industries in California presently generate an estimated 5 million tons of hazardous wastes each year. Approximately 30% of these wastes are organic or partly organic and are amenable to incineration. Most of these wastes are disposed of on the land; however, problems are associated with present disposal practices. First, there is a finite amount of space available in the state for hazardous waste landfills, and this amount of space is being reduced as communities continue to look upon these sites with extreme disfavor. Second, some chemicals have been determined to be carcinogens, mutagens, and teratogens and do not decompose in landfills. Placing these chemicals in a landfill amounts to long-term storage rather than disposal and could result in serious environmental crises long after the site is closed. Third, some organics are quite volatile and are capable of migrating through landfill cover, leading to air pollution problems in the immediate area of the landfill. Incineration is preferable to landfill in many instances.

Historically, California has not been very supportive of incineration as a means of reducing the volume of wastes before disposal or as a means of detoxifying wastes considered to be hazardous. This has been due primarily to the extensive area available for landfill, the low cost of this method of disposal, and continuing problems with air pollution. Recently, however, the state has become more sensitive to problems that may result from disposing of certain hazardous wastes in landfills. Increased attention is now being directed toward incorporating alternative treatment processes, including incineration, into the state's recommended plan for hazardous waste management.

The previously mentioned study by the Office of Appropriate Technology identified 500,000 tons of wastes now being placed in landfills as high priority wastes which, for health reasons, should not be put untreated in landfills. Of these wastes, which include pesticides, polychlorinated biphenyls (PCBs), cyanides, toxic metals, and halogenated and nonhalogenated organics, about 16% were identified as prime candidates for incineration in either specially designed incineration or in cement kilns, an option that is being actively pursued in California. The report identified another 100,000 tons of waste such as "other" organic solids and liquids, and oily sludges that could also be incinerated rather than buried. The combination of these two categories of wastes represent somewhat less than 15% of the total stream. It should be pointed out that the amounts identified as candidates for incineration excludes 10,000 tons of solvent waste identified as candidates for recycling.

SUMMARY

High-temperature incineration is definitely an important aspect of a state's waste management plan for those wastes that otherwise cannot be treated or recycled. The state of the art in incineration is quite advanced, offering several options for achieving destruction efficiencies of greater than 99.99%. Although incineration is currently expensive relative to landfills, this is changing as land disposal costs continue to increase. I believe that the role of incineration will continue to grow as states and communities continue to demand alternatives to landfills.

REFERENCES

Peirce, J. J., and P. A. Vesilind. 1981. *Hazardous Waste Management* (Ann Arbor, MI: Ann Arbor Science Publishers, Inc.).

Scurlock, A., et al. 1975. "Incineration in Hazardous Waste Management," PB 216 049, U.S. EPA.

Stoddard, S. K., G. A. Davis, and H. M. Freeman. 1981. "Alternatives to the Land Disposal of Hazardous Wastes: An Assessment for California," Toxic Waste Assessment Group, Governor's Office of Appropriate Technology, Sacramento, CA.

DISCUSSION

***Ralph Franklin,* DOE:** How would you compare the incineration of hazardous wastes with other competitive methods for disposal such as deep ocean dumping, detoxication through some chemical reaction, or whatever other competitive processes may exist? Do you see incineration as the best option?

H. Freeman: I think it is probably the most here-and-now technology. Some deep wells are an example of something that is technically acceptable but probably a political disaster. I do not believe we could ever conduct deep-well injection in a serious way in California. I have a predisposition for incineration because I know more about that, but from all I read on biological treatment, I think it is coming forward. I think we are going to see more of it. If industry will treat more wastes onsite, it sure will solve a lot of headaches, and I believe they are starting that, especially as the landfills get more expensive.

***Ray Blanco,* ORNL:** It is not deep-well disposal; it is hydrofracture that we practice at Oak Ridge for getting rid of some low-level radioactive waste. We have had an environmental impact statement, and it has been reviewed by EPA, NRC, DOE, and so forth. We mix cement and other additives, which complex heavy metals with the evaporator bottoms, and inject them 1000 ft into the ground where they spread out in a hydrofracture, which is about 0.5 in. thick. The hydrofracture spreads out over several hectares of area. We can inject 100,000 gal at a shot, and we think this technology, which has been developed and used for 15 years successfully, could be applied to hazardous wastes. You mentioned that when you burn hazardous wastes your product is a hazardous waste, and at present you have to put it into a surface landfill. I should think the political furor about a surface landfill with hazardous waste in it would be much greater than any furor about a solidified concrete or grout product that is 1000 ft down in bedrock and 1000 ft under the water table. I think it is something this conference should consider as a new technique, a kind of intermediate level of disposal for hazardous wastes, because it has been successfully used for radioactive wastes.

***David Wilson*, Harwell:** My feeling is that of all these technologies there are very few that are competitive for the same types of wastes. Incineration is an excellent technology for organic wastes, but many of the chemical treatment methods do not apply to organic waste. Certainly cement solidification is not really applicable to organic materials.

Ray Blanco: What will you do with the ashes from the incinerator?

H. Freeman: We have enough capacity in landfills to handle ashes from incinerators. What eats up landfills are drums of waste placed there in tank trucks and so forth. But if we could restrict it to the material that we could do nothing else with, I think we could continue to use them.

***A. H. Purcell*, Technical Information Project:** The work that 3M has been doing in incineration—do you find that of particular merit or potential in terms of general capability?

H. Freeman: Well, 3M's work is always cited as state-of-the-art technology in industrial incineration. I've heard nothing bad about that, but that whole approach to waste destruction has not really caught on in the West. Just let me say one thing in passing; I did not really touch on how cheap land disposal has been in California. Some of the landfills out there are unrestricted or are certified to take anything. In the past this hasn't put a lot of restrictions on their cost, so we ended up with costs like $30–50/ton for getting rid of waste. You can imagine the competition between that and an alternative technology in which we talk about $250 or $300/ton. We are hoping with the new fee schedules that are coming out of the California Department of Health Services, RCRA, and what have you, that the game will start to change, but it has a long way to go.

***P. A. Vesilind*, Duke University:** One last comment about hazardous waste disposal is that we shouldn't forget that technology is always changing and that what we know today is going to be primitive compared to what we should know in 20 years. If we are going to dispose of materials, we ought to put them in places where we can get at them. If we can do a better job of reclaiming them or treating so that they are less hazardous, we ought to do it. I agree with Harry very personally about deep-well injection of any kind, as well as deep ocean disposal, because we can't get at them at that point.

CHAPTER 35

REMEDIAL ACTIONS AT SOLID WASTE LANDFILLS

Robert D. Mutch, Jr.
Wehran Engineering
Middletown, New York 10940

William J. Siok
Wehran Engineering
Hampton, New Hampshire 03942

ABSTRACT

Remedial or mitigative action programs at abandoned hazardous waste disposal sites encompass a wide spectrum of specific remedial measures. Past experiences have clearly demonstrated the site-specificity of remedial measures. Because no two sites are alike in their hydrogeologic environment or their impact on the environment, no two site-restoration projects are identical. Judgment must be exercised in evaluating the environmental impacts and weighing these impacts against the cost-effectiveness of various remedial action alternatives. In most cases, the overall site restoration will encompass several individual remedial measures, each addressing one or more mechanisms of leachate generation or migration.

INTRODUCTION

Not until a deliberative, environmentally sound approach to solid waste landfill design began to emerge approximately 10 years ago was any serious effort made to control the effects of solid waste disposal. Many dumps and landfills that were established 15 or more years ago have been the center of various levels of remedial activity to reduce or eliminate deleterious impacts on the environment, particularly on groundwater quality. Remedial and mitigative measures at abandoned and active solid waste landfills encompass a wide range of specific alternatives among the general categories of those measures that (1) reduce or eliminate leachate generation, (2) contain and collect leachate, and (3) provide for waste excavation and redisposal.

The cost-effectiveness of a particular approach is a critical factor in selecting the most appropriate remedial action approach. Depending on the nature and severity of a problem, costs may be entirely capital-related or may be most heavily associated with system operation and maintenance.

Remedial or mitigative action programs at solid waste disposal sites that contribute to groundwater and surface water degradation encompass a broad spectrum of specific measures. Past experiences have clearly demonstrated the site-specificity of remedial measures. Because no two sites are alike in their hydrogeologic environment or their impact on the environment, no two site-restoration projects are identical. A great deal of judgment must be exercised in evaluating the environmental impacts and in weighing these impacts against the cost-effectiveness of various remedial action alternatives. Solutions involving 100% reductions in contaminant migration are not always practical or possible. They also may not be justifiable in light of real or potential environmental and public health impacts.

The key to making informed decisions as to the nature and comprehensiveness of a site restoration project lies in developing a full understanding of the hydrogeologic conditions and environmental impacts. This understanding must include definition of the present and likely ultimate extent of contamination. It should also quantify to the degree practical the various mechanisms of leachate formation and migration from the solid waste landfill site. Only when these factors are understood can a rational remedial action strategy be formulated.

Measures for the in situ management of solid waste disposal sites generally fall into two broad categories. (Excluded from these categories is the case of excavation and removal of the waste.) These categories include:

1. measures that reduce or eliminate the generation of leachate by minimizing the infiltration of moisture into the landfill, and
2. measures that contain and collect generated leachate.

MEASURES THAT REDUCE OR ELIMINATE LEACHATE GENERATION

Minimization of leachate generation is an integral step in landfill remedial action. Reductions in inflow of moisture to the landfill are repaid handsomely in corresponding decreases in leachate generation.

Additionally, if a high degree of leachate volume reduction is accomplished, the rate of groundwater restoration beneath and downgradient of the facility will be accelerated.

Surface Water Runoff Diversion

A method of effecting immediate results in the reduction of leachate volume involves diversion of surficial runoff patterns, which cause excessive volumes of offsite runoff to be directed over the surface of a disposal site. In all but the most difficult terrain, it is usually possible to divert a major portion of runoff that is directed toward the landfill from offsite areas. Means of accomplishing this may include shallow ditches uphill from the disposal site or deeper trenches, which also deal with infiltration of groundwater.

Diversion of Groundwater Inflow

The hydrogeologic conditions at and around the disposal site dictate the extent to which groundwater inflow contributes to leachate generation. If a problem site is located in a hydrogeologic setting in which the predominant component of groundwater flow is horizontal and there is a water table/solid waste intersection, significant quantities of leachate generation can result.

Diversion or interception of groundwater before inflow to the landfill can be accomplished through cutoff walls, open drainage channels, or subsurface French drains. In any case, groundwater is collected hydraulically upgradient of the landfill and diverted to a discharge point downgradient of the fill. Unless otherwise desired, the system design should preclude the reversal of hydraulic gradient to the extent that leachate from the landfill itself is drawn into the groundwater interception system.

Facilitated Drainage of the Landfill Surface

Because of the usual heterogeneity of landfill waste streams in conjunction with a frequent lack of compaction during placement, most landfills undergo a significant degree of differential settling. This usually results in an irregular landfill surface characterized by poor drainage or an absence of drainage in many areas. The resulting restricted landfill drainage serves to aggravate leachate generation by promoting infiltration of moisture into the landfill. Regrading of the landfill surface to enhance surface water runoff can be accomplished by cut or fill techniques, depending on the magnitude of the differential settling and the practicality of excavation. In many cases, a more attractive and beneficial grading plan can be achieved consisting of meandering drainage swales and knolls where the landfill surface is considerably steepened to promote runoff.

Landfill Capping

Where greater reductions in infiltration of moisture are warranted, capping of the landfill by either natural or synthetic materials can minimize or effectively prevent further leachate generation. Although there are many techniques to cap a landfill, certainly the most common is the use of compacted clay. Other methods include the use of geomembranes, soil cement, bentonite/soil admixtures, and various low-permeability waste materials such as fly ash or cement fines. For the purpose of reviewing the design considerations involved in capping a landfill, we have selected two techniques for discussion: compacted clay and a synthetic membrane.

Compacted Clay

Compacted clay to cap landfills and reduce leachate generation is a commonly used and often very effective technique. To effect a high degree of infiltration reduction, it is usually necessary for the compacted clay to consistently achieve a maximum permeability of 1×10^{-7} cm/s, or better. Depending on the character of the clay selected or available for use in the compacted clay cap, this may be a difficult standard to meet.

With rare exception, the clay cap should be overlain by a suitable thickness of soil capable of supporting vegetation. The depth of this soil will depend on its moisture-holding capacity and local climatic conditions as well as the intended nature of the landfill vegetation. The overlying soil also serves to minimize desiccation of the clay.

Geomembranes

Synthetic membranes may offer substantial cost benefits over the use of compacted clay in some applications. This is particularly true where suitable clays are not available for the proposed project. If properly placed, a geomembrane should offer an even higher degree of infiltration reduction, providing its integrity is maintained through proper construction and maintenance.

The positive aspects of geomembranes to essentially eliminate infiltration into a landfill are most obvious. The limitations are primarily related to the practical considerations and limitations associated with actual placement of the membrane over the landfilled area of concern. These include

1. the ability of the geomembrane to withstand puncturing and tearing because of the soils that will underlie and overlie it, and because of construction practices;
2. the chemical compatibility of the geomembrane with its intended environment, including the possible impact of landfill-generated gases;
3. the resistance of the geomembrane to anticipated root growth in the overlying soil;
4. the stability of the overlying soils and the geomembrane for installations on slopes; and
5. the extent of settling anticipated and its potential effect on the geomembrane.

LEACHATE CONFINEMENT AND COLLECTION

Innumerable techniques and environmental control measures can be used to partially or fully confine leachate migration and, having restrained its further migration, collect the leachate for treatment and ultimate disposal. Containment techniques run the gamut from impermeable perimeter dikes to slurry-trench cutoff walls. In many applications, the leachate collection system also affords a degree (and often a very high degree) of leachate migration containment by manipulation of groundwater flow patterns. Leachate collection systems vary from such simple devices as perimeter French-drain collection systems to deep groundwater recovery wells.

Some of the more commonly employed techniques for confinement and collection of leachate from program landfills are discussed below. Where possible, the benefits and design considerations and limitations associated with the individual remedial measures have been grouped together to minimize repetition.

Perimeter Dikes

Perimeter dikes are often an effective means by which to retain lateral migration of leachate from the landfill where, as a result of hydrogeologic factors, lateral surficial migration of leachate is a primary mechanism of leachate discharge. In the absence of hydraulic gradient reversal techniques or other groundwater manipulation efforts, a perimeter dike must "key into" a confining or semiconfining geologic stratum to be effective. It is implicit in this statement that the confining or semiconfining stratum be continuous beneath the landfill.

Where total confinement of leachate is not the intent of the perimeter dikes, the importance of keying into an underlying confining bed may be overshadowed by other considerations. For example, the intent of a perimeter dike may be to prevent direct emergence of leachate at the landfill perimeter by forcing it to pass beneath the dike through the underlying soils. This might be a desirable effort where the leachate was amenable to soil attenuation processes. A dike may also be constructed to dampen the influence of tidal fluctuations on leachate generation from landfills constructed in tidal regions or in areas subject to frequent flooding.

Dikes designed and constructed for any of the purposes discussed previously are normally intended to be relatively impervious. This impermeability can be achieved by use of compacted clays, geomembranes, bentonitic materials, or other techniques. Depending on the application, the entire dike might be constructed of the impermeable material, or the impermeable material can be restricted to an interior core. In addition to achieving the desired degree of impermeability, a dike should also be structurally stable and capable of withstanding erosional impacts, particularly in riparian applications.

Subsurface Cutoff Walls

In essence, a subsurface cutoff wall represents an extension of the perimeter dike concept discussed previously. The difference is that, because of geologic conditions, the impermeable barrier must extend below grade so as to key into an underlying confining or semiconfining stratum. A typical cutoff wall installation in conjunction with a surficial dike is depicted in Figure 1. Depending on the geologic conditions, the depth of penetration can vary from as little as 1.5 or 3 m to in excess of 33 m. To a large extent, the requisite depth of penetration will dictate the technique employed in the cutoff wall construction. The other major factor affecting cutoff wall design and construction is the consideration of geotechnical conditions.

The principal factor affecting the effectiveness of a subsurface cutoff wall is the depth of penetration with respect to an underlying confining

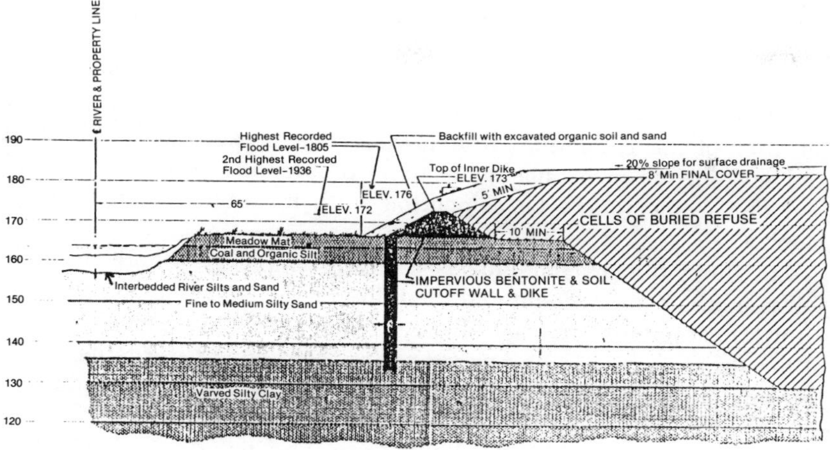

Figure 1. Cutoff wall and dike.

bed. Groundwater flow hydraulics dictate that effectiveness of cutoff walls is severely compromised unless the cutoff wall very nearly or, preferably, fully penetrates the water-bearing zone.

Other factors affecting cutoff wall performance include the permeability of the underlying confining bed, the stability of the soil through which the cutoff wall is to be constructed, and the ease of soil excavation. The degree of soil stability will dictate whether techniques such as the use of compacted clay or synthetic membranes that rely on open trench conditions are feasible. Where soils are unstable, the slurry-trench excavation technique can be used to maintain trench stability during construction and provide the necessary degree of impermeability as well.

Some of the most commonly used techniques of cutoff wall construction are compacted clay, geomembrane, and slurry trench. A compacted clay cutoff wall necessitates an open trench in which placement and compaction of the clay can be undertaken. The ability to work effectively in an open trench may be precluded by many factors, including the required depth of the trench, the extent of groundwater inflow, the stability of the soils, and space limitations. Assuming that the compacted clay technique is applicable, it will usually be necessary for the compacted clay to achieve some minimum degree of impermeability. A maximum permeability of 1×10^{-7} cm/s is often applied as the standard.

Under certain geologic conditions and for certain applications, a geomembrane may be used as a subsurface cutoff wall. As in the case of the compacted clay cutoff wall, for construction of a geomembrane cutoff wall, the soil must be stable. Major design considerations for application of this technique are that the geomembrane can be installed and backfilled in such a manner as to prevent tearing or puncturing and that the geomembrane be compatible with its subterranean environment.

The slurry-trench technique is typically employed where unstable soil conditions are encountered. Stability of the trench is achieved by a thick bentonitic slurry that fills the trench at all times. The elevation of the slurry surface is maintained above the groundwater table in the surrounding soils. This factor, together with the high relative density of the slurry, allows for excavation of a vertical walled trench in most soils with a minimum likelihood of cave-in. The impermeable character of the cutoff wall can be achieved by backfilling with an admixture of bentonitic slurry and native soils or by the addition of cement to the slurry such that it sets up.

Leachate Collection Systems

A variety of leachate collection techniques are available. The technique most suited to the specific landfill under consideration will be dependent on the objectives of the program and the site-specific hydrogeologic conditions. Whether the objective is collection of the entire volume of leachate or interception of a particular segment of the leachate flow that is particularly problematic, an accurate estimate of leachate volume generation rates is requisite.

Some of the more commonly applied leachate collection systems are listed. Each of these concepts has in common the requirements for basic data essential to proper design. These include, but are not necessarily limited to,

1. accurate estimates of anticipated leachate generation rates;
2. knowledge of the chemical compatibility of the system components with the leachate being located;
3. accurate prediction of the required functioning life of the system based on rate of leaching and mass of leachate material;
4. the constructability of the system with particular respect to safety and public health considerations where excavation or drilling within the landfill itself is necessitated; and
5. the hydraulic capacity and costs of the leachate/groundwater con-

veyance and handling facilities, particularly those associated with leachate treatment and discharge.

French Drains

A French-drain groundwater interceptor is essentially a manmade conduit of high permeability sand or gravel designed to collect and conduct groundwaters. The capacity of the French drain can be enhanced significantly by inclusion of a perforated drainage pipe within the drain. The drain is typically sloped to a low point where flow discharges either under the action of gravity or by means of dewatering pumps.

In the design of a French-drain groundwater interceptor, two conditions are requisite to proper long-term functioning of the facility: (1) each member of the French drain should have sufficient hydraulic capacity to handle anticipated design flows, and (2) the French drain should be designed to minimize or preclude piping of soil particles through the system.

Collection Sumps, Wells, and Well Points

In some instances, it may be possible to collect leachate from one or more collection sumps, wells, or by a well point system constructed within the landfill. This approach may be feasible where lateral confinement has been provided by perimeter dikes or cutoff walls and the hydraulic capacity of the collection system is equivalent to the rate of leachate generation. Since a well point system depends on suction lift, such a system is only applicable in the case where the landfill and leachate depths are shallow. The well points themselves can be installed by either driving them into place or by drilling and construction within the complete borehole. Since the hydraulic capacity of each well point is usually quite small, a large number of well points would typically be necessary to withdraw generated leachate from most landfill sites. Systems of this nature are more applicable to wastes possessing high permeabilities or to situations where leachate generation is minimal.

COST/BENEFIT ANALYSES

In all but the most straightforward site cleanups, consideration should be given to the anticipated benefit of each remedial action alternative vs the attendant costs. The resulting cost/benefit analysis is an excellent tool by which to weigh the cost-effectiveness of any number of alternative approaches for mitigating or arresting the impacts of an abandoned landfill or other source of pollution. The measure of the benefit of a proposed remedial action program can be expressed in many forms, depending on site conditions and the nature of the site's environmental or public health impacts. Measurements of anticipated abatement can be expressed as reduction in total release of chlorinated hydrocarbons in pounds per year, in predicted incidence of cancer per 100,000 persons, or in total gallons of leachate released per year.

Whatever units are chosen to evaluate the effectiveness of various alternatives, invariably a region of diminishing return is encountered as one strives for progressively higher degrees of abatement. This is true in most areas of environmental control, whether it be wastewater treatment, air pollution control, or remediation of landfill sites.

A typical cost/benefit analysis is shown in Table I. In this case, the site is a small, inactive, chemical waste dump in a rural setting (actual site). In this instance, it was decided to evaluate the benefit of each of ten alternatives based on their ability to prevent and otherwise manage leachate discharge from the landfill. To account for both initial construction costs and long-term operating costs, a study period of 30 years was selected. As presented in Table I, each alternative was evaluated over the 30-year evaluation period. Present-worth values were used to assess long-term operational costs. (Alternative 4 was ultimately selected as representing the best mix of environmental control and cost in this particular case.)

DISCUSSION

***Robert Saar*, Geraghty and Miller:** I have one observation about capping and also one question. There is usually quite a bit of water in a landfill even at the point that you put a cap on, and you may not have observed a stopping of leachate generation for quite awhile after the cap is in place. In fact, with less water going into the landfill when the cap is put into place, you may have a temporary rise in concentration of some chemicals in the groundwater until the water in the landfill has drained out.

R. Mutch: That is very much the case, and that is why I stress the use of water balance in coming up with an overall water balance of the landfill so that we can know what effect the cap will really have. It is going to cut off future inflow of precipitation, hopefully, if it is designed and built properly, but if there is 10 million gallons of leachate in storage in the landfill, it does nothing for that. That will slowly continue to bleed out. If the landfill is actually below the water table and has inflow of groundwater, capping the

Table I. Evaluation of Remedial Action Alternatives

Alternatives	Anticipated Degree of Abatement (%)	Estimated Construction Cost ($)	Estimated Operational Costs; 30-yr Duration (Present Worth) ($)	Estimated Total Cost ($)	Cost Per 1000-gal Leachate Managed[a] ($/1000 gal)
Alternative 1: No Action	0	0	188,600	188,600	
Alternative 2: Regrading; Surface Water Diversion; Cover; Revegetation	44	100,000	207,460	307,460	15.92
Alternative 3: Impermeable Cap	87.5	200,000	207,460	407,460	10.61
Alternative 4: Impermeable Cap; Southern Cutoff Wall	+99	250,000	207,460	457,460	10.42
Alternative 5: Circumferential Collection System; Leachate Treatment	+99	150,000	2,751,674	2,901,674	66.11
Alternative 6: Circumferential Collection System; Southern Cutoff Wall; Leachate Treatment	+99	200,000	2,434,826	2,634,826	60.03
Alternative 7: Circumferential Cutoff Wall and Collection System; Leachate Treatment	+99	225,000	1,484,282	1,709,282	38.94
Alternative 8: Alternative 6; Impermeable Cap	+99	350,000	850,586	1,200,586	27.35
Alternative 9: Alternative 7; Impermeable Cap	+99	375,000	229,564	604,564	13.77
Alternative 10: Exhumation	+99	12,500,000	0	12,500,000	284.80

[a]Estimated total cost × 1000, divided by anticipated degree of abatement × 43,890,000.

landfill is only attacking one mechanism of leachate generation; the other mechanism is the groundwater inflow. I think it's important that we recognize that capping a landfill is only attacking that one mechanism. If groundwater inflow is 20% of the problem, you're still going to have 20% of the impact when you're done.

Robert Saar: Then I have one question. Is there available or is there being developed a kind of cap that is permeable to gas and not to water along the lines of the cortex kind of fabric that is now available for clothing?

R. Mutch: I don't know of anything which in and of itself has both those properties, but it is not a major problem in terms of capping landfills to provide for the ventilation of gases through gas ventilation systems. These can be nothing more than trenches of gravel beneath the cap, venting at one or more locations to allow the accumulation of gas under the liner to be conducted and vented to the atmosphere. In the case of some of the chemical waste landfills, that gas is a potential problem and has to be scrubbed before it is allowed to be released into the environment. That has not really been a major problem other than there is nothing uglier than the sight of a landfill with 100 gas vents poking through the surface, all at different angles. We see a lot of that, but there are better ways of doing it than simply putting PVC gas vents at 100-ft intervals through the cap. There are some better techniques.

***Dev Sachdev,* Envirosphere Company:** I have two questions. One, after you collect the leachate, do you treat it or spray it on the completed cell to lose it to evaporation? What have you recommended on this thing?

R. Mutch: There are a number of different options. One of the interim things is done if you have an immediate problem where you have to stop the leachate, like in the case of Monroe township landfill; the leachate can be collected within a matter of a couple of weeks by some open collection systems followed by a more permanent system. But that leachate in this case was simply recirculated back to the landfill. So that is a temporary measure. What is being done in terms of long-term treatment, one of the most viable options I would think, is sending it to a municipal sewage treatment plant where it represents a very small percentage of the overall flow from the treatment plant. In addition to that, pretreatment facilities are being built to pretreat and remove certain harmful constituents before the leachate goes into the sewage treatment plant or full-scale treatment plants onsite.

(***Questioner***): What you have recommended in these couple of landfills that you have designed—you have recommended to send it to the municipal treatment system?

R. Mutch: Well, there are a lot of different things. In some cases, as in the Monroe Township landfill, treatability studies have been completed. Simulating the effects of the leachate on the municipal sewage treatment plant and gaining acceptance from the sewage treatment plant to accept the leachate have been done there. In other cases, leachate is going to be treated by building full-scale treatment plants onsite which will treat it and discharge it to a receiving body. In three cases that I know of in New Jersey, municipal sewage treatment plants will be the ultimate recipient of this material, either in a pretreated form or, in many cases, untreated prior to discharge to the plant.

(***Questioner***): You said you had to dig and install groundwater wells to bring back the aquifer. These are groundwater wells only on the downgradient side, and how many were there?

R. Mutch: Well, every situation is a little bit different. In the case I was speaking of, there were 13 recovery wells, and they were on the downgrade inside of the landfill. In most cases, the recovery wells are usually on the downgradient side of the source of contamination, the objective being to contain further spread of the plume and eventually to recover that plume and remove it from the aquifer. But depending on the hydrogeologic conditions, it could be one well or it could be 100 wells depending on the extent of the problem and how wide the plume has spread before you catch it.

(***Questioner***): How deep had you gone there? How deep below the bottom of the landfill?

R. Mutch: I think the maximum depth of the wells is about 60 ft in most cases you are dealing with; in most cases that I have been involved with it is shallower than that. You are dealing with an upper groundwater table aquifer that has been contaminated. Very often, in the case of industrial plants in conjunction with the collection of contaminated groundwater, you may also be skimming off layers of floating organics on the water table with scavenger systems and treating that as a different waste thing, hopefully recovering it.

***Doug Burgess,* Stone & Webster Engineering:** Somebody suggested to us recently that an alternative method of constructing a slurry trench to the methods that you described would be to use an advancing beam which is pile-driven into the ground. It has the advantage of being able to go deeper, maybe as deep as 200 ft, does not produce any spoil that has to be disposed of, etc. There were several other advantages cited. Could you comment on that method, please?

R. Mutch: Yes. That technique, I think, is called a vibratory beam concept in which they are injecting a beam down into the ground, pounding it down into the ground, and then they withdraw it while they are injecting a slurry into the space, so they end up with an imprint of a beam in the ground filled with the slurry. They then move over two-thirds of a beam and do it again, and one ends up with sort of an interlocking pattern of beams. The wall is only about 3 in. thick, which is a disadvantage in terms of the groundwater hydraulics of the situation, because even if it achieves the same permeability, which I understand it does, you are dealing now with only a 3-in.-wide wall. If used in an application where you are going to have a steep hydraulic gradient between inside and outside of the wall, you are going to have much facter rates of penetration. Not only will the velocity be faster, but now you've only got a 3-in.-wide wall, and leachate does get through a little quicker. I think some of the advantages are that it requires less space: it may very well be able to go deeper in some cases, but it is a very confined type thing, so you can sneak it into spaces where you might not be able to sneak in slurry trench equipment. A slurry trench operation is a major construction activity. It takes a little wider space than I believe this vibrating beam technique does.

***Jens Hansen,* Denmark:** Traditional solid waste landfill could be perceived as a digestion chamber where you stabilize the solid waste over time. In having water percolating through, you can control the time it takes, say, from

100 years to several thousand years. Why would you like to cap it so as to prevent water from coming into it? I have some doubts whether it is really a reasonable thing to create this kind of time-delayed bomb.

R. Mutch: I agree with your reasoning completely. I think a lot of thought has to be given before the decision is made to cap a landfill. I think too often we are jumping right in to putting a cap on that landfill. In the case of the Monroe Township landfill, we were collecting leachate. Then the decision came up, because the regulatory officials and some consultants for an opposition group wanted a clay cap on it as well. And when you look at the benefits of putting a clay cap on it at that point, I did not think they looked adequate for the reasons you were saying. This was mainly a solid waste landfill with some component of hazardous waste. We wanted to promote the leaching out and stabilization of that waste. We had a viable approach for collecting and treating it. We could cap it, and we figure we'd be treating 5000 gal/d for who knows how long, 50 years or more perhaps, if we capped it. If it was uncapped, we generated about 60,000 gal/d and we would be treating it for a lot less time. I do not think anybody knows the answer to how long you are going to treat it, but it would be a lot less. I think that is clear. The economics sway heavily in favor of not capping it, but I'm not sure that applies to all landfills. When you are dealing with a site that is mostly chemical waste, I don't think that rationale applies. There you are better off capping it and just minimizing the discharge period because there is such an amount of leachable material in there that that approach may not be viable. But a solid waste landfill is another sitution. It can reach a stage where it is not leaching contaminants to the extent that it does initially.

CHAPTER 36

LAND TREATMENT OF HAZARDOUS WASTES

K. W. Brown

Texas A&M University
Soil and Crop Sciences Department
College Station, Texas 77843

ABSTRACT

Hazardous industrial wastes, including effluents and sludges, are being disposed of by land treatment in many areas of the country. These wastes typically contain inorganic and organic constituents that could adversely affect human health or the environment if they migrate from the site or accumulate to unacceptable levels. A considerable amount of information on the fate and transport of nutrients, common salt, and metals in the soil environment is available; however, much less is known about the behavior of the complex organic mixtures found in wastes. Laboratory, greenhouse, and lysimeter studies of the fate and mobility of two industrial wastes were conducted. The residual toxicity and mutagenicity in the soil and runoff decreased with time after application, whereas the leachate water collected 1.5 m below the subsurface was essentially free of toxic and mutagenic components.

INTRODUCTION

Land treatment is now being used to dispose of hazardous industrial wastes at about 250 facilities throughout the United States. The interest in land treatment has been increasing, perhaps because of the relatively low cost of this method of disposal compared with landfilling and incineration, which typically cost 2.5 and 10 times more than land treatment respectively. Another possible advantage of land treatment is that if properly managed, it should result in degradation of the toxic organic constituents of the waste and sorption and retention of undesirable inorganic constituents in the soil environment.

Although a reasonable amount of data are available on the fate and mobility of nutrients, salts, and metallic constituent of waste applied to the soil, much less is known about the fate of the organic constituents of these wastes in the soil. Some data are available on the rates of degradation of oil and grease and the bulk biodegradation of organic fractions. However, very little is known about degradation, movement via runoff or leachate, and plant uptake of waste constituents that are mutagenic, carcinogenic, or teratogenic.

There is a long history of land treatment of waste materials. Organized land treatment of municipal wastes began in the last century. More recently, soft industrial wastes (including the residue of food processing and paper pulp plants) and hard industrial wastes (including residues from oil refineries, petrochemical plants, and pharmaceutical manufacture) have been disposed of on land. Many of these modern industrial wastes contain constituents that are toxic or that could result in genetic damage to plants or animals.

Genetic toxicity should be determined by using a series of biological systems that will accurately predict the potential of waste constituents to cause gene mutations and other types of genetic damage. Biological tests predict most accurately the genetic toxicity of a sample, whereas chemical analysis of the components of a complex mixture may fail to account for transformations in the soil or various interactions that can occur between chemicals in a complex mixture. Four groups of biological test methods may be used to evaluate the genetic toxicity of a sample: (1) epidemiologic; (2) long-term animal; (3) long-term plant and animal cell culture; and (4) simple microbial, plant, or animal cell culture assays. The only method that is not restricted by time and cost is the short-term assay that uses plant, microbial, or mammalian cell cultures. Some possible short-term systems and the genetic events they can detect are listed in Table I.

A complete set of test systems should be capable of detecting gene mutations, compounds that cause various types of chromosome damage, and compounds that inhibit DNA repair. This project used several test systems to (1) respond to the types of genetic damage previously described; (2) detect the anticipated compounds of the waste; and (3) incorporate metabolic activation into their testing protocol. These systems include provisions for solvent and positive controls, which demonstrate the sensitivity of the test system and functioning of the metabolic

Table I. Biological Systems That May Be Used to Detect Genetic Toxicity of a Hazardous Waste

Organism	Genetic Event Detected		Metabolic Activation	References
	Gene Mutation	Other Genetic Damage		
Prokaryotes				
Bacillus subtilis	Forward, reverse	DNA repair	Mammalian	Felkner et al., 1979; Kada et al., 1977; Tanooka et al., 1978
Escherichia coli	Forward, reverse	DNA repair	Mammalian, plant	Green et al., 1976; Mohn et al., 1974; Scott et al., 1978; Slater et al., 1971; Speck et al., 1978
Salmonella typhimurium	Forward, reverse	DNA repair	Mammalian, plant	Ames et al., 1975; Plewa and Gentile, 1976; Skopek et al., 1978
Streptomyces coelicolor	Forward	DNA repair	Not developed	Carere et al., 1975
Eukaryotes				
Aspergillus nidulans	Forward, reverse	DNA repair, chromosome aberrations	Mammalian, plant	Bigani et al., 1974; Roper, 1971; Scott et al., 1978, 1980
Neurospora crassa	Forward	Not developed	Mammalian	Deserres and Malling, 1971; Ong, 1978; Tomlinson, 1980
Saccharomyces cervisiae	Forward	Mitotic gene conversion	Mammalian	Brusick, 1972; Loprieno, et al., 1974; Mortimer and Manney, 1971; Parry, 1977
Schizosaccharomyces pombe	Forward	Mitotic gene conversion	Mammalian	Brusick, 1972; Loprieno et al., 1974; Mortimer and Manney, 1971; Parry, 1977
Plants				
Tradescantia	Forward	Chromosome aberrations	Plant	Nauman et al., 1976; Underbrink et al., 1973
Arabidopsis thaliana	Chlorophyll mutation	Chromosome aberrations	Plant	Redei, 1975
Hordeum vulgare	Chlorophyll mutation	Chromosome aberrations	Plant	Kumar and Chauhan, 1979; Nicoloff et al., 1979
Pisum sativum	Chlorophyll mutation	Chromosome aberrations	Plant	Ehrenburg, 1971
Triticum	Morphological mutation	Chromosome aberrations	Plant	Ehrenburg, 1971
Gylcine max	Chlorophyll mutation	Chromosome aberrations	Plant	Vig, 1975

Vicia faba	Morphological mutation	Chromosome aberrations	Plant	Kihlman, 1977
Allium cepa	Morphological mutation	Chromosome aberrations	Plant	Marimuthu et al., 1970
Insects				
Drosophila melanogaster	Recessive lethals	Nondisjunction, deletions	Insect	Wurgler and Vogel, 1977
Habrobracon	None developed	Dominant lethals	Insect	Von Borstel and Smith, 1977
Mammalian Cells in Culture				
Chinese hamster ovaries	Forward, reverse	Chromosome aberrations	Mammalian	Beek et al., 1980; Neill et al., 1977
V79 Chinese hamster cells	Forward, reverse	Chromosome aberrations	Mammalian	Artlett, 1977; Soderberg et al., 1979
Chinese hamster lung cells	Forward	Chromosome aberrations	Mammalian	Dean and Senner, 1977
Human fibroblasts	Forward	DNA repair	Mammalian	Jacobs and DeMars, 1977
Human lymphoblasts	Forward	DNA repair	Mammalian	Thilly et al., 1976
L5178Y mouse lymphoma cells	Forward	Chromosome aberrations	Mammalian	Clive, 1973; Clive and Spector, 1975; Clive et al., 1972
P388 mouse lymphoma cells	Forward	Chromosome aberrations	Mammalian	Anderson, 1975
Human peripheral blood lymphocytes	Forward	Chromosome aberrations	Mammalian	Evans and O'Riordan, 1975
Various organisms	None developed	Sister chromatid exchange	Mammalian	Perry and Evans, 1975; Stetka and Wolff, 1976

activation system, and act as an internal control for the biological system. Thus far, we have examined two biological test systems to detect compounds that can cause point mutations and produce increased lethal damage in DNA-repair-deficient organisms. The microbial mutagenicity assay using *Salmonella typhimurium* as an indicator organism is 80–95% efficient for detecting certain classes of organic chemical carcinogens (McCann et al., 1975; Purchase et al., 1976). Categories of carcinogens that are poorly detected by *Salmonella* assay included chlorinated hydrocarbons, chemicals that are bactericidal or volatile, and chemicals that crosslink DNA (Rinkus and Legator, 1979). The *Bacillus* DNA repair assay is sensitive to pesticides (Shiau et al., 1980) and bactericides (Brown and Donnelly, 1981) that fail to induce a clear response in *Salmonella*. The two biological test systems were selected to efficiently detect mutagens and potential carcinogens; the DNA repair assay was used to enhance the efficiency of the mutagenicity assay.

The test systems used in this project also include biological systems using *Aspergillus nidulans* (Scott et al., 1980), *Glycine max* (Vig, 1975) and *Tradescantia* (Sparrow et al., 1974) as indicator organisms. The various types of chromosome damage that can be indicated by these test systems include nondisjunction, mitotic crossing over, major chromosomal damage, recessive lethals, translocation, and duplication (Nilan and Vig, 1976; Scott et al., 1980).

MATERIALS AND METHODS

Two hazardous wastes were collected for study. After the wastes were characterized, extracts were subjected to bioassays to detect mutagenic activity. The wastes were also incubated with soil to evaluate the fate of the mutagenic compounds when exposed to the soil environment. The same wastes were applied to the large undisturbed lysimeters from which leachate and runoff were collected. These samples were concentrated on XAD resin. Organic residue was also evaluated for mutagenic activity.

Waste

The two wastes used in this study were collected from American Petroleum Institute (API) oil-water separators, one from a petroleum refinery and one from a petrochemical plant. Two 50-gal samples of each waste were collected. An aliquot of each waste was removed from the drum for the laboratory study, and the rest was stored for use in the field study. Physical and chemical characteristics of each waste are given in Table II. Dichloromethane was selected from a group of agents to extract the organic fractions of the waste and the soil because it consistently provided the greatest extraction efficiency for the wastes used. The residue from this extraction was partitioned into acid, base, and neutral fractions by liquid-liquid extraction following the scheme outlined in Figure 1.

Table II. Sludge Characteristics

Oil Sludge	Extractable Carbon (g/100 g)	Ash	Water	Total Hydrocarbon (%) Pentane	Benzene	Dichloro-methane
Refinery	10	41	46	72	22	6
Petrochemical	62	25	12	63	36	1

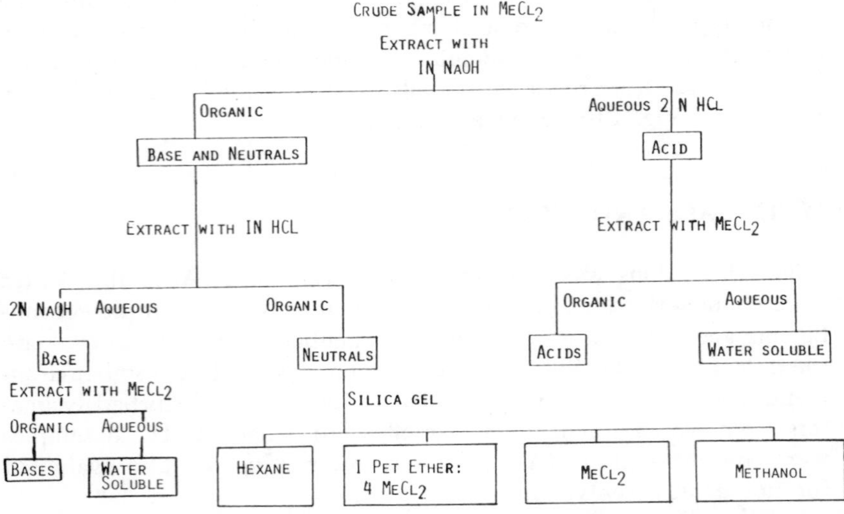

Figure 1. Schematic diagram for extraction procedure used to separate hazardous waste fractions.

Soil

Two soils were selected to represent a range of soil textures: a Norwood sandy clay (Typic Udifluvent) and a Bastrop clay (Udic Paleustalf). Characteristics of these soils are given in Table III. Soils were analyzed for mutagenic activity using *Salmonella* only.

Table III. Characteristics of Unamended Soils Used in the Greenhouse and Lysimeter Study

	Soil Series	
	Norwood	Bastrop
Sand (%)	48.2	60.3
Silt (%)	15.2	10.0
Clay (%)	36.6	29.7
Organic Matter (%)	1.4	1.0
pH	7.69	6.86
Moisture (%)		
Wilting Point	12	6
Field Capacity	18	22
Saturated	33	25
Mutagenic Activity		
Revertants/mg Organic Extract of Soil	47	365
Revertants/mg Soil	0.028	0.292

Biodegradation

The effect of land application on the toxic potential of the petrochemical waste was evaluated in a soil respirometer (Figure 2). The

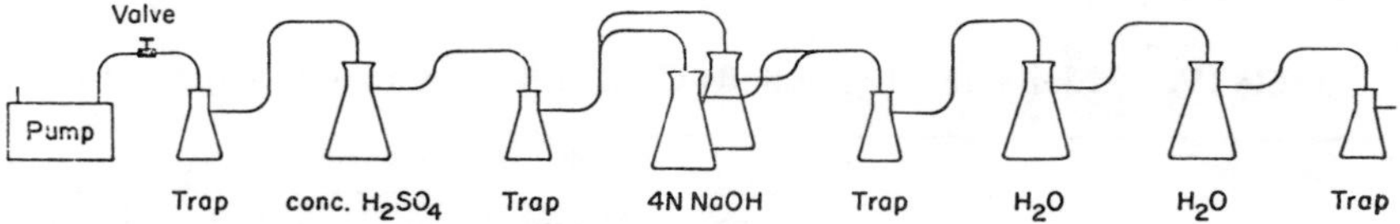

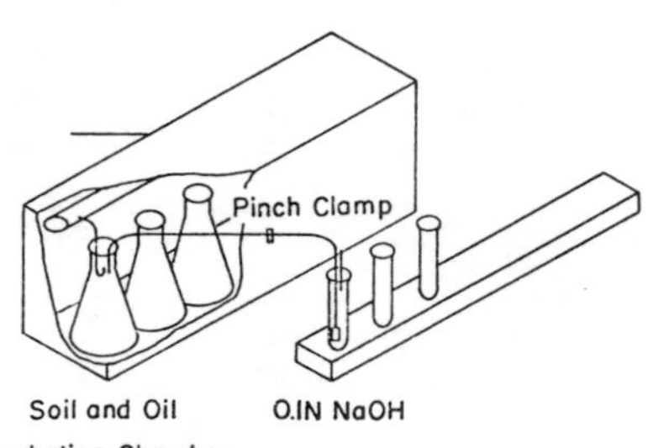

Figure 2. Schematic diagram of a respirometer.

waste was mixed with a Norwood sandy clay soil at two treatment rates and incubated at 30°C for 180 d at a moisture content of 18%. One set of flasks received one 5% waste application; one set received four 5% applications on days 0, 45, 90, and 135; one set, which served as a control, received no application of waste. Biodegradation was monitored by adsorption of evolved CO_2 onto 0.1 *N* NaOH during the incubation period. At the end of the incubation period, 25 g of the waste-soil mixture was removed and extracted with six volumes of dichloromethane before separation on a silica gel column. The dichloromethane extract was separated into aromatic, aliphatic, and condensed ring fractions with petroleum ether, dichloromethane in petroleum ether (1:4) and dichloromethane respectively. The effect of biodegradation on toxicity was determined by comparing mutagenicity of the soil extract towards *Salmonella typhimurium* of a fresh sample of soil-waste mixture with one that had been incubated for 180 d.

Leachate and Runoff

Leachate and runoff samples were collected from large, undisturbed field lysimeters containing four soils. Soil characteristics are given in Table IV. Each waste was incorporated into the top 15 cm of two lysimeters of each soil. A second and third waste application was made to one lysimeter of each soil, and one lysimeter to which no waste was applied served as a control. Five monoliths of each soil were encased in

Table IV. Characteristics of Soils Used in Biodegradation Experiments

	Soil Series			
	Norwood	Nacogdoches	Lakeland	Bastrop
Sand (%)	48.2	41.8	81.1	60.3
Silt (%)	15.2	12.9	4.5	10.0
Clay (%)	36.6	45.3	14.4	29.7
Organic Matter (%)	1.4	1.3	0.7	1.0
CEC (meq/100 g)	19.6	17.2	0.3	27.4
pH	7.69	5.95	6..45	6.86
Nitrogen (ppm)	70	47	42	160
Phosphorus (ppm)	94	3	20	25
Potassium (ppm)	312	164	200	200
Calcium (ppm)	4000	1280	600	2920
Magnesium (ppm)	485	400	100	385
Moisture (%)				
Wilting Point	12	8	1	6
Field Capacity	18	20	5	22
Saturated	33	36	23	25

the 150-cm-deep lysimeters and brought to a central location. A detailed description for the installation of lysimeters is given by Brown et al. (1974). The leachate was collected from porous ceramic suction cups (Coors Type No. 7001-P-6-C), which were installed in the soil at the bottom of the profile. The suction cups were connected by nylon tubing to a 20-L glass bottle, which was connected to a vacuum manifold. Leachate samples were collected once a week unless the rate of water flow required more frequent sampling.

Runoff samples were collected by a peristaltic pump. Glass beakers were placed in a small hole in the corner of each lysimeter; the soil surface in the lysimeter was sloped so that runoff water could collect in the beaker. A rainfall of greater than 1.27 cm activated a switch, which supplied power to the pumps. The collected runoff was pumped through nylon tubing to a 20-L glass bottle in which the water was stored in an air-conditioned shed until collected. Water samples were collected in amber glass bottles and stored at 4°C until they were processed.

Two methods were used to extract water-bound organic compounds. Initially, a 20-cm^3 bed was made of nonpolar XAD-2 resin (Applied Science Laboratory, State College, Pennsylvania) by following the methods of Hooper et al. (1978). Later in the study, the XAD-2 resin was supplemented with a moderately polar resin because of the complex nature of the samples being analyzed. Thus, samples that were extracted near the end of the study were passed through a mixed bed of 4.0 g XAD-2 and 6.3 g XAD-7, or about 20 cm^3 of each resin, as suggested by Rappaport et al. (1979). The resins were washed before use by swirling and decanting them three times with 10 volumes each of acetone, methanol, and distilled water. Washed resins were stored at 4°C before use. Glass econo-columns (Bio-Rad, Richmond, California), 1.5×50 cm^2, were packed with 20 cm^3 of XAD-2 resin, and then with 20 cm^3 of XAD-7 resin. A glass-wool plug was placed above the resin to trap soil particles. The columns were flushed with 1200 mL distilled water before loading the water sample.

Leachate or runoff water was placed in a reservoir and allowed to pass through the column by gravity flow at about 50 mL/min. After loading the water sample, dry nitrogen was introduced into the column to remove the residual aqueous phase, and the column was washed with 120 mL distilled water to remove residual histidine. The adsorbed organic compounds were then eluted with 160 mL acetone. The acetone extract was filtered through 30 g of anhydrous sodium sulfate and Whatman No. 42 filter paper into a flat-bottom flask. Extracts were reduced $<$10 mL on a Brinkman-Bucci Model R roto-evaporator and taken to dryness in a screw-capped glass culture tube under a stream of nitrogen. Dimethyl sulfoxide (Sigma) was added to the dried extract at a rate of 0.5 mL DMSO/L unconcentrated water; the resulting solution was passed through a 0.2-μm-average-pore-diam Teflon® filter (Millipore-Fluoropore, Bedford, Massachusetts). Samples were stored at 4°C before use.

Bioassay

The genetic toxicity of the extracted samples was measured with two microbial systems capable of detecting compounds that produced point mutations or increased lethal damage in DNA-repair-deficient strains. The *Salmonella*/microsome assay, as described by Ames et al. (1975), was used to measure the ability of a sample to revert strains of bacteria to histidine prototrophy. Samples were tested in two strains at a minimum of three dose levels with and without microsome activation from Aroclor-1254-induced rat liver from Litton Bionetics (Kensington, Maryland). The procedures for the S-9 mix and plate incorporation assay were the same as those used by Ames et al. (1975), except that overnight cultures were grown in 5 mL of Oxoid broth in a 125-mL Erlenmeyer flask. Positive controls and solvent and sterility controls were run. The control used to verify metabolic activation was 10-μg 2-acetylaminofluorene, whereas positive controls for strain TA98 and TA100 were 25-μg 2-nitrofluorine and 2-μg N-methyl-N′-nitro-N-nitrosoguanidine, respectively. Genetic characteristics of the *Salmonella* and *Bacillus* strains used in this study are given in Table V.

The *Bacillus subtilis* DNA repair assay was used to evaluate samples that produced a toxic response in *Salmonella.* The toxicity of the samples was compared in six strains of *Bacillus subtilis* (Table V). The procedures used were the same as in Felkner et al. (1979). Overnight cultures grown in Brain Heart Infusion Broth (Difco) were streaked radially from a centered sensitivity disk, and the agar plates were grown for 24 h at 37°C. After the incubation period, the toxicity of the sample was evaluated by comparing the zone of inhibition produced. Controls included ultraviolet (UV) sensitivity, 2-μg mitomycin C and 2-μL methylmethane sulfonate.

RESULTS AND DISCUSSION

Waste Characterization

Distribution of mutagenic activity in the acid, base, and neutral fractions of the refinery and petrochemical waste is given in Table VI.

Table V. Microbial Systems Used to Detect Genetic Damage

Bacteria	Strain	Genetic Event Detected	Mutation That Enhances Sensitivity[a]	Reference
S. typhimurium	TA1538	Frameshift mutation	hisD3052, rfa, △uvrB	Ames et al., 1975
	TA98	Frameshift mutation	hisD3052, rfa, △uvrB, pkM101	
	TA1535	Base-pair mutation	hisG46, rfa, △urvB	
	TA100	Base-pair mutation	hisG46, rfa, △urvB, pkM101	
B. subtilis	168 wt	Inhibition of DNA repair	Wild type	Felkner et
	rec A8	Inhibition of DNA repair	rec−, DRC−, trans−	al., 1979
	rec E4	Inhibition of DNA repair	rec−	
	mc-1	Inhibition of DNA repair	mitomycin C sensitive	
	hcr-9	Inhibition of DNA repair	△uvrB	
	fh2006-7	Inhibition of DNA repair	△uvrB, rec−	

[a]hisD3052, hisG46: mutation in histidine operon; rfa: deep rough characteristic—mutants lack lipopolysaccharide cell wall, enhances permeability; △uvrB: deficient in excision repair, sensitive to UV light; pkM101: contain R factor, plasmid which enhances sensitivity; rec−: deficient in recombinant repair; DRC−: blocked in donor-recipient complex formation; trans−: deficient in transformation and transduction.

Additional information is presented on the activity of crude oil (Epler et al., 1978) and cigarette smoke condensate (Kier et al., 1976) for comparison purposes. Cumulative mutagenic activity of both the refinery and the petrochemical waste appears to be greater than that of unrefined crude oil (437 revertants/mg) but considerably less than for cigarette smoke condensate (7820 revertants/mg). For the two wastes studied, the basic fraction of refinery waste induced the greatest response, producing more than 2000 revertants/mg residue. The mutagenic activity of the petrochemical waste was distributed more evenly among the acid, base, and neutral fractions; the greatest activity was detected in the acid fraction (498 revertants/mg).

Table VI. Distribution of Mutagenic Activity in Fractions of Several Complex Mixtures

Sample	Revertants/mg			Reference
	Acid	Base	Neutral	
Refinery Waste	813	2308	33	
Petrochemical Waste	498	278	246	
Crude Oil	10	277	150	Epler et al., 1978
Cigarette Smoke	5450	430	1940	Kier et al., 1976

Biodegradation

Table VII gives the mutagenic activity in (1) unamended soil; (2) soil with 5% waste that had not been incubated; and (3) soil with 5% waste and soil with 20% waste, both of which had been incubated for 180 d at 30°C. No activity was evident in any fractions from the unamended soils. At the highest application rate, the number of revertants induced by the saturate and aromatic fractions was reduced by 21 and 26%, respectively. The toxic potential of the saturate and aromatic fractions was reduced to a greater extent at the lowest application rate, although the dichloromethane fraction representing condensed ring compounds failed to exhibit a similar reduction at either application rate. Dibble and Bartha (1979) reported an increase in the concentration of residual condensed ring fractions after soil incubation. This increase, which may be caused by accumulation of metabolites from

Table VII. Effect of Soil Incubation for 180 d on Mutagenic Activity of Petrochemical Waste

Sample	Revertant/mg solvent extract		
	Petroleum Ether Extract	Petroleum Ether: Dichloro-methane	Dichloro-methane
Soil	0	0	0
5% Waste and Soil, Not Incubated	33	479	100
5% Waste and Soil, Incubated	24	107	112
20% Waste and Soil, Incubated	26	354	111

other fractions, could explain the increase in observed mutagenic activity in this fraction. With adequate time, the condensed ring fractions should also degrade, reducing the activity of the residual materials.

When the mutagenic activity of incubated and nonincubated soil is expressed as revertants per gram of soil, the reduction produced by chemical and biological activity becomes more evident. Results in Table VIII show a significant reduction in mutagenic activity of both the saturate and aromatic fractions. As discussed earlier, the dichloromethane extract did not show a reduction in mutagenic activity for 6 months of incubation.

Table VIII. Mutagenic Activity Extracted from Soil and Waste-Amended Soil

	Revertant/g soil		
Sample	Petroleum Ether Extract	Petroleum Ether: Dichloromethane	Dichloromethane
Soil	0	0	0
5% Waste and Soil, Not Incubated	1.3	1.9	3.8
5% Waste and Soil, Incubated	0.3	1.8	3.9

The extraction efficiency of these procedures for hydrocarbons and mutagenic activity from soil are given in Table IX. These results indicate that less than 5% of the applied hydrocarbons remained in the soil after solvent extraction. The efficiency of the recovery of mutagenic activity from soil was significantly less than that for hydrocarbons. The dichloromethane extract of the soil-waste mixture recovered only 67% of the mutagenic activity extracted from the waste alone with dichloromethane. Whether the reduction was a result of soil binding or an artifact created by the solvent extraction is unknown. Work conducted in the future will attempt to better define the cause of this deficiency so that the extraction efficiency can be improved.

Table IX. Efficiency of Extraction of Hydrocarbons and Mutagenic Activity from Soil Extraction Efficiency (%)

Sample	Petroleum Ether Extract	Petroleum Ether: Dichloromethane	Dichloromethane
Hydrocarbon	99	98	95
Mutagenic Activity	85	95	67

Leachate and Runoff Water Samples

The mutation frequency of the leachate and runoff water collected from field lysimeters was calculated by using a method (Commoner, 1976) in which a mutation frequency >1 indicates a doubling of revertant colonies. The mutation frequency was calculated as

$$MF = \frac{E - C}{C_{avg}}$$

where E = number of revertant colonies on an experimental plate with the residue from 200 mL of water,
C = number of revertant colonies on the solvent control plates from the same day with the same tissue preparation,
C_{avg} = overall average of revertant colonies for solvent control plate for the entire study

When using this scheme, a sample that induces a mutation frequency between 1 and 1.5 is considered to have a borderline mutagenic potential, whereas a compound that induces a mutation frequency >1.5 is considered to be a positive mutagen.

Analysis of water collected from the refinery-waste-amended soil (Table X) indicates that the leachate samples having the greatest mutagenic activity were collected within 2 months after the first waste application. These data represent a summary of the mutation frequencies calculated from dose-response curves where the leachate or runoff water was tested at a minimum of three dose levels. The first leachate water, collected from the Bastrop clay 47 d after application and from the Norwood soil 17 d after application, induced mutation frequencies of 6.7 and 5.1, respectively. All but one of the leachate samples collected later induced a mutation frequency ≤1.5. None of the water samples collected from unamended soils induced a significant increase in the number of revertants at the dose levels of water concentrate tested.

Although the runoff water samples collected from the Bastrop lysimeters exhibited a reduction in toxic potential with time, the potential remained significantly above background levels. Runoff samples collected from the Bastrop soil 40 and 145 d after the second waste application induced mutation frequencies of 9.0 and 4.2, respectively.

After the second waste application on Lakeland sand, eight runoff samples were collected over 138 d. Five of the first six samples pro-

Table X. Mutation Frequency of Leachate and Runoff Water Collected from Refinery Waste Amended Soils (waste was applied on days 0, 202, and 403)

	Time (d)										
	0	**17**	**47**	**67**	**155**	**228**	**250**	**260**	**275**	**293**	**318**
Leachate											
Bastrop			6.7	0.9	0.2	0.4		0.3		0.2	
Norwood		5.1	0.7								
Nacogdoches								0.5		0.3	
Unamended soil			0.5	0	0.1			0.1		0.8	0.1
Runoff											
Bastrop				1.4			9.0				
Lakeland								1.6	0.8	1.9	5.3
Unamended soil						0	0.3		0.2		

	Time (d)											
	322	**330**	**335**	**345**	**355**	**375**	**388**	**395**	**410**	**420**	**465**	**475**
Leachate												
Bastrop					0.6		1.6		0.1	1.1		
Norwood		0.1				0.7				1.5		0.1
Nacogdoches		0.3				0.1			1.5	0.2	0.1	1.4
Unamended soil									0.12	0.8		
Runoff												
Bastrop			0.3	0.4	4.2			2.6	2.7			
Lakeland	1.8	2.5	0.3	0.3								
Unamended soil												

duced a clear, positive response in the mutagenicity assay. The dose-response curves for these samples, presented in Figure 3, indicate that samples collected 131 and 138 d after waste application failed to induce a doubling of revertant colonies. Thus, the toxic potential of runoff water from refinery-waste-amended Lakeland sand appears to have returned to background levels within 5 months after the second waste application. The dose-response curves for runoff water collected from unamended soil (Figure 4) indicate that, even at higher water concentrate dose levels, there was no mutagenic response.

Analysis results of the leachate and runoff water collected from petrochemical waste–amended soils are presented in Table XI. Only three of the leachate samples collected induced a positive response in the mutagenicity assay. Two of these were collected from the Norwood soil 38 and 52 d after the first and third waste application respectively. One

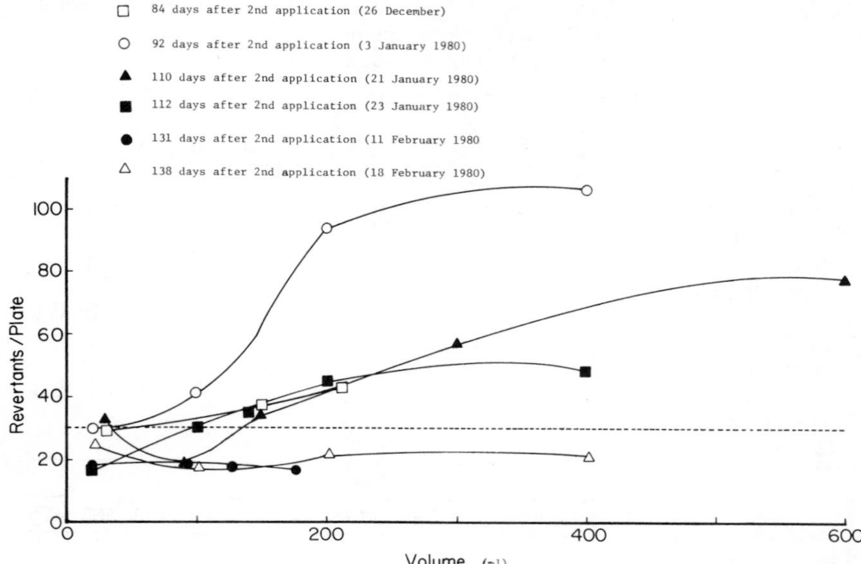

Figure 3. Mutagenicity, as measured with S. *typhimurium* strain TA98 without enzyme activation of runoff water from refinery waste–amended Lakeland sand.

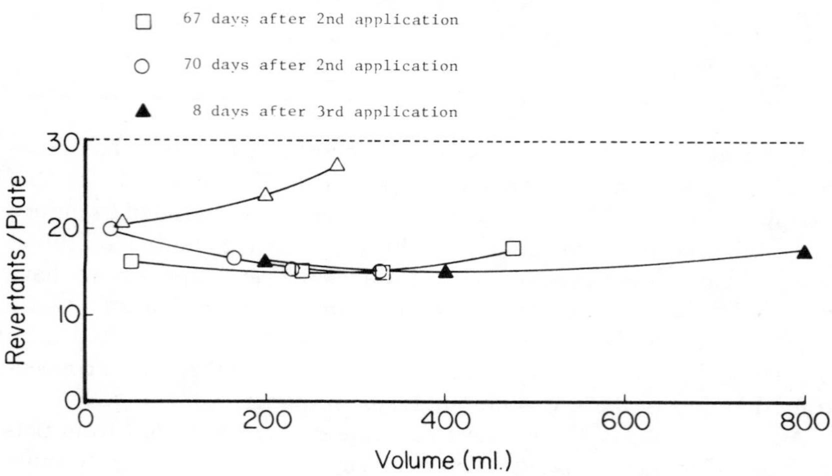

Figure 4. Mutagenicity, as measured with S. *typhimurium* TA98, of runoff water from unamended soils. Line of significance (---) is equal to twice the number of spontaneous revertants.

Table XI. Mutation Frequency of Leachate and Runoff Water Collected from Petrochemical Waste–Amended Soils (waste was applied on days 0, 202, and 403)

	Time (d)														
	0	38	53	60	70	148	153	183	190	202	245	255	263	268	280
Leachate															
Bastrop		0.3	1.4		0.7	1.2	0	0.3							
Nacogdoches													1.2	1.3	
Norwood		5.8			1.1										
Unamended soil			0.5		0		0.1					0.1			
Runoff															
Bastrop		14.3		0.5								1.1			
Norwood											1.0				1.3
Lakeland				1.1					1.5						0.7
Unamended soil												0.3			0.2

	Time (d)													
	295	305	315	325	332	340	345	383	410	420	430	455	468	495
Leachate														
Bastrop										0.5				
Nacogdoches		0.8						3.3		0.9	1.0	0.7		
Norwood									0.6		0.1	3.1	1.5	0.5
Unamended soil	0.8			0.1					0.2	0.8				
Runoff														
Bastrop			8.4								0.6			
Nacogdoches	0.7		1.5	0.7	1.6		0.9							
Norwood	1.8		4.2	2.8	0.6	0.7	0.9							
Unamended soil														

sample, collected more than 6 months after the second waste application from the Nacogdoches soils, induced a mutation frequency of 3.3. No leachate samples collected from the Bastrop soil induced a mutation frequency >1.5.

Analysis of runoff water from petrochemical waste-amended soil indicates that large quantities of mutagenic compounds are released at the soil surface. Runoff water collected from the Bastrop soil 38 d after the first waste application and 113 d after the second waste application induced mutation frequencies of 14.3 and 8.4, respectively. One runoff sample from the Norwood soil and two from the Lakeland sand, all of which were collected during the winter, had mutation frequencies >1.5.

Results of the analysis of leachate and runoff water by the *Bacillus subtilis* DNA repair assay are given in Table XII. These results indicate that a negative correlation exists between mutagenicity in *Salmonella* and toxicity to *Bacillus subtilis*. No runoff samples that were mutagenic were also toxic in the DNA repair assay, whereas most samples that were toxic to *S. typhimurium* at higher dose levels produced increased lethal damage in repair-deficient strains of *B. subtilis*. A runoff sample from petrochemical waste-amended soil collected 130 d after the second waste application from the Lakeland sand produced a clear positive response in the DNA repair assay and was not mutagenic in the *Salmonella* assay. Thus, these results indicate that the DNA repair assay complements the mutagenicity assay by responding to mutagenic compounds not detected by a *Salmonella* assay and by detecting a second type of genetic damage.

CONCLUSIONS

Results obtained from testing with two petroleum sludges indicate that short-term bioassays can be a valuable tool for evaluating the mutagenic activity of a hazardous waste. Analysis of the refinery and petrochemical waste indicates that they are more mutagenic than crude oil and that they possess less mutagenic activity on a per-weight basis than cigarette smoke condensate.

These procedures also seem adaptable for determining the treatment potential of a hazardous waste. Although mutagenic materials can be extracted from soil, methods for extraction will require further development to determine the extraction efficiency of various solvents. Determination of the extent of reduction of the hazardous characteristics of a land-applied waste is essential to the success of the land treatment method of waste disposal. Incubation studies revealed that

Table XII. Comparison of Lethal Effect of Water Extracts on DNA-Repair-Deficient and -Proficient Strains of *B. Subtilis*

Soil	Sample: Volume (mL eq)	Sample: Day	Inhibition Radius (mm): 168 wt	recA8	mc-1	recE4	hcr-9	fh2006-7	Response
Runoff Refinery									
Bastrop	400	335	3					3	−
		345	7	9	9	7	9	8	−
		355	6	5	6		6	4	−
Lakeland	400	275	0	0	0	0	0	0	−
		293	8	5	4	3	8	6	−
		318	3	5	5	3	4	4	−
Runoff Control									
Lakeland	400	315	4	3	4	4	4	4	−
Positive Control									
Methylmethane sulfonate	2 μL		8	33	26	21	17	20	+++
Runoff Petrochemical									
Bastrop	400	255	6	10	9		7	9	+
		315	8	7	10		9	5	±
		430	9	5	5	8	9	10	−
Lakeland	400	280	2	3	2	5	4	4	±
		295	5	6	4		5	5	−
		315	4	4	3	4	4	4	−
		325	5	6	5	5	5	7	−
		332	3	4	4	5	8	10	++
		340	1	1	1		1	1	−
Norwood	400	280	2	5	5	5	3	3	±
		295	4	5	4	5	5	4	−
		315	4	7	5	5	4	4	±
		325	4	4	3	8	5	6	±
		332	7	7	6	8	6	9	−
		345	5	3	4	5	8	10	++
		352	4	3	2		4	3	−

the mutagenic activity of waste added to soils decreased over a 6-month period.

Bioassays have also been effective for evaluating the level of mutagenic activity in the effluent from a land treatment facility. Only low levels of mutagenic activity were detected in a few samples of leachate water collected shortly after waste application. This result indicates that the soil retained most of the mutagenic constituents and prevented them from entering the groundwater. The runoff water collected during the year after waste application exhibited decreasing amounts of activity. The most active sample of runoff water was 14.3 times the background level. At this level, 3 L of water could contain the activity of one cigarette.

As these systems are further developed and as the testing protocol becomes better defined, the utility of short-term bioassays for detecting and characterizing mutagens and potential carcinogens will increase. Test systems that are used as part of a monitoring system should be selected on the basis of the types of compounds in the waste and the types of damage caused by these compounds. Additional work to be performed as part of this project will include complete characterization of three additional hazardous wastes and a comprehensive greenhouse and lysimeter study to determine the treatment potential and environmental fate of hazardous waste constituents. Although additional work is needed to define more clearly the advantages and limitations of short-term testing, preliminary results indicate that short-term tests can be a valuable tool for defining the mutagenic characteristics of a waste, evaluating their treatment potential, and evaluating their fate and mobility from land treatment disposal sites.

REFERENCES

Ames, B. N., J. McCann and E. Yamasaki. 1975. "Methods for Detecting Carcinogens and Mutagens with the *Salmonella*/Mammalian Microsome Mutagenicity Test," *Mutat. Res.* 31:347.

Anderson, D. 1975. "The Selection and Induction of 5-Iodo-2-deoxyuridine and Thymidine Variants of P 388 Mouse Lymphoma Cells with Agents Which Are Used for Selection," *Mutat. Res.* 33:399.

Arlett, C. F. 1977. "Mutagenicity Testing with V79 Chinese Hamster Cells," in *Handbook of Mutagenicity Test Procedures,* B. J. Kilbey, M. Legator, W. Nichols, and C. Ramel, Eds. (Amsterdam: Elsevier Biomedical Press), pp. 171-191.

Beek, B., G. Klein, and G. Obe. 1980. "The Fate of Chromosomal Aberrations in a Proliferating Cells System," *Biol. Zentralbl.* 99(1):73-84.

Bigani, M., G. Morpurgo, R. Pagliani, A. Carare, G. Conte, and G. DiGuideppe. 1974. "Non-disjunction and Crossing Over Induced by Pharmaceutical Drugs in *Aspergillus nidulans,*" *Mutat. Res.* 26:159.

Brown, K. W., and K. C. Donnelly. 1981. "Genetic Toxicity of Water Concentrates from the Effluent of a Waste Oil Storage Pond."

Brown, K. W., C. J. Gerard, B. W. Hipp, and J. T. Ritchie. 1974. "A Procedure for Placing Large Undisturbed Monoliths in Lysimeters," *Soil Sci. Soc. Am.* 38:981-983.

Brusick, D. J. 1972. "Induction of Cyclohexamide Resistant Mutants in *Saccharomyces cervisiae* with N-Methyl-N-nitro-N-nitrosoguanine and ICR-170," *J. Bacteriol.* 109:1134.

Carere, A., G. Morpurgo, G. Cardamone, M. Bignami, F. Aulicino, G. DiGuisepie, and C. Conti. 1975. "Point Mutations Induced by Pharmaceutical Drugs," *Mutat. Res.* 29:235.

Clive, D. W. 1973. "Recent Developments with the L5178Y TK Heterozygote Mutagen Assay System," *Environ. Health Persp.* 6:119.

Clive, D. W., and J. F. S. Spector. 1975. "Laboratory Procedure for Assessing Specific Locus Mutations at the TK Locus in Cultured L5178Y Mouse Lymphoma Cells," *Mutat. Res.* 31:17.

Clive, D. W., W. G. Flamm, M. R. Machesko, and N. H. Bernheim. 1972. "Mutational Assay System Using the Thymidine Kinase Locus in Mouse Lymphoma Cells," *Mutat. Res.* 16:77.

Commoner, B. 1976. "Reliability of Bacterial Mutagenesis Techniques to Distinguish Carcinogenic and Noncarcinogenic Chemicals," EPA-600/1 76-022.

Dean, B. J., and K. R. Senner. 1977. "Detection of Chemically Induced Mutation in Chinese Hamsters," *Mutat. Res.* 46:403.

Deserres, F. J., and H. V. Malling. 1971. "Measurement of Recessive Lethal Damage over the Entire Genome and at Two Specific loci in the ad-3 Region of a Two Component Heterokaryon of *Neurosporo crassa,*" in *Chemical Mutagens, Principles and Methods for Their Detection, Vol. 2,* A. Hollaender, Ed., (New York: Plenum Publishing Corporation), pp. 311-341.

Dibble, J. T., and R. Bartha. 1979. "Effect of Environmental Parameters on the Biodegradation of Oil Sludge," *Appl. Environ. Microbiol.* 37:729-738.

Ehrenburg, L. 1971. "Higher Plants," in *Chemical Mutagens, Principles and Methods for Their Detection, Vol. 2,* A. Hollaender, Ed. (New York: Plenum Publishing Corporation), pp. 365-386.

Epler, J. L., J. A. Young, A. A. Hardigree, T. K. Rao, M. R. Guerin, I. B. Rubin, C. H. Ho, and B. R. Clark. 1978. "Analytical and Biological Analyses of Test Materials from the Synthetic Fuel Technologies," *Mutat. Res.* 57:265-276.

Evans, H. J., and M. L. O'Riordan. 1975. "Human Peripheral Blood Lymphocytes for Analysis of Chromosome Aberrations in Mutagen Tests," *Mutat. Res.* 31:135.

Felkner, I. C., K. M. Hoffman, and B. C. Wells. 1979. "DNA-Damaging and Mutagenic Effect of 1,2-Dimethylhydrazine on *Bacillus subtilis* Repair-Deficient Mutants," *Mutat. Res.* 28:31-40.

Green, M. H. L., W. J. Muriel, and B. A. Bridges. 1976. "Use of a Simplified Fluctuation Test to Detect Low Levels of Mutagagens," *Mutat. Res.* 38:33.

Hooper, K., C. Gold, and B. N. Ames. 1978. "Development of Methods for Mutagenicity Testing of Wastewater and Drinking Water Samples," report to Water Resources Control Board, State of California, Sacramento, CA.

Jacobs, L., and R. DeMars. 1977. "Chemical Mutagenesis with Diploid Human Fibroblasts," in *Handbook of Mutagenicity Test Procedures,* B. J. Kilbey, M. Legator, W. Nichols, and C. Ramel, Eds. (Amsterdam: Elsevier Biomedical Press), pp. 193-220.

Kada, T., M. Morija, and Y. Shirasu. 1974. "Screening of Pesticides for DNA Interactions by REC-Assay and Mutagenic Testing and Frameshift Mutagens Detected," *Mutat. Res.* 26:243.

Kier, L. D., E. Yamasaki, and B. N. Ames. 1974. "Detection of Mutagenic Activity in Cigarette Smoke Condensates," *Proc. Nat. Acad. Sci., U.S.* 71(10):4159-4163.

Kihlman, B. A. 1977. "Root Tips of *Vicia fava* for the Study of the Induction of Chromosomal Aberrations," in *Handbook of Mutagenicity Test Procedures,* B. J. Kilbey, M. Legator, W. Nichols and C. Ramel, Eds. (Amsterdam: Elsevier Biomedical Press), pp. 389-410.

Kumar, R., and S. Chauhan. 1979. "Frequency and Spectrum of Chlorophyll Mutations in a 6-Rowed Barley *Hordeum-vulgare,*" *Indian J. Agri. Sci.* 49(11):831-834.

Loprieno, N., R. Barale, C. Bauer, S. Baroncelli, G. Bronzetti, A. Cammellini, A. Cinci, G. Corsi, C. Leporini, R. Nieri, M. Mozzolini, and C. Serra. 1974. "The Use of Different Test Systems with Yeasts for the Evaluation of Chemically Induced Gene Conversions and Gene Mutations," *Mutat. Res.* 25:197.

Marimuthu, K. M., A. H. Sparrow, and L. A. Schairer. 1970. "The Cytological Effects of Space Flight Factors, Vibration, Clinostate, and Radiation on Root Tip Cells of *Tradescantia* and *Allium cepa,*" *Radiat. Res.* 42:105.

McCann, J., E. Choi, E. Yamasaki, and B. N. Ames. 1975. "Detection of Carcinogens as Mutagens in *Salmonella*/Microsome Test: Assay of 300 Chemicals," *Proc. Nat. Acad. Sci., U.S.* 72:5135-5139.

Mohn, G., J. Eltenberger, and D. B. McGregor. 1974. "Development of Mutagenicity Tests Using *Escherichia coli* K-12 as an Indicator Organisms," *Mutat. Res.* 23:187.

Mortimer, R. K., and T. R. Manney. 1971. "Mutation Induction in Yeast," in *Chemical Mutagens, Principles and Methods for Their Detection, Vol. 1,* A. Hollaender, Ed. (New York: Plenum Publishing Corporation), pp. 289-310.

Nauman, C. H., A. H. Sparrow, and L. A. Schairer. 1976. "Comparative Effects of Ionizing Radiation and Two Gaseous Chemical Mutagens on Somatic Mutation Induction in One Mutable and Two Non-mutable Clones of *Tradescantia*," *Mutat. Res.* 38:53-70.

Neill, J. P., P. A. Brimer, R. Machanoff, G. P. Hirseh, and A. W. Hsie. 1977. "A Quantitative Assay of Mutation Induction of the HGPRT Locus in Chinese Hamster Ovary Cells: Development and Definition of the System," *Mutat. Res.* 45:91.

Nicoloff, H., K. I. Gecheff, and L. Stoilov. 1979. "Effects of Caffeine on the Frequencies and Location of Chemically Induced Chromatid Aberrations in Barley," *Mutat. Res.* 70:193-201.

Nilan, R. A., and B. K. Vig. 1976. "Plant Test Systems for Detection of Chemical Mutagens," in *Chemical Mutagens, Principles and Methods for Their Detection, Vol. 1,* A. Hollaender, Ed. (New York: Plenum Publishing Corporation), pp. 143-170.

Ong, T. N. 1978. "Use of the Spot, Plate and Suspension Test Systems for the Detection of the Mutagenicity of Environmental Agents and Chemical Carcinogens in *Neurospora crassa*," *Mutat. Res.* 54:121-129.

Parry, J. M. 1977. "The Use of Yeast Cultures for the Detection of Environmental Mutagens Using a Fluctuation Test," *Mutat. Res.* 46:165.

Perry, P., and H. J. Evans. 1975. "Cytological Detection of Mutagen-Carcinogen Exposure by Sister Chromatid Exchange," *Nature* 258:121.

Plewa, M. J., and J. M. Gentile. 1976. "The Mutagenicity of Atrazine: A Maize-Microbe Bioassay," *Mutat. Res.* 38:287-292.

Purchase, I. F. H., E. Longstaff, J. Ashby, J. A. Styles, D. Anderson, P. A. Lefevre, and F. R. Westwood. 1976. "Evaluation of Six Short-Term Tests for Detecting Organic Chemical Carcinogens and Recommendations for Their Use," *Nature* 264:624-627.

Rappaport, S. M., M. G. Richard, M. C. Hollstein, and R. E. Talcott. 1979. "Mutagenic Activity in Organic Wastewater Concentrates," *Environ. Sci. Technol.* 13(8):957-961.

Redei, G. P. 1975. "*Arabadopsis* as a Genetic Tool," *Ann. Rev. Genet.* 9: 111-125.

Rinkus, S. J., and M. S. Legator. 1979. "Chemical Characterization of 465 Known or Suspected Carcinoges and Their Mutagenic Activity in the *Salmonella typhimurium* System," *Cancer Res.* 39(9):3289-3804.

Roper, J. A. 1971. "*Aspergillus*," in *Chemical Mutagens, Principles and Methods for Their Detection,* A. Hollaender, Ed. (New York: Plenum Publishing Corporation).

Scott, B. R., A. H. Sparrow, S. S. Lamm, and L. Schairer. 1978. "Plant Metabolic Activation of EDB to a Mutagen of Greater Potency," *Mutat. Res.* 49:203-212.

Scott, B. R., E. Kafer, G. L. Dorn, and R. Stafford. 1982. "*Asperigillus nidulans:* Systems and Results of Test for Induction of Mutation and Mitotic Segregation," *Mutat. Res.* 98:49-94.

Shiau, S. Y., R. A. Huff, B. C. Wells, and I. C. Felkner. 1980. "Mutagenicity and DNA-Damaging Activity for Several Pesticides Tested with *Bacillus* Mutants," *Mutat. Res.* 71:169-179.

Skopek, T. R., J. L. Liber, J. J. Krowleski, and W. G. Thilly. 1978. "Quantitative Forward Mutation Assay in *Salamonella typhimurium* Using 8-Azaguanine Resistance as a Genetic Marker," *Proc. Nat. Acad. Sci. U.S.* 75:410.

Slater, E., M. D. Anderson, and H. S. Rosenkranz. 1971. "Rapid Detection of Mutagens and Carcinogens," *Cancer Res.* 31:970.

Soderberg, K., J. T. Mascarello, G. Breen, and I. E. Scheffler. 1979. "Respiration Deficient Chinese Hamster Cell Mutants Genetic Characterization," *Somatic Cell Genet.* 5(2):225-240.

Sparrow, A. H., L. A. Schairer, and R. Villalobos-Pietrini. 1974. "Comparison of Somatic Mutation Rates Included in *Tradescantia* by Chemical and Physical Mutagens," *Mutat. Res.* 26:265-276.

Speck, W. T., R. M. Santella, and H. S. Rosenkranz. 1978. "An Evaluation of the Prophage Induction (Inductest) for the Detection of Potential Carcinogens," *Mutat. Res.* 54:101.

Stetka, D. G., and S. Wolff. 1976. "Sister Chromatid Exchange as an Assay for Genetic Damage Induced by Mutagen-Carcinogens," *Mutat. Res.* 41:333.

Tanooka, H., N. Munakata, and S. Kitahara. 1978. "Mutation Induction with UV- and X-Radiation in Spores and Vegetative Cells of *Bacillus Subtilis*," *Mutat. Res.* 49:179-186.

Thilly, W. G., J. G. DeLuca, I. V. Hoppe, and B. W. Penmann. 1976. "Mutation of Human Lymphoblasts by Methylnitrosourea," *Chem.-Biol. Interact.* 15:33.

Tomlinson, C. R. 1980. "Effects of pH on the Mutagenicity of Sodium Azide in *Neurospora crassa* and *Salmonella typhimurium*," *Mutat. Res.* 70(2): 179-192.

Underbrink, A. G., L. A. Schairer, and A. H. Sparrow. 1973. "*Tradescantia* Stamen Hairs: A Radiobiological Test System Applicable to Chemical Mutagenesis," in *Chemical Mutagens, Principles and Methods for Their Detection, Vol. 3,* A Hollaender, Ed. (New York: Plenum Publication Corporation), pp. 171-207.

Vig, B. K. 1975. "Soybean (*Glycine max*): A New Test System for Study of Genetic Parameters as Affected by Environmental Mutagens," *Mutat. Res.* 31:49-56.

Von Borstel, R. C., and R. H. Smith. 1977. "Measuring Dominant Lethality in Haborbracon," in *Handbook of Mutagenicity Testing Procedures,* B. J. Kilbey, M. Legator, W. Nichols, and C. Ramel, Eds. (Amsterdam: Elsevier Biomedical Press), pp. 375-387.

Wurgler, F. E., and E. Vogel. 1977. "*Drosophila* as an Assay System for Detecting Genetic Changes," in *Handbook of Mutagenicity Test Procedures,* B. J. Kilbey, M. Legator, W. Nichols, and C. Ramel, Eds. (Amsterdam: Elsevier Biomedical Press), pp. 335-373.

DISCUSSION

***Ralph Franklin*, DOE:** Is this the kind of disposal strategy that you had in mind when you were commenting on the value of a solid waste extraction procedure vs a total analysis?

K. Brown: I think that if we are going to characterize waste and design a land treatment facility, the first thing we need is a total analysis. We need to know how much copper and how much generic material is in the waste.

Ralph Franklin: You have to recognize that you are looking at a totally different disposal strategy, and I couldn't agree with you more. Your strategy is to try to recycle the material in a more-or-less natural way back into the environment, and I would agree with you that an extraction procedure type test is going to give you almost meaningless information. Clearly, what is called for is some kind of test that would involve an incubation. The important questions are what is the capacity of the soil to accommodate this material, and what would allow it to be recycled into the environment in an acceptable way. What is going to be mobilized although its importance becomes much less significant. On the other hand, the intent of the other disposal strategy, where you are putting it into a landfill in a concentrated form, is basically isolation of the waste from the environment; some kind of test for mobilization would be extremely important. My point is, you have to recognize there are two completely different disposal strategies and each calls for different kinds of tests in evaluating the hazard and risk.

K. Brown: Yes, I can appreciate that. I would contend though that we really need to know what we are putting in a landfill. In fact, the components of the waste that go into hazardous waste landfills may adversely affect the permeability of the clay liner.

***T. Tamura*, ORNL:** Could you tell me what you think is the possibility of shortening the time so that you can get it back to an innocuous level by identifying and thereby cultivating the microbes that attack your pollutant?

K. Brown: There has been very limited work done on that. I think the best thing that we could go to is the situation in which some people have looked at cultivating microbes for cleaning up high concentrations of pesticides that were spilled in the soil, and they found they could speed up the system. However, I think that for many of the wastes that we have dealt with there has been such a diversity of compounds that at this point I do not see that as a possibility. However, we have taken extracts from incubated soils with wastes and applied those extracts as a wetting medium to a fresh soil. We have also applied a waste to a soil to see whether we could get increased microbial respiration, but we have not seen it. We have tried to backspike these things, but we haven't seen an advantage of backspiking them with the wastes we have looked at.

Harry Freeman: What do you recommend not putting on the land treatment sites? You said some classes could and some classes could not.

K. Brown: We would not want to see high concentrations of chlorinated hydrocarbons go on, nor would we want materials that contain radioactive wastes to go on the land. There are certain wastes that would not be very amenable, such as something that has a long half-life. Completely chlorinated PCB is something else that you would not want to be putting out there.

***Rufus Chaney*, USDA:** What do you think about our arguments about composting as a pretreatment for wastes that contain materials that will be toxic to the microbes and retard treatment in a land treatment setup?

K. Brown: I think there is a lot of potential for pretreatment of wastes, and composting may well be one of them. There are many other situations where chlorinated hydrocarbons could possibly be broken down by some pretreatment and make the metabolites more amenable to digestion by the soil microorganisms. I think there is a lot of potential in that area.

CHAPTER 37

SETTING PRIORITIES IN HAZARDOUS WASTE MANAGEMENT

Arthur H. Purcell

Technical Information Project
1345 Connecticut Avenue, NW, Suite 217
Washington, DC 20036

ABSTRACT

Management of hazardous wastes covers a wide spectrum of activities and disciplines: source reduction, materials handling, toxicity evaluation, transportation, use, disposal, storage, repositing and recycling. Each activity encompasses many considerations and requires strict priority setting to optimize the effectiveness of the activity in overall management programs. A few priority areas that are of immediate attention in the management of hazardous wastes are identified.

INTRODUCTION

Managing the flow of hazardous* wastes in our society includes a complex series of tasks, beginning with source control and continuing through final disposal and recycling. Each activity entails considerations in policy, administration, science and technology, and information flow areas and requires the integration of expertise from many disciplines in the private and public sectors.

As the problems of hazardous waste proliferation mount, developing

*"Hazardous" and "toxic" are used here interchangeably to underscore both the safety and health aspects of hazardous wastes.

and setting priority actions that will help minimize these problems have become increasingly urgent. Two distinct challenges must be met: to prevent or minimize (1) the actual flow of hazardous substances into the environment, and (2) the adverse biological impacts of those substances that do enter the environment as a result of both planned processes and accidental releases. The phrase "prevent or minimize" is used to indicate that, in many instances, the total prevention of toxic effluents of impacts may be possible, whereas in other cases toxics can only be controlled, not eliminated. Until recently, the prevailing belief was that the effects of few hazardous effluents could be prevented through any means other than direct elimination of production. Recent technological advances have, however, shown that the prevention of toxic effects is a viable and growing field.

TOXIC AND HAZARDOUS SUBSTANCES MANAGEMENT

Management of toxic and hazardous substances falls into four basic stages:

1. *Source flow:* The source of the flow of toxic substances is a function of both consumer and producer actions. Toxic substances are produced because of (1) demand for products that may entail the use of such substances in processing and manufacturing and (2) decisions at the production level to use toxic substances in manufacture of demanded goods. Changes in demand and production can alter the source flow. The magnitude of the source flow of toxic substances determines, in turn, the potential magnitude of toxic substance impacts all the way downstream—from worker exposure during manufacturing, through consumer exposure during use, to public exposure during and after disposal.
2. *Processing and manufacturing:* After decisions have been made to use toxic substances in processing and manufacturing, the possibility arises of damaging exposure to workers. Many management options exist for preventing or minimizing such exposure.
3. *Use:* Goods and products made with hazardous substances—ranging from aerosol sprays to industrial cleaning fluids—pose potential health threats to users. Effective management of hazardous and toxic materials becomes difficult at the use stage because of the wide dispersion of the substances among the population.
4. *Disposal:* Toxic substances enter the effluent stream as by-products and disposables in significant quantities at both the processing and

manufacturing and the use stages. The scandals of Love Canal and Valley of the Drums are two dramatic illustrations of the adverse impacts of inadequate hazardous waste management (HWM) practices at the disposal stage.

For the past few years we have been engaged in a difficult and expensive struggle to upgrade overall HWM. Most efforts have been concentrated on the fourth HWM stage—disposal. However, all four stages entail high HWM priorities.

Source Flow and Processing and Manufacturing

Coordination of Toxic and Nontoxic Materials Management Efforts

A major problem in controlling the flow of toxic substances is that the intimate relationship between the problems of toxic and nontoxic materials flow is not recognized. As a result, little coordination exists between overall resource conservation efforts and the activities aimed at minimizing the flow of toxic substances. For example, little effort is made to demonstrate to policymakers the link between reduction and recycling of overall materials flow and the flow of toxic materials. Consumers have little idea of the toxic substances that are produced in the manufacture of the products they buy and, hence, have little power to buy the "least toxic" product.

The federal government has established a large-scale hazardous waste program but has effectively downgraded its nonhazardous waste program to a minimal level, despite the fact that much of the "nonhazardous" waste generated in the postconsumer stream has toxic and hazardous components upstream. The government also has a large-scale toxic substances program aimed at minimizing the threats of toxic substances that enter the marketplace. To date, however, there has been little substantive communication and coordination between these two significant programs.

A top priority in HWM, at both the government and private sector level, is to integrate materials flow programs so that the opportunities to minimize or eliminate the toxic and hazardous components of materials flows can be enhanced.

Source Reduction

The most obvious way to minimize the risks of exposure to toxic substances is to minimize the physical flow of these substances. Source reduction can be accomplished through several strategies, which have varying economic impacts on producers. Across-the-board cutback or elimination, which generally comes to mind when source reduction is discussed, is only one of six major materials conservation strategies (Purcell, 1980). Alternative technologies, using nontoxic materials, improved design, recycling, and cooperation from consumers in maximizing materials life are all viable mechanisms for reducing the magnitude of the source flow of toxic materials.

Source reduction must become a top priority in our efforts to manage toxic and hazardous materials. The fewer toxics we release into the environment, the less risk they pose and the less it will cost to dispose of them. The field of source reduction, however, is in its infancy. Incentives for producers to radically change processing designs to reduce the quantity and harmful quality of toxic materials are lacking.

To date, few substantive efforts in HWM have been geared toward genuine source reduction. This is particularly true at the federal regulatory level. The Resource Conservation and Recovery Act (RCRA) and Superfund, both multibillion-dollar hazardous and toxic substances management programs, are directed almost solely toward disposal and destruction of toxic substances released as waste. The Toxic Substances Control Act (TSCA), administered through EPA, seeks to characterize substances in terms of their health-damaging potential and prevent manufacture or to strictly control handling of certain substances. However, TSCA is not concerned, per se, with questions of how to minimize the flow of toxic substances or how to develop nontoxic alternatives.

The costs of following TSCA and RCRA regulations provide some indirect incentives, of course, for source reduction, but direct incentives are needed. Such incentives can be established through industry and government efforts aimed at upgrading source reduction priorities.

A few industries have initiated active source reduction programs. For example, 3M has created a photographic developing process that uses water instead of toxic chemicals. To demonstrate the nontoxicity of the process, 3M demonstrated it in a fish tank at the Shedd Aquarium in Chicago. Also, AT&T has developed dry processing techniques to replace wet processes in microcircuitry printing. The result has been a significant reduction in hazardous manufacturing effluents. Arm and Hammer baking soda, a widely distributed household product, is packaged in recycled cardboard, with a resulting reduction in hazardous effluents of virgin board manufacture. These types of source reduction concepts can be greatly expanded by other industries.

Use

Consumers of goods and products made with toxic substances have certain rights that must be recognized in the overall management of toxic and hazardous substances. These include the right to know the toxicity of processes involved in developing products and the implications of waste generation entailed in product development and disposal. This right is controversial and even borders on such difficult subjects as a producer's right to trade secrecy; consequently, it has been given a very low priority. Consumers cannot effectively influence the source flow of hazardous substances, however, unless they know the full "hazardous history" of products. Most goods and products on the marketplace—whether or not they contain hazardous substances—entail the dispersion of hazardous materials somewhere in the processing chain. Through development of a system of environmental health product codes, it would be possible for consumers to know with some certainty the total hazardous implications of products available to them and to have the opportunity to buy the "least hazardous" product.

Disposal

Tens of millions of tons of toxic and hazardous substances enter the environment every year as unwanted wastes. Until recently, no standards or uniform regulations existed for disposing of these wastes. As a consequence, critical situations such as Love Canal have built up over the years. Considerable attention and money, particularly at the federal level, have been directed toward achieving safe disposal of hazardous wastes.

Siting

Very high priority has been placed on the difficult task of siting hazardous waste disposal facilities. Geological and political constraints have made finding and developing adequate repositories for our growing hazardous waste extremely difficult. Siting is a complex problem, involving public education and political maneuvering as well as sophisticated engineering. Many similarities exist between the problems of siting hazardous waste facilities and those of siting nuclear waste repositories. In the latter case, however, priorities dictate an overriding federal role in siting. No such agreement has been achieved with hazardous waste; like nonhazardous municipal waste, hazardous waste disposal is still generally considered a local matter.

To ease siting problems, greater attention must be given to methods that will reduce the volume of wastes. The first priority, as discussed above, is source reduction. The second is the physical reduction of waste volumes, which can be achieved through alternative disposal systems that do not require long-term burial and recycling hazardous wastes.

Several alternatives to burial of hazardous wastes have been developed (Powers, 1976). These range from incineration to landfarming to biological treatment. Although estimates vary on how much of the total hazardous waste stream could be treated through these systems, it is becoming increasingly clear that more attention must be given to these alternatives.

Monitoring

Recycling of hazardous wastes traditionally has been a low-priority item, both in industry and government programs. As a recycler complained at an EPA-sponsored meeting on hazardous waste last year, "The government spends millions of dollars aimed at destroying materials and virtually nothing on reusing them" (EPA, 1980). As with nonhazardous wastes, some hazardous effluents have been considered unrecyclable, mainly because industrial processes have not been designed with hazardous waste recycling in mind. Therefore, hazardous effluent streams are generally complex, heterogeneous chemical mixes.

One system of incentives developed by the private sector, which has been receiving growing support at the state government level, is the waste exchange. A waste exchange is simply a listing of waste materials and substances, often hazardous, generated by exchange members and available for sale or trade. Waste exchanges are one way to promote hazardous waste recycling.

The so-called product charge system that has been proposed to promote recycling of basic materials could also be applied to recycling of hazardous materials. In such a system, products that are made only from virgin materials are taxed on a weight or unit basis, whereas those using recycled materials are exempted. Developing incentives for improved hazardous waste recycling is a clear priority in overall HWM that must be upgraded.

Compensation

The third priority category in the disposal stage of toxic and hazardous materials is compensation. Establishment of equitable compensation mechanisms for damage from these materials becomes increasingly difficult as they are dispersed further into the human environment. The hardships suffered by some of the Love Canal victims, however, and the possibility that more victims of many more Love Canals may emerge, make the high-priority nature of development of such systems evident.

The problems of compensation can, however, be substantially minimized or even eliminated through development of comprehensive emergency prepardedness and response programs. These programs must have two essential objectives: (1) containing hazardous waste effluent flow into the ambient environment and (2) rapidly and efficiently moving people and other endangered organisms away from areas where hazardous materials threaten to harm them. In line with assigning emergency preparedness and response a high priority in hazardous waste management, the EPA renamed its Office of Solid Waste, the Office of Solid Waste and Emergency Response. Also, the establishment of the Superfund Office, as mandated by Congress, is an indication that emergency planning and response—the main function of Superfund—is gaining priority in hazardous waste management.

CONCLUSION

Priority, by its very definition, connotes an urgency. The fact that several high-priority areas characterize the management of hazardous wastes is indicative of the set of urgencies that surround our hazardous and toxic waste problems. The priority areas we have outlined will require cooperation and interaction among several disciplines if they are to be met. For example, source reduction is as much a management and policy problem as a technological one. Compensation, at first glance, seems to be of only a legal nature, but it actually requires significant scientific input to have a meaningful basis, and competent administrative effort to put systems into practice. Similarly, other priority areas are of equally interdisciplinary natures.

With resources limited and time running short, it is becoming ever clearer that these few areas must receive greater attention in programs to manage our toxic and hazardous substances. None of these areas is small or simple to tackle, but with proper efforts, each can be more effectively addressed.

REFERENCES

EPA. 1980. "Waste Alert," unpublished proceedings, Region III Conference.

Powers, P. W. 1976. *How to Dispose of Toxic Substances and Industrial Wastes* (Park Ridge, NJ: Noyes Data Corp.).

Purcell, A. H. 1980. *The Waste Watchers* (New York: Anchor Press/Doubleday).

DISCUSSION

***Jens Hansen*, Aalborg University, Denmark:** I appreciate this talk on source management. I think it's really a strong point that wasn't made too well before. I would comment in two respects. One is that I think that we are not really creating hazardous waste anymore. We do have solvents as left-overs, and as long as we keep them as solvents in the containers where they used to be, it's not a hazardous waste problem. It's a problem of taking care of some chemicals. That can be done and is done. We do have hazardous waste problems, and they are rather severe, but these hazardous waste problems are all the drums and barrels and everything else we have buried in the ground throughout the country. That was done a long time ago because we didn't really pay attention to the fact that if dumped in the wrong place it would create a hazard in the environment. Along these lines, I think a lot of the so-called hazardous waste products could be taken care of.

INDEX

Fourth Annual Life Sciences Symposium

Attendee List

Abbott, D.T.
Monsanto Research Corp.
Mound Facility
P.O. Box 32
Miamisburg, OH 45342

Albert, R.E.
Department of Environmental Medicine
New York University Medical Center
New York, NY 10016

Anderson, C.H.
U.S. Environmental Protection Agency
505 Cedar Creek Drive
Athens, GA 30605

Auerbach, S.I.
Environmental Sciences Div.
Oak Ridge National Laboratory
Oak Ridge, TN 37830

Bain, G.L.
Rt. 4, P.O. Box 226
Asheboro, NC 27203

Basinski, R.R.
J&L Steel
900 Agnew Road
Pittsburgh, PA 15229

Blanco, R.E.
Oak Ridge National Laboratory
P.O. Box X
Oak Ridge, TN 37830

Blaunstein, R.
U.S. Department of Energy
Washington, DC 20545

Boegly, W.J., Jr.
Environmental Sciences Division
Oak Ridge National Laboratory
Oak Ridge, TN 37830

Bond, A.D.
EG&G–No. 13
1343 Headlee Street
Morgantown, WV 26505

Boothby, A.H.
Mineral By-Products, Inc.
6370 Hillside Drive
Cincinnati, OH 45233

Boyenga, D.B.
Mineral By-Products, Inc.
8070 Condor Court
Centerville, OH 45459

Brown, D.K.
Environmental Sciences Division
Oak Ridge National Laboratory
Oak Ridge, TN 37830

*Brown, J.
Scientific Services Department
Midlands Region of Central Electricity Generating Board
Nottingham, ENGLAND

*Brown, K.W.
Soils and Crop Sciences Dept.
Texas A&M University
College Station, TX 77843

Burchard, J.K.
Research Triangle Institute
P.O. Box 12194
Research Triangle Park, NC 27709

Burgess, P.D.
Stone & Webster Engineering Corp.
245 Summer Street
Boston, MA 02107

Carden, D.M.
Tennessee Valley Authority
Forestry Building
Norris, TN 37828

Carter, L.
Soil & Material Engineers
Marjan Drive
Atlanta, Georgia

*Chaney, R.L.
USDA-SEA-AR
Room 124, Building 007
BARC-West
Beltsville, MD 20705

Cherry, J.A.
Dept. of Earth Sciences
University of Waterloo
Waterloo, Ontario
Canada N2L 3G1

Cho, P.
U.S. Department of Energy
ER-73
Washington, DC 20545

Cope, D.R.
International Energy Agency
14-15 Lower Grosvenor Place
London, ENGLAND SWIW OEX

Cordle, S.R.
Office of Environment Processes and Effects Research
RD-682
Environmental Protection Agency
Washington, DC 20460

Cowherd, D.C.
Bowser-Morner
420 Davis Avenue
Dayton, OH 45401
Cowser, K.E.
Oak Ridge National Laboratory
P.O. Box X
Oak Ridge, TN 37830
Dahlberg, M.
U.S. Department of Energy
Washington, PA 15301
Dailey, Nancy
Oak Ridge National Laboratory
P.O. Box X
Oak Ridge, TN 37830
Davis, E.C.
Oak Ridge National Laboratory
P.O. Box X
Oak Ridge, TN 37830
Dole, L.R.
Oak Ridge National Laboratory
P.O. Box X
Oak Ridge, TN 37830
Ensminger, J.T.
Oak Ridge National Laboratory
P.O. Box X
Oak Ridge, TN 37830
Epler, J.L.
Biology Division, Bldg. 9769
Oak Ridge National Laboratory
Oak Ridge, TN 37830
*Falco, J.W.
Exposure Assessment Group (RD–689)
Office of Research and Development
Environmental Protection Agency
Washington, DC 20460
Fisher, G.E.
E.I. Du Pont de Nemours & Company
Elastomers Division
3245 Sunnyside Avenue
Brookfield, IL 60513
*Florence, L.
Environmental Action Foundation
724 DuPont Circle Building
Washington, DC 20036
Fore, C.D.
Oak Ridge National Laboratory
P.O. Box X
Oak Ridge, TN 37830
*Francis, C.W.
Environmental Sciences Division
Oak Ridge National Laboratory
Oak Ridge, TN 37830
*Franklin, R.E.
Project Officer, EV-34
F-224, GTN
U.S. Department of Energy
Washington, DC 20545
*Freeman, H.M.
Office of Appropriate Technology
State of California
1322 O Street
Sacramento, CA 95814
Friedman, David
U.S. Environmental Protection Agency
401 M. Street
Washington, DC 20460
Gerstner, H.
Oak Ridge National Laboratory
P.O. Box X
Oak Ridge, TN 37830
Getz, P.A.
Union Carbide Corporation
P.O. Box 1410
Paducah, KY 42001
*Giddings, J.M.
Environmental Sciences Division
Oak Ridge National Laboratory
Oak Ridge, TN 37830
Glover, E.W.
Soil & Material Engineers, Inc.
P.O. Box 58069
Raleigh, NC 27658
Gorman, C.
Soil & Material Engineers, Inc.
3025 McNaughton Ave.
Columbia, SC 29206
*Gratt, L.B.
IWG Corporation
975 Hornblend Street Suite "C"
San Diego, CA 92109
*Griffin, R.A.
Illinois State Geological Survey
615 E. Peabody Drive
Champaign, IL 61820
Guerin, M.R.
Oak Ridge National Laboratory
P.O. Box X
Oak Ridge, TN 37830
Gulledge, W.P.
Tennessee Valley Authority
1140 Chestnut St., Tower II
Chattanooga, TN 37415
*Hansen, J.A.
Institute for Water, Soil, and
Environmental Engineering
Aalborg University
Sohngaardsholmsvej 57, DK–9000
Aalborg, DENMARK
*Hasselriis, F.
52 Seasongood Road
Forest Hills Gardens
New York, NY 11375
*Hassett, J.J.
Department of Agronomy
University of Illinois
Urbana, IL 61801

Henderson, Rogene F.
Lovelace ITRI
P.O. Box 5890
Albuquerque, NM 87115

Homerosky, Frank
Goodyear Atomic Corp.
Box 628
Piketon, OH 45661

*Hounslow, A.W.
Department of Geology
Oklahoma State University
Stillwater, OK 74074

Hunter, D.R.
U.S. Environmental Protection Agency
345 Courtland Street, N.E.
Atlanta, GA 30365

*Hupe, D.W.
Michael Baker, Jr., Inc.
4301 Dutch Ridge Road
Beaver, PA 15009

*Kenaga, E.E.
Dow Chemical Company
1803 Building
Health and Environmental Sciences
Midland, MI 48640

*Kingsbury, G.L.
Environmental Assessment Methods Section
Research Triangle Institute
P.O. Box 12194
Research Triangle Park, NC 27709

Kirchgessner, David
U.S. Environmental Protection Agency
125 Ridgewood Road
Chapel Hill, NC 27514

Knaver, Peter
Federal Environmental Agency
Bismarckplatz 1
1 Berlin 33, WEST GERMANY

*Kronenberger, L.
Regulatory Affairs Coordinator
Synthetic Fuels Department
Exxon Company, USA
P.O. Box 2180
Houston, TX 77001

*Logan, T.J.
Department of Agronomy
Ohio State University
Columbus, OH 43210

Mann, J.H.
U.S. Environmental Protection Agency
345 Courtland Street, N.E.
Atlanta, GA 30365

*Mason, B.J.
Ethura
7606 Long Pine Drive
Springfield, VA 22151

*Maulbetsch, J.S.
Electric Power Research Institute
3412 Hillview Avenue
Palo Alto, CA 94303

*Mayhew, J.J.
Chemical Manufacturing Association
2501 M Street, N.W.
Washington, DC 20037

*Meglen, R.R.
Director, Analytical Laboratory
University of Colorado, Denver
Denver, CO 80302

Mezga, L.J.
Oak Ridge National Laboratory
P.O. Box X
Oak Ridge, TN 37830

*Murarka, I.P.
Electric Power Research Inst.
P.O. Box 10412
3412 Hillview Avenue
Palo Alto, CA 95120

*Mutch, R.D.
Wehran Engineering
666 E. Main Street
Middletown, NY 10940

*Nelson, R.W.
Battelle-Pacific Northwest Laboratories
P.O. Box 999
Richland, WA 99352

O'Sullivan, K.M.
FMC Corporation
9 Aspen Court
Buffalo Grove, IL 60090

Paidoussis, Olga
Power Authority of the State of New York
10 Columbus Circle
New York, NY 10019

Portner, E.M.
Johns Hopkins University
Applied Physics Lab.
Johns Hopkins Road
Laurel, MD 20707

Poulson, Eric S.
IU Conversion Systems, Inc.
115 Gibraltar Road
Horsham, PA 19055

*Purcell, A.
Technical Information Project, Inc.
1346 Connecticut Avenue, N.W.
Suite 217
Washington, DC 20036

Quinn, H.B.
U.S. Environmental Protection Agency
401 M. Street (RD–682)
Washington, DC 20460

Richmond, C.R.
Oak Ridge National Laboratory
P.O. Box X
Oak Ridge, TN 37830

Rickert, L.W.
Oak Ridge National Laboratory
P.O. Box X
Oak Ridge, TN 37830
*Saar, R.A.
Geraghty & Miller, Inc.
6800 Jericho Turnpike
Syosset, NY 11791
Sachdev, Dev R.
Envirosphere Company
Two World Trade Center, Fl. 90
New York, NY 10048
Samsel, G.L.
Dames & Moore
3399 Tates Creek
Lexington, KY 40502
*Santhanam, C.J.
Arthur D. Little, Inc.
20 Acorn Park
Cambridge, MA 02140
Schloss, M.
Bureau of Land Management
U.S. Department of the Interior
18th and C Streets, N.W.
Washington, DC 20240
*Simmons, B.P.
California Department of Health Services
2151 Berkeley Way
Berkeley, CA 94704
*Springer, C.
Department of Chemical Engineering
University of Arkansas
Fayetteville, AR 72701
Stief, Klaus
Federal Environmental Agency
Bismarckplatx 1
D 1000 Berlin 33, WEST GERMANY
Stow, S.H.
Environmental Sciences Division
Oak Ridge National Laboratory
Oak Ridge, TN 37830
Suloway, J.J.
Charles T. Main, Inc.
Southeast Tower, Prudential Center
Boston, MA 02199
Tamura, T.
Environmental Sciences Division
Oak Ridge National Laboratory
Oak Ridge, TN 37830
Taylor, E.A.
Tennessee Valley Authority
Forestry Building
Norris, TN 37828
Theis, T.L.
Dept. of Civil Engineering
University of Notre Dame
Notre Dame, IN 46556
Thomas, S.J.
NUS Corporation–Government Operations
4 Research Place
Rockville, MD 20850
Tiegs, S.M.
Oak Ridge National Laboratory
P.O. Box X
Oak Ridge, TN 37830
Turner, R.R.
Environmental Sciences Division
Oak Ridge National Laboratory
Oak Ridge, TN 37830
Udwari, J.J.
Dames & Moore
7101 Wisconsin Avenue, Ste. 700
Washington, DC 20014
Vavruska, J.S.
Los Alamos National Laboratory
P.O. Box 1663
Los Alamos, NM 87545
*Vesilind, A.
Dept. of Civil Engineering
Duke University
Durham, NC 27706
Walia, D.S.
Research and Development Department
United Coal Company
P.O. Box 1280
Bristol, VA 24201
Weeter, D.W.
59 Perkins Hall
University of Tennessee
Knoxville, TN 37916
Wilcox, J.R.
Florida Power & Light Company
P.O. Box 529100
Miami, FL 33157
Willis, B.C.
U.S. Environmental Protection Agency
345 Courtland Street
Atlanta, GA 30365
*Wilson, D.C.
Waste Research Unit
Harwell Laboratory, Bldg. 146.3
Oxfordshire OX11 ORA, ENGLAND
Yamada, M.J.K.
Los Angeles Dept. of Water & Power
P.O. Box 111
Los Angeles, CA 90051
*Zenz, D.R.
Metropolitan Sanitary District of Greater Chicago
100 E. Erie Street
Chicago, IL 60611
Zittel, H.E.
Oak Ridge National Laboratory
P.O. Box X
Oak Ridge, TN 37830